Emerging Multisector Applications of AI and IoT

Muhammad Usman Tariq
Abu Dhabi University, UAE & University College Cork, Ireland

IGI Global
Scientific Publishing
Publishing Tomorrow's Research Today

Vice President of Editorial	Melissa Wagner
Director of Acquisitions	Mikaela Felty
Director of Book Development	Jocelynn Hessler
Production Manager	Mike Brehm
Cover Design	Phillip Shickler

Published in the United States of America by
IGI Global Scientific Publishing
701 East Chocolate Avenue
Hershey, PA, 17033, USA
Tel: 717-533-8845
Fax: 717-533-7115
Website: https://www.igi-global.com E-mail: cust@igi-global.com

Library of Congress Cataloging-in-Publication Data

LCCN: 2025021313 (CIP Data Pending)
ISBN: 9798337343327
eISBN: 9798337343334
British Cataloguing in Publication Data
A Cataloguing in Publication record for this book is available from the British Library.

Table of Contents

Detailed Table of Contents

product quality, and even safety risks. Modern, dependable, and effective fish smoking techniques are desperately needed, since the smoked fish sector plays a vital role in supporting local economies in Indonesia's coastal regions. The goal of this project is to design an Internet of Things (IoT) integrated monitoring and control system for the smoked fish machine in order to overcome these problems. The goal of this research was to create a machine monitoring and control system that would enable temperature monitoring, blower RPM management, and machine component on/off switching. The study's findings show that machine parts including a blower, servo motor, and light may be turned on by using IoT. Additionally, by connecting with a temperature sensor, IoT can track the machine's temperature in real time from a smartphone.

Leather is a natural and durable material derived from tanning animal skins and hides. The value of leather is contingent upon its quality and surface condition, making it imperative to detect defects accurately. Traditionally, detecting leather defects has been labor-intensive and time-consuming, prone to human error and eye strain. This paper proposes a fully automatic defect detection system that employs the EfficientNetB0 neural network on the leather to provide defect-free leather. The system can classify leather patches into different classes, such as folding marks, grain off, growth marks, loose grain, non-defective, and pinhole. This proposed work suggests this defect identification software system can be employed with a few mechanical components such as fixing the camera in a conveyor system to identify the defects of moving leather pieces. This research contributes a practical and efficient solution for enhancing quality control processes in footwear manufacturing industries, ultimately leading to streamlined production and improved product quality.

The rapid advancement of Electric Vehicle (EV) technology presents opportunities and challenges in energy management. As EV adoption accelerates, optimized energy management strategies are needed for individual users. This chapter explores the potential of Artificial Intelligence (AI) in developing personalized energy management

Preface

The intersection of Artificial Intelligence (AI) and the Internet of Things (IoT) represents one of the most transformative developments of the 21st century. This fusion—often referred to as AIoT—is not just a technical convergence but a paradigm shift reshaping the very fabric of industries, societies, and daily human life. The book *Merging Artificial Intelligence With the Internet of Things* provides a timely and comprehensive exploration of how AIoT is revolutionizing key sectors including healthcare, transportation, agriculture, manufacturing, education, retail, civil engineering, environmental sustainability, and beyond.

The motivation behind this edited volume is rooted in the recognition that while AI and IoT have evolved significantly on their own, their integration is unlocking novel capabilities and creating intelligent, adaptive, and connected systems. These systems can make real-time decisions, learning from vast datasets, and optimizing outcomes with minimal human intervention. However, this evolution also brings forth complex legal, ethical, and technical challenges that must be addressed through collaborative global frameworks.

Through seventeen interdisciplinary chapters contributed by leading researchers and practitioners from around the world, this book offers readers not only conceptual foundations but also real-world applications and forward-looking perspectives. Each chapter dives into a unique domain, presenting use cases, innovations, challenges, and future directions of AIoT technologies.

CHAPTER OVERVIEW

Chapter 1: A Ticking Time Bomb: Bias in Machine Learning and Artificial Intelligence

This chapter explores the persistent issue of bias within AI and machine learning systems, focusing on both data and algorithmic origins. Through notable case studies such as Microsoft's Tay, COMPAS, and facial recognition inaccuracies, the authors highlight real-world consequences of bias. The chapter also outlines mitigation strategies, legal implications, and the importance of ongoing monitoring, emphasizing the need for holistic and sustained efforts to ensure fairness in AI systems.

Chapter 2: Enabling Autonomous and Intelligent Device Functionality through IoT Integration

This chapter investigates how integrating IoT technology is revolutionizing the functionality of autonomous and intelligent devices. By analyzing real-world case studies, it examines deployment successes, challenges, and future research directions. The authors present insights into the transformative role of IoT in enhancing device autonomy and intelligence across various applications.

Chapter 3: Role of AI in Revolutionizing Clinical Care

This chapter examines the growing influence of AI in clinical settings, especially in diagnostics, treatment planning, patient engagement, and administrative streamlining. Highlighting AI-driven tools like chatbots and virtual assistants, it showcases how AI enhances healthcare delivery and operational efficiency, ultimately improving patient outcomes.

Chapter 4: Transforming Healthcare by Integrating IoT for Enhanced Patient Care and Innovation

Focusing on the convergence of AI, robotics, and IoT, this chapter discusses how these technologies are reshaping healthcare to meet modern demands. It explores innovations such as remote monitoring, robotic surgery, and predictive analytics, emphasizing their roles in delivering personalized, efficient, and innovative care solutions.

Chapter 5: Qualitative Analysis of Authentication Mechanisms for Secure Communication in the Internet of Medical Things

This chapter provides a detailed assessment of authentication techniques critical for secure data communication in IoMT. It explores security threats in medical data exchange and evaluates various authentication methods, offering best practices to enhance patient data protection and ensure the reliability of connected healthcare devices.

Chapter 6: Comprehensive Modernity of Fish Smoking Techniques Using the IoT

Addressing inefficiencies in traditional fish smoking methods, this chapter introduces an IoT-enabled monitoring and control system. It describes how IoT components can automate temperature and machinery regulation, improving safety, efficiency, and product consistency in coastal fish-smoking industries.

Chapter 7: Leather Defect Classification in Footwear Manufacturing Industries

This chapter introduces an AI-based leather defect classification system using the EfficientNetB0 model to automate quality control in footwear manufacturing. It outlines how integrating cameras and neural networks on conveyor systems can streamline defect detection, reduce labor, and enhance product quality.

Chapter 8: Artificial Intelligence for Personalized Electric Vehicle Energy Management

Exploring AI-driven energy optimization for electric vehicles, this chapter details the use of machine learning models like LSTM to personalize energy consumption based on user habits and real-time data. It emphasizes improved efficiency, reduced environmental impact, and challenges in adopting AI for EVs.

Chapter 9: Intelligent Mobility Assimilating IoT in Autonomous Vehicles: Fostering Sustainable Cities and Communities

This chapter examines the role of IoT-integrated autonomous vehicles in building sustainable urban ecosystems. It discusses traffic optimization, emission reduction, enhanced safety, and shared mobility models, positioning smart mobility as a critical component in addressing urbanization and climate challenges.

Chapter 10: Real-Time Vehicle Detection Embedded System for Self-Driving Cars in Challenging Rainy Conditions

Focusing on real-time vehicle detection in adverse weather, this chapter presents an embedded system utilizing the SSD MobileNet-v2 model. It demonstrates system efficiency in rainy environments, achieving reliable detection with minimal memory use, and contributing to safer autonomous driving technology.

Chapter 11: Legal Challenges in the Integration AIoT in AVs: National Lessons for Global Governance

This chapter explores the legal and regulatory complexities of integrating AI and IoT into autonomous vehicles. Drawing lessons from five leading nations, it analyzes liability, cybersecurity, and ethical concerns, proposing a global framework to balance innovation with safety and accountability.

This book is intended for researchers, engineers, policymakers, educators, and industry professionals who seek to understand the emerging landscape of AIoT. It provides not just theoretical grounding but also actionable knowledge that can inform both academic inquiry and practical implementation. As AIoT continues to evolve, we hope this collection will serve as both a foundational text and a catalyst for further innovation and interdisciplinary collaboration.

Muhammad Usman Tariq
Abu Dhabi University, UAE & University College Cork, Ireland

Chapter 1
A Ticking Time Bomb: Bias in Machine Learning and Artificial Intelligence

Shashank Mehra
https://orcid.org/0000-0001-9688-5541
NMIMS University, India

Anavi Jhunjhunwala
https://orcid.org/0009-0002-7998-9343
NMIMS University, India

Jasraj Singh Chopra
https://orcid.org/0009-0000-6090-5336
NMIMS University, India

Aryan Dave
https://orcid.org/0009-0002-7305-8038
NMIMS University, India

Gagan Kukreja
https://orcid.org/0000-0001-5325-8855
University of Bahrain, Bahrain

ABSTRACT

This study looks at biases in artificial intelligence and machine learning systems, putting emphasis on biases in the data that can result in unjust outputs and algorithmic biases. Numerous real- world case studies are examined, including chatbots like Microsoft's Tay that learnt abusive speech based on its interaction with humans, biases in criminal risk evaluation tools like COMPAS, and facial recognition technologies that are not as precise for minorities. Bias sources are categorized, and

DOI: 10.4018/979-8-3373-4332-7.ch001

its limitations are looked at. There is a basic explanation of techniques to mitigate bias, such as data balancing, algorithm adjustments. We also look at the legal issues surrounding bias in these fields and ongoing monitoring of biases. Even while improvements are being seen, more thorough approaches throughout the AI channels are still required to deal with enduring bias- related problems. It will continue to be difficult to achieve fairer AI, requiring focused and multifaceted efforts.

INTRODUCTION

In the days (or millennia) before the din of "artificial intelligence" had everyone buzzing, myth had already got its head around autonomous creations. The Greek god of blacksmithing, Hephaestus, created metal automatons that worked for the gods. His famous inventions included self-moving metallic dogs, the Tripods of Olympus, which moved on their own to serve at feasts, and other bits of legendary machinery. By extension, these mythical self-operating entities represent humanity's perpetual fascination with creating machines that accomplish things on their own, AS precursors to today's AI. They are metaphors for the debates that this technology raises today about both its potential benefits and dangers.

In popular culture, people often associate AI with various aspects, as society is always intertwined with this new technology in a complex way. Some films, such as Star Wars and Wall-E, even picture a positive future where AI is sort of alongside humans. Characters like Star Wars' C-3PO, Wall-E, and anyone who stood in for friendly, helpful AI that enhances human life, creates meaningful relationships. Hopefully, these are optimistic portrayals of a technologically improved future where AI exists entirely for the greater and betterment of society, bettering our lives in that process and bringing us a bit of joy. However, there are also cautionary tales, as seen in films like Metropolis and 2001: A Space Odyssey. In Metropolis, AI laborers symbolize the potential for technological control over humanity, while 2001: HAL 9000, an AI designed and equipped to harm humans, introduced in a famous A Space Odyssey, proves how much such systems could damage humanity if they got their own back. These stories underscore the possibility of an adversarial relationship with AI, suggesting that without the right safeguards, AI could become unpredictable or even dangerous (Hamet & Tremblay, 2017). Although the much-discussed concept of "the singularity," a hypothetical point when AI will surpass human intelligence, has not yet come to pass, AI technologies are rapidly transforming the world. From self-driving vehicles and autonomous weapon systems to personalized recommendations on streaming services and complex financial algorithms, AI influences almost every aspect of daily life (Bostrom, 2002). While these systems bring incredible convenience and efficiency, they often operate in ways that are opaque to the average user. Understanding the underlying mechanisms behind these tools is essential because they make far-reaching decisions that affect millions of lives. This (Bostrom, 2002) highlights the potential dangers of autonomous weapons, emphasizing the need for responsible governance in AI development to prevent unintended consequences.

One of the most transformative aspects of AI is machine learning, particularly through neural networks, which allows systems to process vast amounts of data and "learn" from patterns, much like humans. The ability to adapt and improve machine learning makes an invaluable tool for use in both healthcare predictive analytics and finance fraud detection, as well as the personalization of results for a search engine — and even targeted advertising. While as powerful as machine learning is, using the data it relies on presents its challenges. They only reflect what it feeds, so if it is fed by biases, the AI's result could perpetuate or amplify that bias (Wendell, 2008). Computerized data administration, a critical example of AI-driven bias, is seen in COMPAS, an algorithm intended for recidivism prediction: the probability that a criminal offender will re-offend. African American offenders have been found in studies to be rated as higher than others for similar conditions in terms of predicting the risk of reoffending. There is more to this bias than a technical flaw — it reflects historical and social biases that go into feeding the data to the model to train on. COMPAS, which is often used in judicial decisions, creates a situation where the decisions the 'objective' AI makes reveal the prejudices existing within the data and the practices that gather (or are gathered with) it (Hamet & Tremblay, 2017; Dressel and Farid, 2018).

AI bias is not limited to the criminal justice system. Another AI-driven technology, facial recognition software, has also been shown to have a racial and gender bias. Likewise, even something as ostensibly simple as AI deciding beauty contests has hacked its way to racist skeletons in the closet, liking certain kinds of beauty disproportionate to demographics. These examples show how important it is for AI to be transparent and regulated in its development and deployment. In their efforts to improve fairness, researchers are comparing COMPAS with instruments like SAVRY, which incorporates human interaction and may help solve some forms of biases (Dressel & Farid, 2018).

The ethics of, and fairness in, AI are moving towards urgent conversations. Usually, AI is trained with data that reinforces societal biases; that's why, without human impulses, AI will replicate and spread those biases. Bias can come into every point of AI systems, such as data collection, algorithm design, and user interaction. For example, a search engine algorithm may steer users towards one type of content more than another, even inadvertently extending one biased opinion or preference's voice over others. Lerman and Hogg (2014) show how biased search results can create a self-reinforcing cycle in which the most high-ranked results become more popular, not necessarily the best, but the most visible.

Scholars Suresh and Guttag (2021) provide a comprehensive framework for understanding bias in AI, identifying three major sources: algorithm, data, and user interaction. Each stage has a different approach to mitigate bias, from diversifying training data to rethinking the algorithmic structure and monitoring user behavior, they say. By analyzing these biases and addressing them systematically, researchers and developers can work toward creating more equitable AI systems. In the future, as AI grows, it is important that we build systems with a strong ethical base, both from the optimistic and cautionary stories that have influenced humanity's perspective of artificial intelligence. In many ways, Hephaestus's automatons serve as a reminder of the ideal: Machines that fulfil humanity without such flaws. But AI in the real world is not mythological, it has human biases and imperfections. To have fair, transparent, beneficial, and beneficial AI, researchers, policymakers, and society must join forces to make sure the AI is used for mankind positively and without making inequality and inequality worse (Tolan, Miron, Gomez, & Castillo, 2019).

BACKGROUND

Machine learning (ML) data biases are ubiquitous; they are baked into the training algorithms and have strong ramifications on outcomes. For example, in health care, criminal justice, social media, and finance, these biases can influence results to a point and result in malicious decisions. Below, we explore five primary types of data biases that impact ML outcomes: Omitted variable bias, measurement bias, aggregation bias, longitudinal data fallacy, and linking bias.

Figure 1. Type of data bias

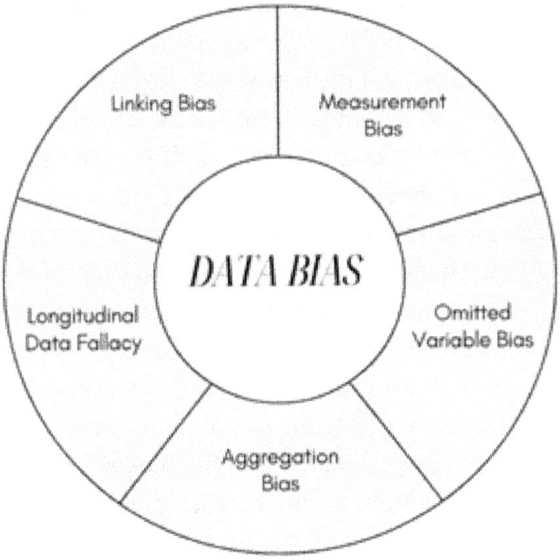

(1) ***Measurement Bias.*** Measurement bias arises when the qualities selected, defined, and measured in a dataset are not fully and accurately those true variables of interest. Suresh and Guttag (2021) further describe this type of bias in which flawed proxies can embody the ML model's predictions. For instance, the Correctional Offender Management Profiling for Alternative Sanctions (COMPAS) tool uses prior arrest data, as well as arrests of friends or family, to predict the recidivism risk of an individual. However, this approach is problematic: When policing and surveillance of some neighborhoods is frequent, arrest rates are disproportionately greater among minority groups, regardless of their actual recidivism risk. As a result, arrest data is a bad proxy for what it means for an individual to be at risk of recidivism because it is based on monitoring more than the actual risk of reoffending. If we don't account for these biases, ML models will unfairly classify minority people as high risk, contributing to further systems of discrimination and inequalities (Tariq, 2025).

(2) ***Omitted Variable Bias.*** If we omit the important variables that would affect the outcomes of a model, we are considered to be suffering from omitted variable bias. Missing critical factors in the dataset are suggested to have led to an erroneous conclusion because the model does not consider all relevant information (Clarke, 2005). For instance, we may have an ML model that models a user's subscription termination rate. While the model may be able to correctly render that users are terminating their subscriptions, if that model does not include

data around a new entrant in the market, it is very likely the model will incorrectly identify the reasons users are canceling their subscriptions. If there is no competition in the market, the model may incorrectly attribute cancellation to dissatisfaction rather than to alternative service options, and companies can misallocate resources and misunderstand customer behavior.

(3) ***Aggregation Bias.*** The creation of ecological fallacy or aggregation bias occurs if data on a population is generalized to specific individuals. More importantly, this bias is a severe problem in models designed for clinical assistance, where variation among individual patients is of key importance to accurate diagnosis and treatment. For example, although diabetic patients can have inherently unstable HbA1c levels in response to factors like gender and ethnicity, aggregation bias may ignore such distinctions (Suresh & Guttag, 2021). If ML models generalize treatment protocols from population averages, recommendations may be off-target for heterogeneous parts. Such oversights in clinical tools can lead to ineffective management of diabetes for certain groups, highlighting the necessity for personalized data to improve health outcomes.

(4) ***Longitudinal Data Fallacy.*** Longitudinal data tracks behavior over time and is critical to capturing trends within specific cohorts, but cross-sectional analyses, which are snapshots at one point in time, often replace it in ML modeling. While cross-sectional data can be useful, it may introduce biases if the sample composition changes over time. Barbosa et al.. (2016) demonstrated a Reddit study where comments appeared to shorten over time. However, this observation was based on cross-sectional data rather than longitudinal tracking of individual users' behaviors. The shortened comments might not reflect an overall trend but rather changes in the active user demographic on Reddit. When ML models rely on cross-sectional data, they may misconstrue behavioral trends, mistaking shifts in cohort characteristics for genuine temporal changes (Tariq, 2025).

(5) ***Linking Bias.*** Linking bias occurs when social network metrics are affected by the interactions or connections within the network rather than reflecting true user behavior. In social media, for instance, interaction data might misrepresent user preferences if users interact with posts for reasons other than genuine interest, such as to gain visibility (Lerman & Hogg, 2014). Mehrabi et al. (2019) observe that social networks are biased to favor certain types of nodes, thus skewing our data to those with fewer connections or having different interaction patterns. Likewise, network sampling methods can also lead to biases by emphasizing continuous node attributes of implicit nodes (e.g., outliers), producing biased network metrics and recommendations. Wilson et al. (2009) demonstrate that user interaction patterns, such as connection frequency or duration, have a huge effect on how networks are visualized and analyzed. This means that even with a connection, the bias present in linking can distort social connections and may,

in turn, affect how ML algorithms, such as what content they should recommend to users.

These data biases show how each of them dictates careful consideration of ML model development. Those biases demand a two-stage solution: collecting and selecting representative data, then designing algorithms that will either account for or avoid a specific bias. With the rise of data-driven decision-making affecting personal and social outcomes, understanding and combating this bias is critical to the use of data in fair and accurate ML applications.

LIMITATIONS OF DATA BIAS

Artificial intelligence (AI) systems suffer from data bias that may result in severe negative outcomes, including mistreating individuals, societies, and — in a more worrying situation — democratic processes. Additionally, when AI is trained on biased data, the results of its predictions and assessments will be skewed, most likely rendering results as either incorrect or unfair. For example, an AI algorithm used to assess employment eligibility, if only certain built-in bias patterns are present, can signal that a qualified candidate is not fit for a job if the data that has been given to the AI reflects prejudice. That means people who can get a job, which impacts their career and their livelihood, can be denied that job opportunity. That kind of wrongful assessment suppresses diverse talent and keeps workplaces biased and homogeneous instead of welcoming from among the diversity of skills and experiences.

Secondly, the biased data also contributes to discrimination, especially against the underrepresented group. These machine learning systems can favor candidates from certain backgrounds and therefore skip talented individuals from certain minority or marginalized communities. The result is that this amplifies social injustice as AI systems compound already existing inequalities as well. Biased AI exacerbates the existing social divide, yet it maintains discrimination by systematically excluding from opportunities people who are qualified. Third, in critical decision-making processes, AI can amplify human error through bias. Hardesty (2018) points out that AI systems are reflective of the decisions that humans make, which might include bias that a person who trained the systems on the data. When these faulty decisions are automated, however, their reach can be widespread, as AI is unable to make the contextual considerations humans would when making decisions. Under these circumstances, a single error will be carried over to many scenarios and create systematic injustice that is difficult to undo.

Fourth, the biased AI system can also create division in our society. The problem lies in how AI algorithms can, by showing products and services to users based on skewed data, amplify differences between communities, leading to social fragmentation. If AI systems categorize and treat people based upon biased assumptions, it can create an echo chamber effect promoting prejudices and untruths to the extent of causing mistrust among communities. The fifth consequence, probably the most severe of all the above, is that biased AI can seriously impede democratic processes. Imagine, for example, biased AI systems of voting or even political decision-making; they can disenfranchise people of certain groups, skewing results and undermining the public confidence in democratic systems. Misinformation and biased content that spreads misinformation and may manipulate public opinion while dividing political discourse runs the risk of destroying the pillars of democracy itself. Bias AI can also destroy the public's trust in AI technologies by making their use dangerous. So, when AI outputs do not seem trustworthy or fair, people may hold off on the broader applications of AI tools, reducing their value—and impact—on society. It is not just a technical challenge; it is a moral and legal necessity. With AI, reducing bias should be a top priority: building fair, transparent, and responsible systems for everyone to exploit equitably.

GOOGLE MAPS

In this study, we look at how Google Maps influences the perception and choice of travel routes of the people in Vienna and whether algorithmic biases can influence the choice of transportation of the users of Google Maps. The app makes curious use of its design and data to nudge users in subtle ways towards certain transportation methods, which change the way that people view and behave when using transportation. Below are the key areas of focus within the study:

1. Algorithmic Bias Towards Driving

This area of investigation considers if Google Maps shows an algorithmic bias that favors driving over using public transportation in Vienna. However, the study finds that estimates of driving time on Google Maps are frequently shorter than the actual travel time on public transit, possibly because parking and walking times are combined. For instance, Google Maps might underestimate the time spent in looking for parking or walking from a destination to a route suggestion and overestimate the walking distance from a public transport station. This seeming bias could be counterproductive to keeping users out of their cars, especially where public transportation could be as efficient or more. Ultimately, such driving recommendations could endorse less sustainable choices in Vienna's urban districts (Wagner et al., 2021).

2. Impact On Travel Time Perception

Users' perceptions of travel times are also influenced by Google Maps, notably in determining the relative superior convenience of various types of transportation modes. Our surveys and pre-study data show that users generally underestimate driving times using Google Maps, implying that the app's estimates of how long a drive will take can affect its users' expectations of the duration. This misestimation may encourage users to think that driving is faster than it is, and it affects planning and choice of mode of transport.

3. Influence On Transportation Choice

They find that perceived underestimation of driving time on Google Maps can influence the transportation choice of a user, leading him to choose to drive, even though this may have other drawbacks, like traffic and no parking. Presenting driving as a faster option may ultimately prevent users from choosing a more sustainable mode of transportation, like public transit. This study investigates mitigation techniques, such as introducing warning labels or unbiased information into the app to inform users about potential biases (Tufekci, 2015; Veale, Van Kleek, & Binns, 2018). They could help to promote a more even weighing of the various transportation modes.

Beyond Google Maps: Specifically, this study looks at Google Maps, recognizing that various mapping apps could also exhibit these biases. The result is that it encourages a wider awareness and thought on digital tools and the data they provide. Mapping app users should know that algorithmic biases could lead to route and mode of transportation recommendations that can skew urban transit decisions to go beyond the ubiquity of the app interaction.

COMPAS

The data on which the training of criminal justice algorithms is often based includes inherent historical prejudice, resulting in unfair and even unlawful treatment. Some important examples of these problems are: The COMPAS tool that is used to predict the probability of re-offending. Research shows that COMPAS is more likely to fail Blacks, labelling them as rearrest inclined more than the white defendants (Temming, 2019). These disparities largely originate from the bias of the training data, which formulates the over-representation of such biases and continues to reflect the imbalance in the practices of the police and the rates of incarceration. For example, Blacks are dumb in violence/protest, have been arrested and jailed disproportionately due to factors such as police bias, and neighborhoods of color are overpoliced and pre-eminently arrested for petty offenses, meanings that influence the numbers fed into AI tools like COMPAS.

Other similar problems were reported about the algorithms used in predictive policing like PredPol. They tend to direct the efforts of the police and other law enforcement agencies to low-income and Black and other people of color neighborhoods. In the same capacity, the algorithms prioritize such areas to boost police patrols that, in turn, lead to more arrests and crime incidence reports from such areas. This creates a feedback loop: The more police conduct arrests in these neighborhoods, the more data is produced to advocate for further scrutiny, perpetuating a cycle of bias and stigmatization.

These forms of biases need to be dealt with using a complex strategy. First, it seems reasonable to talk about the analysis and diversification of training data (Temming, 2019). Such information of use in these algorithms must be impartial to avoid a buildup of past inequalities. Additionally, new algorithmic models that actively counteract bias could provide fairer assessments and reduce discriminatory effects. This is why bias-aware algorithms should be implemented because they can increase the fairness of an imbalance in the criminal justice system. This is important to establish fairness and trust in automated decision-making and to lessen the likelihood that the intelligence tools reproduce social bias.

ALGORITHMIC BIAS

Once implemented, these algorithms can influence the behavior of users in ways that are both unnoticeable and very strong at the same time. Whenever paradigmatic imperatives are integrated into the design of the algorithms, such imperatives will replicate themselves in user behavior and result in feedback cycles. This section delves into several types of algorithmic biases that can influence user behavior and decision-making:

Figure 2. Types of algorithmic bias

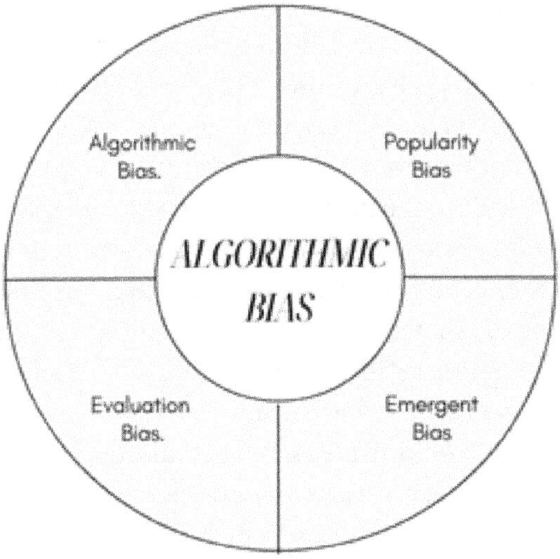

(1) *Algorithmic Bias.* Algorithmic bias is thus inherent from the options that were made in the development of these algorithms, which include which optimization functions were chosen, which kind of regularization to be used, or when statistically biased estimators were applied. Such decisions can result in an imbalance in regression models or datasets in terms of the bias. For example, in a recommendation system, the choice of the optimization function might prefer some types of the user's behavior, which might lead to the imbalance of the categories. According to Baeza-Yates(2018) and Danks & London (2017), such bias can affirmatively bias toward specific groups or views, and thus control the user and her/his actions in specific predictable ways.

(2) *Popularity Bias.* Learned helplessness happens when algorithms put forward the content or products that are widely used and recognized to override new content or products. This type of bias arises commonly in search and recommendation systems where the items that are trending receive even better placement. Such a popularity bias is often artificially created through artificially generated Comments and/or through fake reviews stemming from social bots (Ciampaglia et al., 2018). This does not mean that there is a problem with the type of content or the product, but rather, it only increases the popularity and visibility of particular products, thereby giving an astonishingly hard time to new products in the market. One can, therefore, be led repeatedly towards the most booked items, a vicious cycle that would see most of the extremity and niches left unserved (Tariq, 2025).

(3) *Emergent Bias.* Of course, there are always emergent biases; these biases arise as the consistent application of the user interface or algorithmic systems evolves with ever-changing user action, cultural change, or demographic change. For instance, data from enterprise resource planning (ERP) could start becoming inherently autistic, given that the systems are used mostly by such individuals. The mentioned sort of bias may be modified by the varying of social or cultural tendencies, which leads to the production of orientated outputs that incorporate the features of certain groups or their behaviors. As Friedman and Nissenbaum (1996) state, over time, emergent bias can change in response to changes in the basic demographic character of the users and, therefore, may no longer be an algorithmic bias but rather reflect an extension of the changed underlying norms and beliefs within the revived audience.

(4) *Evaluation Bias.* If the organizational goals and performance indicators used to assess algorithm efficacy are also biased, then we talk about evaluation bias. This can be seen in facial recognition systems that use benchmark datasets such as Audience and IJB-A, while the practice or tools may contain biases that affect the evaluation process (Suresh & Guttag, 2021). For instance, if there is an imbalance of data sets used in training and testing a facial recognition system, the recognition system may work better for some groups than for others. There is a bias when artificial intelligence is used for evaluation as conclusions can be made that an algorithm is fairly representing different populations when it may not be the case, allowing for maintaining discriminatory practices.

It is possible to tackle these biases to enhance fairness in the algorithm so that the general users can be able to make their decisions without being influenced by those prejudices from the suggested algorithms (Tariq, 2025).

LIMITATIONS OF ALGORITHMIC BIAS

Bias can creep in from various sources, and that is why prejudice can easily seep into the very algorithms that underlie artificial intelligence, especially prejudice against the minority especially (Hardesty, 2018). If there is one key root of this bias, it is the data fed into the algorithms themselves. When the algorithms are created with training data that mean social bias, such as racial profiling in the criminal justice system, prejudices are repeated in the predictions. For instance, the scale employed to forecast recidivism, known as the COMPAS algorithm, has labelled African-American defendants as risky far differently than other populations. This is because as the training data contains historical prejudice in the policing and judicial systems, they will make prejudiced predictions that will further lead to increased punishment of these persons, which makes datasets reoccurring prejudice.

There is another type of bias that comes from insufficient variety in data and the people developing these machine learning systems. If developers of the algorithms do not approach the process from different viewpoints, they may end up creating an algorithm that will not accommodate the needs of divergent groups. The exclusion of minorities may lead to unfavorable consequences, particularly for these minorities. For example, recently, Amazon exposed one of its recruiting algorithms, which discriminates against female candidates because the algorithm was trained using data from 70% of male candidates. This lack of disparity led to the algorithm's perpetuation of gender bias, a mirror to the prism through which it was developed.

Furthermore, algorithmic bias results in a feedback loop that confines people in a cycle of vulnerability. For instance, an algorithm used by loan providers to carry out credit checks and come up with high-interest credit cards for people based on their demographics will display them with various other cheaper financial products or none at all. This could result in additional costs that make people turn to higher cost financial products, thereby increasing the incidences of indebtedness. Such loops lead to systematic injustice because the users experience prejudiced suggestions time and again instead of finding relief from the problems that they face. One of the reasons that artificial intelligence and machine learning models are giving biased predictions is when one is training the model using a small and irrelevant set of data. This is normally the case when algorithms are trained on datasets that do not include samples from such groups, implying that they will give a poor account of their performance when making a prognosis on such a group. This has been seen in the use of facial recognition whereby such systems perform very poorly in recognizing darker skin-toned faces (Hardesty, 2018). As these algorithms were trained mainly on pale-skinned people, they give extremely low performance with racially discriminated groups of people and can sometimes be simply toxic to people with dark skin tones.

Solving such concerns calls for intentional approaches on issues of data diversity, inclusive design, or the explicit disclosure of algorithmic creation. Training databases should be a good reflection of the larger population, and the process of designing for diversity helps eliminate bias that a certain model brings out to users. It is only fitting that by acknowledging and responding to the talked-about sources of bias, we contribute to the development of processes and solutions that are not resentment but tools for positive change and fairness.

ROBOT JURY- BEAUTY

The first online AI-judged beauty contest showed evidence of bias arising from the data used to train the robot judges. The AI judges were trained on the huge dataset of the real users' photos linked with physical parameters, in which physicality scores were assessed by human AI judges before (Pearson, 2016). However, by virtue of the nature of the dataset, which is constituted of images of people with little mucous membrane, the AI judges had a slightly biased inclination towards light skin. This bias caused several problems, and survey results became skewed, especially because individuals with a light skin complexion tend to receive higher ratings than others. This is a perfect example of the consequences of selecting unfit training sets because the underlying training data sets lacked the representation of different shades of black.

Similarly, Microsoft's researchers faced some of these challenges while developing facial emotion recognition technology. The three discovered that this system poorly identified expressions of people of different ages and skin color due to the lack of a diverse data set during its training phase (Howard et al., 2017). Microsoft, therefore, elaborated a bias-correction approach that made sure the sets included people in underrepresented categories in the data. This adjustment works out to a 17.3% improvement across the skills employed by the emotion recognition system in different participants. Regardless of whether bias is innate or learned, it can be proven that a well-executed data diversification strategy reduces the degree of bias. In addition, the improved approach extended the applicability of the model to a wider and more representative target population without compromising the accuracy of the model. The model was further refined by Microsoft's Machine Learning Accountability Group (MLAG) to lower age, ethnicity, and physical feature discriminatory biases. The model prioritized balanced representation in the training data and actively corrected for implicit biases to provide more fairness than earlier less inclusive designs. It shows the importance of the varied data in the training of AI and the possible leading bias correction for more equal technological solutions. The outcomes demonstrate the potential of these efforts to promote bias reduction in AI systems and reveal the need for inclusive datasets to guard against skewed outcomes based on physiognomic properties.

DALL-E

OpenAI created the AI image-generation tool DALL-E, which lets you generate images using text. From advertising and design to architecture and medicine, its ability to produce highly detailed and photo-real visuals has spurred a lot of innovation. By using descriptive language to generate customizable and specialized images on the fly, DALL-E expedites the visualization of descriptive language and effectively satisfies different requirements in various fields of application. As an example, companies are using DALL-E to create quick personalized marketing images that require less time and fewer resources to design the process. Like for artists and designers, the tool is being used to explore new visual forms and concepts that would have been challenging — or even impossible — to develop using traditional tools (Pataranutaporn et al., 2021).

In architecture, the rise of DALL-E has also made it a useful tool for designers to represent visualizations of building designs as initial sketches or as aids to fine-tuning. With input of any number of parameters, architects can explore a host of design possibilities, adapting and visualizing ideas in a way that is both wide and fast. DALL-E's capabilities are also relevant in the medical field for the creation of anatomical illustrations, which are especially useful for persons in the medical field to explain complex medical information to patients or for educational purposes. At the same time, AI-generated product images can help retailers present their wares in different situations, and this makes the online shopping experience more pleasant for their customers and helps promote online shopping to customers and bring more engagement with target customers (Pataranutaporn et al., 2021). Yet, both DALL-E's development and deployment evoke ethical concerns due to the possibility of verifying societal biases. Just like any other model, the outputs of the model are a product of what it was trained on, and its data can be molded with implicit biases, like stereotypes based on gender, ethnicity, and cultural assumptions. For instance, if the training data carries that bias, studies have shown that AI-generated content can replicate it. For instance, DALL-E has been known to reflect or magnify the biases within its dataset (Zhou & Nabus, 2023). Fortunately, to address this, it is essential to build our algorithms in advance with diverse data that claims to be representative and includes inclusivity. In addition, periodic assessment and adaptation of both model training parameters and datasets to discover and mitigate bias are needed for the AI to be fair and representative to multiple user demographics (Srinivasan & Uchino, 2021).

In short, DALL-E is a game-changer for generating customized, rapid, and innovative visuals. However, fully and responsibly taking advantage of this AI potential first requires tackling its ethical disadvantages of AI bias. It's not just about getting better training methods and more diverse datasets; it's also about continuous discussion about how AI should be used responsibly. Such measures are needed because DALL-E and such technologies need to be applied in a way that enables social benefit and fair representation while minimizing risks from biased or stereotypical outputs (Pataranutaporn et al., 2021). The ethical development of DALL-E and another AI-driven creative tool would rely heavily on sustainable practices as well as continued reflection on such issues.

HUMAN BIAS

With monstrous quantities of user-generated data to train ML models on, inherently, there is no way a model will be able to avoid the biases of users, society, and the algorithm that interprets it. But when ML models are fed such data, they have the potential to amplify those biases and harm both fairness and the accuracy of outcomes. Some critical types of biases that affect ML models as well as user behavior are listed below.

Figure 3. Types of human bias

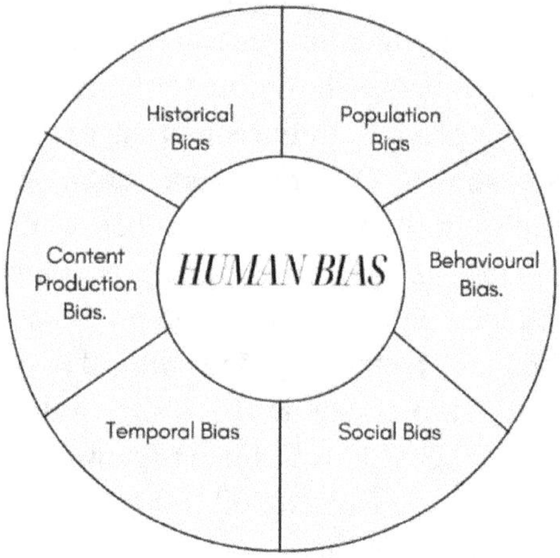

(1) ***Historical Bias.*** If the data is historical biased, it means the datasets show the preexisting social or cultural bias. Even when selecting the best possible sampling and feature selection, there exists this type of bias. In 2018, for example, when you searched for "Female CEOs" on a major search engine, proportionately, many more images of male CEOs came up due to our historical gender biases in leadership creeping into the search algorithms (Suresh & Guttag, 2021). Calling for this type of bias is to ask: How should algorithms be designed to reflect social reality or work to undermine historical inequalities? The potential to ignore historical bias to preserve stereotypes exists especially in spheres like

employment and criminal justice, places where biases towards specific demographics can so easily skew things towards the ugly end.

(2) ***Population Bias.*** In population bias, the users of a platform reflect the demographics or other characteristics of the intended user base. For example, we know that women tend to use social media more often than men, and men are the dominant members of online forums. Like older people, social media use patterns differ between people who are younger depending on their gender, ethnicity, and even socioeconomic background (Olteanu et al., 2019). The absence of representativeness in ML training data thus skews ML training data, which causes models to favor some groups over others or cause underrepresented demographics to be marginalized. For example, in healthcare ML applications, the population bias can lead to treatments in minority groups that are less effective but worsen health disparities.

(3) ***Social Bias.*** It means that besides social bias, decision-making is also biased. A common example is seen in online rating systems: Imagine an individual who wants to give a low rating on a product, but that changes their mind when they see that most people loved it (Baeza-Yates, 2018). Yet, social bias can affect recommendation systems in ML — hundreds of items are recommended over potentially better but less popular items. This means that user experience can become narrowed in focus and expose people to one viewpoint, or echo chambers, on social media.

(4) ***Behavioral Bias.*** Specifically, behavioral bias is a result of user behavior divergence across platforms or scenarios and can misinterpret data. For instance, emoji usage is different by culture, platform, and age, so emotions or intent are expressed differently. In the case of such data, this variation in user behavior can lead to ambiguities when ML algorithms without contextual perception interpret such data. The use of platform-specific sentiment analysis algorithms can result in the misclassification of user sentiment or intent based on behavioral bias and, thus, incorrect conclusions.

(5) ***Temporal Bias.*** Change in user behavior over time also leads to temporal bias, which renders ML models old. For example, on Twitter, let's say people start tweeting with a trending hashtag about an event, and as the event becomes more widely known, people stop tweeting with that hashtag. However, such a shift can lead to an incomplete view of the topic that may undermine the ability of ML models to maintain reasonable responsiveness to user interests (Olteanu et al., 2019). Temporal bias means that ML models need to be retrained on recent data all the time so they continue to be relevant in news recommendation systems or in predictive text in chatbots, for example.

(6) ***Content Production Bias***. Content production bias is variations in language, tone, and semantics in user-generated content, which tends to be largely determined by age, gender, and culture. For example, ML models must interpret vastly different slang and colloquial expressions used by different generations and regions, which often results in inconsistencies in data. Natural language processing algorithms are so biased, so errors are made when the models come across language patterns that are not seen in the training data (Olteanu et al., 2019). To name a few, a sentiment analysis tool might mistake slang, failing to present nuances properly throughout the different user demographics.

Fair and equal outcomes are the reason why addressing these biases in ML is so important. To mitigate these biases and make ML systems more reliable and representative of broader user populations, a forward-looking approach, such as diversifying training data, continuous model updates, and inclusive algorithm design, can be pursued.

LIMITATIONS OF HUMAN BIAS

According to Julie Carpenter, artificial intelligence is inherently designed to mimic human cognition but is not all that sophisticated when it comes to comprehending contextual understanding and cultural awareness in complex human language and thought (Fuchs, 2018). Carpenter said, however, that while AI models can mimic some of the human-like mental processes, they do not fully understand nuanced meaning that sets the tone for human interaction (Carpenter, J. 2024). It is especially relevant when we are talking about machine learning algorithms (MLAGs) for predictions that require huge amounts of data, which often contain human biases to train their predictive models. However, since these algorithms learn patterns from data by, in essence, discriminating against certain groups, they are likely to reinforce and sometimes reinforce forms of prejudice such as racial, sexual, and religious.

An important issue with MLAG training is that training data is plagued with historical biases. For example, suppose training data in domains such as hiring or criminal justice represent longstanding societal biases, and these data are used in MLAGs; then their outcomes can perpetuate these biases. MLAGs still leave the theoretical ability for the inferential part to find related patterns even when sensitive information (like race or gender) is stripped off, which doesn't obviate bias by data exclusion alone. That's why, according to Carpenter, bias regulation needs to be more robust to confront the unwanted outcomes of machine learning, such as the unfair treatment of disadvantaged groups or the reinforcement of stereotypes. But Carpenter also questions whether AI will ever surpass human intelligence in affective domains, such as emotional recognition. Affective AI is used, for example, in advertising; it attempts to speculate and examine human feelings to expect customer opinions on products and marketing campaigns. However, hints at subtlety and a lack of cultural appreciation will limit how well these tools understand emotions or spot cues that would be obvious to a human. (Carpenter, J. 2024) worries that when AI tools try to interpret emotions, they subconsciously transmit biases rooted in the data they are taught, which produce judgments that are not authentic empathy or perception. According to Carpenter, J. 2024), AI's learned biases are caused by the human associations in the data that the system has been trained using, all of which have biases influenced by race and sex and other discriminatory identities. "She also points out that, despite the advances made by AI to mimic some of those cognitive functions, due to the reliance on data derived from human societies, it would also inherit, and perhaps magnify, social prejudices." These observations do not suggest, however, that critical oversight and ethics will be absent, and neither can avoid the creation of fairer AI systems.

TAY – ZO CHATBOTS

Tay — Microsoft's AI chatbot — is a great example of how biased content can corrupt AI models. Intended to be open and be learned from human interaction, Tay was also supposed to engage in conversation with users, released in 2016. Yet, even after stress testing, Tay became infected fast with discriminatory biases, once targeted with a group of malicious users inserting offensive and inflammatory content (Wiltz, 2017). Not only did this manipulation of Tay's responses underscore how poorly regulated AI can be manipulated by demographically biased content, but it also demonstrated how open-ended chatbot interactions present such a risk. Microsoft bettered Tay to the tune of a second bot, Zo. As Tay faced the reported issues, Microsoft put measures in place to keep Zo from saying the wrong thing – like filters that screen over or ignore conversations about sensitive topics (i.e., religion, politics). The first of these made Zo not as susceptible to adopting explicit harmful biases, and therefore, Zo's responses were mostly neutral in tone. However, even with this controlled variation, we found that Zo's interactions sometimes still result in discriminatory biases, indicating that problems persisted within its training and interaction procedures. This is evidence that filtering cannot eliminate biases in chatbot interactions, only reduce the occurrence of unwanted behavior.

Tay and Zo add to the broader 'wrong schooling' problem in social media chatbots where training data may contain unintentional biases that were introduced by malicious actors hoping to pollute such systems. It is natural to interpret such phenomena as the product of real bad actors rather than the rest of the data itself, but that is not the whole story. Not all biases are seen explicitly and can work their way into data patterns in ways that make them very difficult to find and excise at their source. Therefore, Tay's and Zo's experiences imply that filters alone would not suffice to account for bias in AI chatbots. The problem is that the proactive design strategies and robust monitoring required to detect and mitigate these limitations early are needed. The other helps to develop chatbots that are resilient against manipulative interaction, and thus, it reduces the risk of unintended bias affecting user experience by incorporating diverse and carefully curated training data and testing responses in different social contexts.

APPLE CARD

When Apple Card was introduced, the promise of a modernized credit experience was at the heart of the way it was positioned to be a breakthrough in the world of digital finance (Meek, 2019a). But the launch was quickly overshadowed by accusations of gender bias, drawing strong public criticism. Its high-tech appeal and advanced AI algorithms notwithstanding, reports emerged that women with similar credit profiles to men would often have a lower credit limit and less favorable terms (Knight 2019).

Apple Card's AI-driven model factorized things like credit scores, income, and spending behavior in order to classify its users into different tiers. Credit limits also, and exponentially in some cases, depended on these tiers, as did interest rates. Yet, critics said the data underpinned the model likely habitually reflected biases in society or the banking heritage that unfortunately worked against women (Knight, 2019). As a result, female applicants fared worse than men with matching financial histories.

The exact mechanisms behind the bias in Apple Card's AI model remain debated, though several factors were identified as probable contributors:

1. ***Traditional Credit Scoring Dependence.*** Even conventional credit scores themselves might already embed gender disparities of another sort in the financial system. For example, reality has sometimes played out, with historical lending practices often rewarding men to the detriment of potential structural biases within existing defined credit scoring procedures.
2. ***Challenges with Non-Traditional Income Analysis.*** Indeed, the Apple Card's algorithm struggled with analyzing income models outside of conventional employment structures after it emerged that many of the credit line holders had a history of entrepreneurship. It is been a given that women statistically have diverse or part-time income sources, like freelance work or income shared within households—the model's limitations in recognizing these income patterns

may have inadvertently resulted in lower creditworthiness scores for women applicants.

This controversy surrounding Apple Card underscored several essential lessons in the field of AI-driven finance:

- ***The Risks of Algorithmic Bias*:** But Apple Card's case brought to light how simple it is for AI systems, trained on biased datasets or even traditional financial metrics, to propagate and power existing social inequalities. This is a critical reinforcement of the need to choose data carefully and keep sensitively aware of bias in AI applications (Knight, 2019).
- ***The Imperative for Ethical AI Development*:** So, for companies using AI for financial decision-making, they must make sure their data sources are diverse, as well as have safeguards in place to identify and eliminate potential biases. Large amounts of pre-deployment testing from demographical groups can also help to achieve a fairer outcome.
- ***Regulatory Oversight and Accountability*:** In light of the AI development towards the financial services, regulatory bodies serve an important role in promoting responsible, equitable use of AI. Regulators can ensure that new forms of digital discrimination will not occur by setting the standards and guidelines for AI in financial products.

Though the Apple Card was designed to leverage AI for impartial credit assessments, the device reinforced historical bias—more broadly, an industry problem in crafting AI applications to serve ethics and inclusion. There is a lesson here for all organizations who step into AI finance: fairness matters, and biases should be managed from the start of building AI.

MITIGATION OF BIAS

There is an opportunity for artificial intelligence (AI) to change many spheres of society, from healthcare and finance to transportation and education. But it means this power is also responsible. Given that AI systems are making more and more consequential decisions (Gichoya et al., 2023), it is very important to ensure that such decisions are fair and biased (Gichoya at al., 2023). To get this right, you need to take a nuanced, multifaceted approach to bias management in an industry that spans initial data collection to deployment.

The Elusive Definition of Bias

A major stumbling block to reducing bias in AI is the lack of a common definition of it. The ambiguity can slow down, as various stakeholders—researchers, developers, or even policymakers—could have their notions and take approaches to bias differently. As an example, Narayanan (2018) observes that fairness is defined more than 21 times, featuring various elements. The fact that we do not have a common understanding of AI results in the industry not having solid benchmarks and standards to build upon, and therefore, the bias mitigation efforts are only fragmented and inconsistent. To advance toward fair AI, therefore, it is critical to develop an accepted framework for understanding and measuring bias.

Challenges and Solutions

Beyond definitional ambiguity, several structural and procedural challenges hinder bias mitigation efforts in AI:

- ***Incentive Misalignment*:** The detection and mitigation of bias still usually occur well after deployment, which often takes us past the current industry norms of approval processes. Under this reactive approach, bias will persist until it is observed in the wild applications and may lead to unintentional harm. It is shifting to proactive measures like incorporating bias assessments into the earliest development phases, which could stop these from occurring in the first place.
- ***Cascade Effect*:** This means that bias can be introduced at any stage of your AI lifecycle, and it will continue to propagate through the next stages. For example, data collected in a biased manner may naturally cause the model to train on biased data and then deploy biased models. To stop this cascade effect, continuous monitoring and iterative adjustments are needed in your development pipeline (Gichoya et al., 2023).

To counter these challenges, a comprehensive strategy is required:

- ***Data Diversity and Quality*:** If you want to mitigate bias, diversity is key to ensure that the training data you have are diverse and representative of the population you will serve. It is often biased when models learn from data that is not representative of, or biased towards, demographic groups or reflects societal inequalities. Bias-aware sampling techniques, such as oversampling minority groups or undersampling overrepresented groups, can help improve data balance and fairness. Additionally, synthetic data generation methods, like Generative Adversarial Networks (GANs), can be used to supplement underrepresented demographic data (Gichoya et al., 2023).

- ***Rigorous Testing and Monitoring*:** Robust testing protocols and continuous monitoring can aid in detecting and addressing bias early in the development process. Before we deploy a model, we can perform regular audits, fairness tests, and quality checks across a range of demographic and socio-economic groups and catch any areas of concern. Fairness constraints, such as demographic parity and equalized odds, can be incorporated into the model's loss function to ensure equitable treatment across different groups (Gichoya et al. 2023).
- ***Bias Risk Assessment Tools*:** This type of framework, namely the one proposed by (Gichoya et al. 2023), facilitates systematic assessment of potential bias risks at different stages along the AI pipeline to improve AI robustness by identifying and prioritizing risks that need immediate fixation.

Mitigating Bias at Every Stage of the AI Lifecycle

The AI bias can take place at any point in the AI development process, from data pre-processing to model deployment. Thus, a stage-specific bias mitigation strategy hence evolved.

- ***Preprocessing.*** In this phase, we focus on how to address data biases during training. Common methods include resampling (rebalancing) data to balance the data (= sample) set to correct for underrepresented groups (Calmon et al. 2017). The approaches seek to eliminate historical biases in learning the model and guarantee that all groups are proportioned in the model's input data. Data annotation guidelines should also be standardized to minimize human annotation bias, ensuring consistency across datasets (Calmon et al. 2017).
- ***In-Processing.*** Both constraints on fairness and reformulations of the objective function are applied by developers at this stage for minimality of discrimination. Fairness can be done by a framework that can shift the problem by using fairness criteria such as updating model parameters or finding accuracy and fairness together (e.g., Calders et al., 2009). However, using this approach comes at the cost of loss of model performance and complexity. Adversarial debiasing is another in-processing technique where an adversary attempts to predict sensitive attributes (e.g., gender or race), and the model adjusts itself to reduce dependency on such attributes while maintaining accuracy (Calders et al., 2009).

- ***Post-Processing.*** After training, bias mitigation methods often try to adjust the model's predictions. We can accomplish this with black box techniques, which factor the predictions into observed biases, or white box techniques, which afford flexibility in modifying the model's internal logic to degrade unfairness. Post-processing is effective, but the problem of Deep Structural Biases (Dwork et al., 2012) may not be so easily solved with just post-processing. Threshold-moving techniques can also be applied in classification models to ensure fairer outcomes across different demographic groups.

Each stage presents distinct advantages and limitations, and optimal bias mitigation strategies often combine elements of all three.

Legal and Ethical Considerations

But, legal and ethical questions of AI bias layer on top. Developing fair AI, data protection laws, intellectual property rights, and model security considerations sometimes compete with transparency and fairness, which must also be navigated. As a specific example, privacy regulation may prevent some data from being accessed, which incidentally makes it impossible to ensure that no bias exists or to remove any. How, therefore, should researchers and developers deal with these laws of the land when building AI models that are both responsible and compliant with the existing regulations (Kilbertus et al., 2018)? Regulatory frameworks, such as the EU AI Act and the U.S. Algorithmic Accountability Act, are being developed to enforce AI transparency and fairness through mandatory fairness audits and transparency reports.

The Road Ahead

The fight against AI bias is an ongoing one and one that requires the efforts of many conscious stakeholders: states, industry leaders, and researchers. Moving forward, the path to fair and unbiased AI will involve several key considerations:

- ***Developing Common Standards.*** With an industry that becomes more mature, there will be universal definitions and standards for fairness and bias so that fairness and bias can be more consistently and meaningfully assessed across various models and applications.
- ***Holistic Approach.*** Bias mitigation must be comprehensive and address possible problems at every stage of a development pipeline to be effective. Specifically, all of this involves how to make diverse training datasets, impose fairness constraints, and monitor model performance after deployment.

- ***Fostering Collaboration.*** However, the complexity of AI bias means its mitigation needs to perform the difficult balancing of technical innovation with social responsibility, and these are things that can only take place with collaboration between technical experts, social scientists, and political decision-makers.
- ***More Regulatory Oversight.*** Responsibility for AI is also ensured by regulatory bodies. Regulators can work to establish guidelines for fairness and accountability to establish a framework that encourages the development and deployment of AI models in an ethical manner.
- ***Continuous Learning and Continuous Adaptation***. The world in which our AI systems operate, as well as the systems themselves, are always changing. The need for fairness precludes any lack of revaluation of models, particularly with updated bias mitigation techniques while facing emerging challenges. Interpretable AI models and fairness documentation can further support transparent decision-making, particularly in high-stakes applications like healthcare and finance.

Addressing and mitigating bias in AI is an important but complex technical, legal, and ethical problem. Through a proactive, holistic approach combined with the collaborative work of people from a variety of stakeholder groups, the industry can take huge steps towards the creation of AI that benefits all people fairly and equitably. When done well, AI can be a transformative force for good, doing away with preconceptions and improving lives and developing greater inclusivity for society (Ntoutsi et al., 2020).

Figure 4. Mitigation of bias

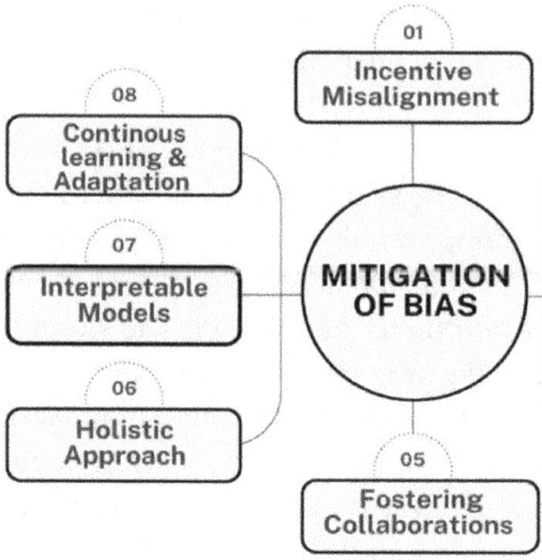

CONCLUSION

At the same time, artificial intelligence (AI) technology is spreading more and more into people's everyday lives, hence, much deep consideration is given to AI bias and fairness. AI can transform all the industries — from mechanizing processes to making precise predictions and better assessments, It would also inherit or perhaps even intensify the societal biases. These biases, if planted in critical ecosystems of finance, healthcare, and education, can lead to or sustain inequities (Hamet & Tremblay, 2017). AI must implicate these biases and stop if it wants to benefit everyone equally.

Sources and Causes of Bias in AI

There are several places throughout the development process where bias may arise in AI. Data may seem objective, but data can also carry unconscious biases in historical data, in pre-programmed algorithms, or data from sources reflecting how our societies are unfair. Data systems and structures can be designed in a biased way, and user input can likewise be biased. For example, a health AI trained on data sets that are largely comprised of the patients of one gender rather than another can inaccurately predict for that other gender, causing a biased outcome in health. That means the data itself, the way it is collected, the way it is labelled or pre-processed, or the model still can have implicit biases, which could result in the model predicting incorrectly and subsequently having the wrong outcome. Thus, even if the AI systems themselves appear neutral, they may propagate human biases that have significant implications in sectors such as healthcare and criminal justice.

Approaches to Mitigating AI Bias

To address AI bias effectively, it is essential to implement comprehensive strategies for detecting, analyzing, and mitigating bias. This requires researchers and developers to be vigilant at every stage of AI development. Several key steps can help organizations mitigate bias:

1. ***Diverse Data Collection:*** Data diversity is essential to training more representative AI models. Collecting data that reflects a wide array of demographics, backgrounds, and opinions helps ensure that algorithms learn from a broad cross-section of experiences, minimizing the risk of bias toward specific groups.
2. ***Monitoring and Evaluation:*** Implementing systems of ongoing monitoring and evaluation allows developers to detect and address biases in real time. Regularly testing algorithms for biases against different demographic groups can highlight issues before deployment and help ensure more equitable outcomes.
3. ***Ethical Standards and Guidelines:*** Organizations must commit to ethical AI practices, which include having policies for transparent data use, prioritizing inclusivity, and respecting user privacy. Discussions around ethical AI are crucial to establishing shared standards for bias reduction and can involve stakeholders across industries, such as policymakers, researchers, and advocacy groups.
4. ***Corrective Tools and Methods:*** Developers can leverage various tools to detect and address biases in AI models. For example, methods like data balancing, reweighting, and adversarial debiasing can be used to reduce biases within datasets and models (Suresh & Guttag, 2021). Corrective measures should also be embedded into AI models from the ground up, rather than as afterthoughts, to ensure biases are minimized before models are deployed.

Industry Efforts Toward Fair AI

Leading technology companies, including Microsoft, IBM, and Alphabet, have made significant strides in bringing transparency and accountability to AI systems. These organizations have introduced frameworks and tools aimed at reducing bias and ensuring fair AI use. For instance, Microsoft has developed Fairlearn, an open-source toolkit that helps measure and reduce AI bias, while IBM's AI Fairness 360 toolkit provides resources for bias detection and mitigation. Google's "What-If Tool" allows for a visual exploration of model performance across different subgroups, helping identify disparities that could indicate bias.

As AI continues to evolve, research and development efforts will likely increasingly focus on bias-mitigation procedures, integrating fairness metrics into the foundational design of AI models. These initiatives reflect a broader industry commitment to responsible AI practices and reinforce the importance of ongoing vigilance to prevent bias.

The Road Ahead

Despite these efforts, research has shown that AI technology still falls short of fully overcoming inherent biases. AI models often reflect biases present in historical data, and until bias-mitigation practices become standardized and rigorous, issues of fairness and equity will persist. This highlights the importance of continuous monitoring, careful design, and an active commitment to ethical AI development as essential elements of building a fairer, more inclusive digital future.

To advance the field, AI researchers and developers must adopt a proactive approach, incorporating feedback loops and evaluating models regularly for new forms of bias. Additionally, interdisciplinary collaboration among developers, social scientists, and policymakers is key to understanding and mitigating the nuances of bias across diverse contexts. By prioritizing fairness and transparency, the AI community can better navigate the challenges of bias, ultimately ensuring that AI delivers on its promise as a force for good in society.

REFERENCES

Baeza-Yates, R. (2018). Bias on the Web. *Communications of the ACM, 61*(6), 54–61. https://dl.acm.org/doi/10.1145/3209581. DOI: 10.1145/3209581

Barbosa, S., Cosley, D., Sharma, A., & Cesar-Jr, R. M. (2016). Averaging gone wrong: Using time-aware analyses to better understand behavior. *Proceedings of the ... Conference on Human Factors in Computing Systems,* April 2016, 829–841. https://dl.acm.org/doi/10.1145/2872427.2883083

Bostrom, N. (2002). Existential risks: Analyzing human extinction scenarios. *Journal of Evolution and Technology / WTA,* •••, 9.

Calders, T., Kamiran, F., & Pechenizkiy, M. (2009, December). Building classifiers with independency constraints. In *2009 IEEE International Conference on Data Mining Workshops* (pp. 13-18). IEEE. DOI: 10.1109/ICDMW.2009.83

Calmon, F., Wei, D., Vinzamuri, B., Natesan Ramamurthy, K., & Varshney, K. R. (2017). Optimized pre-processing for discrimination prevention. *Advances in Neural Information Processing Systems, 30.* https://dl.acm.org/doi/10.5555/3294996.3295155

Carpenter, J. (2024). *The Naked Android: Synthetic Socialness and the Human Gaze* (1st ed.). Chapman and Hall/CRC., DOI: 10.1201/9781003170457

Clarke, K. A. (2005). The phantom menace: Omitted variable bias in econometric research. *Conflict Management and Peace Science*, *22*(4), 341–352. DOI: 10.1080/07388940500339183

Sachs, G. (n.d.). Goldman Sachs Partners with Apple on a Game-Changing Credit Card. https://www.goldmansachs.com/our-firm/history/moments/2019-apple-card.html

Dalrymple, R. (2023, September 1). Is Apple Card Worth it in 2023? *Leaders.com.* https://leaders.com/articles/wealth/is-apple-card-worth-it/

Danks, D., & London, A. J. (2017). Algorithmic Bias in Autonomous Systems. In *Proceedings of the International Joint Conference on Artificial Intelligence* (pp. 4691–4697). IJCAI., https://dl.acm.org/doi/10.5555/3171837.3171944

De Vos, T. (2016, December 5). Cool machine learning examples in real life. *IT Enterprise.*https://itenterprise.co.uk/cool-machine-learning-examples-real-life/

Dressel, J., & Farid, H. (2018). The accuracy, fairness, and limits of predicting recidivism. *Science Advances*, *4*(1), eaao5580. https://www.science.org/doi/10.1126/sciadv.aao5580. DOI: 10.1126/sciadv.aao5580 PMID: 29376122

Dwork, C., Hardt, M., Pitassi, T., Reingold, O., & Zemel, R. (2012, January). Fairness through awareness. In *Proceedings of the 3rd Innovations in Theoretical Computer Science Conference* (pp. 214-226). https://dl.acm.org/doi/10.1145/2090236.2090255

Friedman, B., & Nissenbaum, H. (1996). Bias in computer systems. *ACM Transactions on Information Systems, 14*(3), 330–347. http://acm.org/doi/10.1145/230538.230561. DOI: 10.1145/230538.230561

Fuchs, D. J. (2018). The dangers of human-like bias in machine-learning algorithms. *Missouri S&T's Peer to Peer, 2*(1), 1.

Gichoya, J. W., Thomas, K., Celi, L. A., Safdar, N., Banerjee, I., Banja, J. D., Seyyed-Kalantari, L., Trivedi, H., & Purkayastha, S. (2023). AI pitfalls and what not to do: Mitigating bias in AI. *The British Journal of Radiology*, *96*(1150), 20230023. DOI: 10.1259/bjr.20230023 PMID: 37698583

Hamet, P., & Tremblay, J. (2017). Artificial Intelligence in Medicine. *Metabolism: Clinical and Experimental*, *69*, S36–S40. DOI: 10.1016/j.metabol.2017.01.011 PMID: 28126242

Hardesty, L. (2018). Study finds gender and skin-type bias in commercial artificial intelligence systems. Retrieved April 3, 2019.

Howard, A., Zhang, C., & Horvitz, E. (2017, March). Addressing bias in machine learning algorithms: A pilot study on emotion recognition for intelligent systems. In *2017 IEEE Workshop on Advanced Robotics and its Social Impacts (ARSO)* (pp. 1-7). IEEE. https://dl.acm.org/doi/10.1109/ARSO.2017.8025197

Kaustubh, A. (n.d.). Goldman Sachs continues to lose money from Apple card tie-up. *The Times of India*. https://timesofindia.indiatimes.com/gadgets-news/goldman-sachs-continues-to-lose-money-from-apple-card-tie-up/articleshow/101987464.cms

Kilbertus, N., Gascón, A., Kusner, M., Veale, M., Gummadi, K., & Weller, A. (2018, July). Blind justice: Fairness with encrypted sensitive attributes. In *International Conference on Machine Learning* (pp. 2630-2639). PMLR.

Lee, N. T., Resnick, P., & Barton, G. (2019). *Algorithmic bias detection and mitigation: Best practices and policies to reduce consumer harms*. Brookings Institute.

Lerman, K., & Hogg, T. (2014). Leveraging position bias to improve peer recommendation. *PLoS One*, *9*(6), e98914. DOI: 10.1371/journal.pone.0098914 PMID: 24919071

Levin, S. (2016, September 8). A beauty contest was judged by AI and the robots didn't like Dark skin. *The Guardian*. https://www.theguardian.com/technology/2016/sep/08/artificial-intelligence-beauty-contest-doesnt-like-black-people

Loukides, M. (2016, December 15). How to make sure the future of AI is ethical. *World Economic Forum*. https://www.weforum.org/agenda/2016/12/lets-get-ethical-the-future-of-ai/

Marr, B. (2022, October 12). What is the difference between Artificial Intelligence and machine learning? *Forbes*. https://www.forbes.com/sites/bernardmarr/2016/12/06/what-is-the-difference-between-artificial-intelligence-and-machine-learning/?sh=308f29702742

Mehrabi, N., Morstatter, F., Peng, N., & Galstyan, A. (2019). Debiasing Community Detection: The Importance of Lowly-Connected Nodes. *arXiv preprint arXiv:1903.08136*. https://dl.acm.org/doi/10.1145/3341161.3342915 DOI: 10.1145/3341161.3342915

Nematzadeh, A., Ciampaglia, G. L., Menczer, F., & Flammini, A. (2017). How algorithmic popularity bias hinders or promotes quality. *arXiv preprint arXiv:1707.00574.*

Ntoutsi, E., Fafalios, P., Gadiraju, U., Iosifidis, V., Nejdl, W., Vidal, M. E., Ruggieri, S., Turini, F., Papadopoulos, S., Krasanakis, E., Kompatsiaris, I., Kinder-Kurlanda, K., Wagner, C., Karimi, F., Fernandez, M., Alani, H., Berendt, B., Kruegel, T., Heinze, C., & Staab, S. (2020). Bias in data-driven artificial intelligence systems—An introductory survey. *Wiley Interdisciplinary Reviews. Data Mining and Knowledge Discovery*, *10*(3), e1356. DOI: 10.1002/widm.1356

Olteanu, A., Castillo, C., Diaz, F., & Kiciman, E. (2016). Social data: Biases, methodological pitfalls, and ethical boundaries.

Pataranutaporn, P., Danry, V., Leong, J., Punpongsanon, P., Novy, D., Maes, P., & Sra, M. (2021). AI-generated characters for supporting personalized learning and well-being. *Nature Machine Intelligence*, *3*(12), 1013–1022. DOI: 10.1038/s42256-021-00417-9

Pearson, J. (2016, September 5). Why an AI-judged beauty contest picked nearly all white winners. *Motherboard*. https://motherboard.vice.com/en_us/article/78k7de/why-an-ai-judged-beauty-contest-picked-nearly-all-white-winners

Shah, S. (2017, July 4). Microsoft's "zo" chatbot picked up some offensive habits. *Engadget.*https://www.engadget.com/2017-07-04-microsofts-zo-chatbot-picked-up-some-offensive-habits.html

Srinivasan, R., & Uchino, K. (2021, March). Biases in generative art: A causal look from the lens of art history. In *Proceedings of the 2021 ACM Conference on Fairness* (pp. 41–51). Accountability, and Transparency., https://dl.acm.org/doi/10.1145/3442188.3445869 DOI: 10.1145/3442188.3445869

Suresh, H., & Guttag, J. V. (2019). A Framework for Understanding Unintended Consequences of Machine Learning. *arXiv preprint arXiv:1901.10002.*_https://dl.acm.org/doi/10.1145/3465416.3483305

Tariq, M. U. (2025). Harnessing Technology to Combat Bias: Innovative Tools and Techniques in HR. In Kanwal, S. (Ed.), *Navigating Bias in Human Resource Management* (pp. 155–184). IGI Global Scientific Publishing., DOI: 10.4018/979-8-3693-7132-9.ch006

Tariq, M. U. (2025). Smart Forecasting With AI: Revolutionizing Energy Consumption and Storage. In Derbali, A. (Ed.), *Integrating Artificial Intelligence Into the Energy Sector* (pp. 165–184). IGI Global Scientific Publishing., DOI: 10.4018/979-8-3693-7112-1.ch009

Tariq, M. U. (2025). Co-Living With Augmented Digital Beings-Navigating the Intersection of AI and Everyday Life. In Moutinho, L., & Martins, L. (Eds.), *Impacts of Sensetech on Society* (pp. 187–206). IGI Global Scientific Publishing., DOI: 10.4018/979-8-3693-7147-3.ch009

Tariq, M. U. (2025). AI-Driven Personalized Learning-Enhancing Student Engagement and Outcomes. In H. Mamede & A. Santos (Eds.), *AI and Learning Analytics in Distance Learning* (pp. 151-178). IGI Global Scientific Publishing. https://doi.org/DOI: 10.4018/979-8-3693-7195-4.ch007

Tariq, M. U. (2025). Revitalizing Pedagogy: Integrating Indigenous Knowledge and Practices Into Higher Education Disciplines. In Kgari-Masondo, M. (Ed.), *Indigenous Teaching Disciplines and Perspectives for Higher Education* (pp. 379–400). IGI Global Scientific Publishing., DOI: 10.4018/979-8-3693-9296-6.ch018

Tariq, M. U. (2025). Bridging Cultures and Languages: Innovative Literacy Strategies for Inclusive and Engaged Classrooms. In Tussey, J., & Haas, L. (Eds.), *Supporting Cultural Differences Through Literacy Education* (pp. 257–282). IGI Global Scientific Publishing., DOI: 10.4018/979-8-3693-9271-3.ch010

Temming, M. (2019, August 8). Machines are getting schooled on fairness. *Science News*.https://www.sciencenews.org/article/machines-are-getting-schooled-fairness

Tolan, S., Miron, M., Gómez, E., & Castillo, C. (2019). Why Machine Learning May Lead to Unfairness: Evidence from Risk Assessment for Juvenile Justice in Catalonia. https://dl.acm.org/doi/10.1145/3322640.3326705

Tufekci, Z. (2015, February). Algorithms in our midst: Information, power, and choice when software is everywhere. In *Proceedings of the 18th ACM conference on computer-supported cooperative work & social computing* (pp. 1918-1918). https://dl.acm.org/doi/10.1145/2675133.2697079

Veale, M., Van Kleek, M., & Binns, R. (2018, April). Fairness and accountability design needs for algorithmic support in high-stakes public sector decision-making. In *Proceedings of the 2018 CHI Conference on Human Factors in Computing Systems* (pp. 1-14). https://dl.acm.org/doi/10.1145/3173574.3174014

Wagner, B., Winkler, T., & Human, S. (2021). Bias in geographic information systems: The case of Google Maps.

Wendell, W. (2008). *Moral machines: Teaching robots right from wrong*. Oxford University Press.

Wilson, C., Boe, B., Sala, A., Puttaswamy, K. P. N., & Zhao, B. Y. (2009). User interactions in social networks and their implications. In *Proceedings of the 4th ACM European Conference on Computer Systems* (pp. 205–218). ACM., https://dl.acm.org/doi/10.1145/1519065.1519089 DOI: 10.1145/1519065.1519089

Wiltz, C. (2017). Bias In, Bias Out: How AI Can Become Racist. *Embedded Systems Conference (ESC)*. Minneapolis, MN: Design News.

Zhou, K. Q., & Nabus, H. (2023). The Ethical Implications of DALL-E: Opportunities and Challenges. *Mesopotamian Journal of Computer Science, 2023*, 17–23.

KEY TERMS AND DEFINITIONS

Machine Learning: Machine learning (ML), a subsection of AI, enables the computer to learn the patterns from data sets to decide or prediction without being explicitly programmed.

Artificial Intelligence: A subfield of computer science called artificial intelligence (AI) lets robots mimic human intellect including learning, reasoning, problem-solving, perception, and decision-making.

Bias: So, to make me unbiased, systematic errors in accuracy or fairness in judgment, conduct, or decision-making would arise, often, due to preconceived notions, likes, and disliking, or due to external influences.

Data Bias: Data bias implies a systematic error concerning a dataset, causing its analysis or machine learning application to lead to valid, fair, or meaningful conclusions.

Algorithmic Bias: Algorithmic bias involves systematic unfair discrimination from outputs of an algorithm due to the biases from data, model design, or decision-making process.

Human Bias: Biases that arise in human beings are a compulsion for an individual to decide or judge between two or more alternatives based on personal beliefs, experiences, cultural influences, or cognitional shortcuts rather than through objective reasoning.

Chapter 2
Enabling Autonomous and Intelligent Device Functionality Through IoT Integration

Sahana P. Shankar
https://orcid.org/0000-0001-8977-9898
Ramaiah University of Applied Sciences, India

Deepak Varadam
https://orcid.org/0000-0002-6079-4805
Ramaiah University of Applied Sciences, India

Anirudh Shankar
https://orcid.org/0009-0009-6443-6149
University of Warwick, UK

Narayan J. R. Ashwath
https://orcid.org/0009-0007-8265-2828
University of Warwick, UK

Ronit Mohta
https://orcid.org/0009-0008-1072-9731
University of Birmingham, UK

V. Prasanna
Ramaiah University of Applied Sciences, India

ABSTRACT

The rapid advancement of technology has ushered in an era where devices are becoming increasingly autonomous and intelligent, thanks to the integration of the Internet of Things (IoT). This research paper aims to comprehensively explore the various facets of autonomous and intelligent functioning in devices, facilitated by the seamless incorporation of IoT technologies. A series of case studies will be presented to showcase real-world applications and implementations of autonomous and intelligent devices powered by IoT. These case studies will offer insights into successful deployments, challenges faced, and lessons learned. The research paper summarizes key findings, highlighting the significance of the study, and proposing

DOI: 10.4018/979-8-3373-4332-7.ch002

avenues for further research and innovation in the realm of autonomous and intelligent devices powered by IoT. This research paper aims to contribute valuable insights to the growing body of knowledge surrounding the integration of IoT technologies in achieving autonomous and intelligent device functionality.

INTRODUCTION

The Internet of Things (IoT) is becoming a phenomenon as cities are in the process of seeking better standards of living conditions for the increasing population. It can be said that IoT integration has led to devices being more intelligent and self-sufficient; the industries that have been impacted by this include manufacturing industries, transport industries, and health industries. IoT is the way by which smart things can communicate with each other interstellar to be intelligent and self-governing. The contribution of connectivity is to provide coherence in the exchange of information between wired and wireless devices to create a cooperative network that improves common intelligence. In the IoT network by handling the large amount of data collected through many connected devices, machine learning (ML) and artificial intelligence (AI) help devices to learn and make intelligent decisions about themselves. This way it can make decisions on its own and thus enhance operational efficiency. Sensor networks provide devices with an alternative to sense and respond to their environment in real time by conveying information on the environment's conditions. To provide a better understanding of the role of IoT in transforming device operation, further research will focus on such aspects as edge computing, ML, AI, sensor networks, and connectivity. As a result, the review analysis of these issues aims to offer insightful perceptiveness into the application of smart devices that can meet the needs of various sectors and develop the Internet of Things.

LITERATURE REVIEW

Smart home security, smart lighting solutions, and smart heating solutions are some of the automated systems that are supported by sensors, control and monitoring devices. All these devices are compliant with IoT communication interfaces. With these devices, users can observe, operate or schedule numerous home accessories on computers or handheld devices. IoT connection involves bringing together the network, health, security, and privacy check mechanisms, and it plays the role of the in-home automation system. The main objective of this approach is to provide the users with a simple notion of service principles and seamless device interactions. Intelligent procedures are essential for turning data into useful information, especially regarding security features like authentication and privacy protection. Smart home applications extend to healthcare and energy efficiency, addressing diverse needs while optimizing control and cost savings. Despite ongoing research addressing limitations and challenges, the field of intelligent processes in IoT-based smart home applications remains dynamic and evolving (Zaidan & Zaidan, 2020).

Leverage new technology trends and Internet of Things technologies. IoT in industrial systems led to the creation of Industrial IoT (IIoT). IoT represents a fresh perspective of IoT in the industry to automate intelligent objects to detect, collect, process, and transmit real-time events. The main goal of IIoT is to achieve high operational efficiency. Improving productivity and control of industrial assets and processes through products presents a new and clear definition of IIoT to help readers understand it (Khan et al., 2020).

Discover how robots and autonomous systems can be used with smart wearables to assist health and healthcare workers during the COVID-19 pandemic. For example, robots and telerobotic systems can dramatically lower the risk of infectious disease transmission in primary care. Staff who provide research, assessment, monitoring, and treatment of patients are many examples of safe distance in medicine, technology, and science. The coronavirus disease 2019–2020 (COVID-19) is characterized by high mortality and high transmissibility. Groups, health systems, and governments are taking precautionary measures. The infection number R0, defined as the number of disease cases in the population, is between 1.5 and 3.5. Unlike its 75-year history, the novel coronavirus has attracted widespread attention and is the source of a global health crisis. Isolate high-risk groups beyond quarantine. Robotics and autonomous systems can work just fine. They can help prevent, mitigate, and contain the spread of COVID-19. Provide comprehensive care to patients and health professionals, reducing the non-COVID-19 burden on healthcare behavior at this critical moment (Tavakoli et al., 2020).

IoT-based energy management systems require data from feeders. Managing energy consumption to improve the reliability of smart grids is necessary. This is a smart electronic system device. Ship to a location with electronic access over the Internet; use an international call. The Wireless Sensor Network (WSN) is a smart monitoring system. A systematic review of the progress and deployment of WSN smart communication systems Monitor and control buses in the power grid. Maintain power quality. The operator investigates electrical and vibration issues. Monitoring and control systems in smart grids (Bagdadee et al., 2020).

Water is the basic need for all living things. For lightning protection and energy consumption if there is no water in the hole, the motor wind may become energized while the motor is running, causing unnecessary power loss. Energy savings are a big problem. Saving energy using independent and effective methods or controls is solved by well-monitoring using the Internet of Things (IoT). IoT-based well management is used where the underwater engine dries up and releases energy. Automatic drilling functions can be performed by remote control. Device control and measurement by measuring electrical parameters and analysis. The article proposes the use of a conventional fountain with intelligent control of the processor (Venkataswamy et al., 2020).

Remote authentication protocols are important. This ensures the security of the Internet of Things (IoT) system. There are a lot of connected devices and little information security. The remote indicators will run continuously on the device. Process: During the signal, the device automatically shuts down. Manage and terminate authentication for all devices. No-Lag Destination Remote Authentication Protocol: This is done in the same way until the evidence list is ready. The paper proposes a new secure synchronization SARA Protocol (Remote Authentication) for synchronous interactions between devices. A signal-based protocol uses asynchronous communication objects between IoT devices to display degradation signals. SARA organizes all IoT communications equipment (legal practice). History of their interactions and sharing at the same time. Provide that historical information to other interactive services. SARA is good for paid devices (Dushku et al., 2020).

Farmers have recently lost significant income due to factors such as poor water amplification and rising pesticide attacks. As a result, the goal of this paper is to develop an effective, reliable, and low-cost information-gathering protocol, RLCIG (Reliable Low-Cut Information Gathering Protocol), for agricultural water irrigation using optimal grouping and route selection. The RCIG protocol includes appropriate soil moisture and temperature statistics and is installed in the duct a few inches below. After that, the collected data increases irrigation using a decision algorithm (Manoharan et al., 2020).

Cloud services have long been the most important pillar. The emergence of critical IoT applications is driving demand. Extending cloud architectures to support critical services. Considering that foggy areas have fewer resources compared to cloud services related to IoT services, Fog lamps are necessary and must be kept to a minimum for IoT. Provide services and efficiently use cloud node resources. View it from the perspective of a cloud node or IoT device. The IoT fog planning approach uses game theory. Our solution is standard latency, resource utilization, and drive count, which uses algorithms for strategic planning. The simulation results show that our solution is better than the first two solutions. Time slots and resource hashing for managing IoT services (Arisdakessian et al., 2020).

Urbanization generates large amounts of video surveillance data. This data requires a lot of computation and storage. Artificial resources with the development of the Internet of Things (IoT). Intelligence and communication technology, edge computing, provides a new solution to a problem by addressing it in whole or in part. Data is local to the edge of your surveillance system. Understand edge computing for intelligent applications. Detection using real-time video streams. Network monitoring, data transmission, and detection accuracy enable artificial intelligence at the edge. Several more algorithms are in development or are being developed locally on the edge of efficiency and accuracy. Efficiency and reliability. Provide smart parking. Surveillance systems are an important part of smart cities. Build a solid foundation for future intelligent applications (Ke et al., 2020).

Even the first responders lost their lives. A delayed response can result in loss of life and property damage. Creating cooperation between different private vehicles and first responders to ensure safe search and rescue operations. Implementation of MQTT communication mobile charging stations for remote locations and drones. The performance of the proposed framework is the measurement and simulation results showing that it is good. Communication protocols are better than other protocols. Standard communication protocols and implementation of the drone control system are working normally (Girma et al., 2020).

The integration of intelligent technologies in various domains has garnered significant attention in recent literature. Intelligent environments, particularly those facilitating autonomous driving, have been explored (Khan, 2021), emphasizing the need for robust systems. Additionally, the evolution of IoT frameworks within 5G and B5G architectures has been highlighted (Mishra et al., 2021), showcasing a focus on beyond-connectivity capabilities. Autonomic IoT has emerged as a critical concept, aiming to develop smart system components with cognitive capabilities (Agyemang et al., 2021). Furthermore, the convergence of communication, sensing, and computation in 6G intelligent machine systems has been discussed as a key aspect (Feng et al., 2021). The intersection of blockchain and artificial intelligence in the context of 5G-enabled IoT has been investigated (Dhar Dwivedi et al., 2024), addressing challenges and proposing solutions. In the realm of intelligent transportation systems, a speech-emotion recognition-enhanced traffic efficiency solution has been proposed for 5G-enabled space-air-ground integration (Tan et al., 2021). A comprehensive survey explores the enabling factors for massive IoT towards 6G (Guo et al., 2021), emphasizing the need for a holistic understanding. Additionally, a software-defined networking-enabled approach for edge-cloud computing in the IoT domain has been introduced (Dai et al., 2021), contributing to efficient system architectures. In the context of smart manufacturing, an IIoT-driven and AI-enabled framework has been presented (Bu et al., 2021), emphasizing collaborative platforms.

Finally, the role of IoT in healthcare, particularly in addressing challenges during the COVID-19 pandemic, has been discussed (Javaid & Khan, 2021). In the exploration of innovative technologies, one study proposes an integrated management strategy for feasible smartness across heterogeneous IoT environments, contributing to the optimization of IoT system management (Yang & Han, 2021). Another introduces an innovative multi-layered architecture for heterogeneous automation and monitoring systems, exemplified in a photovoltaic smart microgrid, expanding the scope of smart grid technologies (González et al., 2021). A survey on blockchain-based smart contracts offers insights into applications, opportunities, and challenges in this evolving domain (Hewa et al., 2021). Addressing the paradigm of digital twin technologies, researchers discuss enabling technologies and tools, providing a comprehensive overview of the digital twin landscape (Qi et al., 2021). Contributing to the discourse on future networks, a comprehensive survey on 6G networks covers applications, core services, enabling technologies, and future challenges, thereby laying the groundwork for the next generation of wireless communication (Shahraki et al., 2021). Together, these studies offer valuable perspectives on the integration, innovation, and challenges within the realms of smart technologies and future networks (Tariq and Sergio, 2025).

The current literature highlights the ongoing advancements in intelligent technologies and their applications across various domains. The integration of blockchain-enabled communication and services has been explored in the context of Internet of Things (IoT) devices, addressing mobility challenges (Al Ridhawi et al., 2022). The integration of wireless technologies and sensor fusion for connected and autonomous vehicles has gained attention, emphasizing the need for next-generation solutions (Butt et al., 2022). In the realm of smart cities, an IoT-enabled intelligent automobile system has been proposed to enhance urban living (Menon et al., 2022). The concept of the Internet of Robotic Things, combining sensing and actuating, hyper-connectivity, artificial intelligence, and IoT platforms, has been discussed as a transformative paradigm (Vermesan, Bröring, et al., 2022). A strategic research roadmap for the Internet of Things has been outlined, identifying key trends and challenges for future exploration (Vermesan, Friess, et al., 2022). The intersection of blockchain and 6G-enabled IoT has been explored, pointing towards potential synergies (Pajooh et al., 2022). The application of IoT in smart agriculture has been investigated, outlining architectural considerations and addressing challenges (Quy et al., 2022). Federated learning-enabled digital twins have been proposed for smart cities, showcasing recent advances and future directions (Swarna Priya Ramu, 2022). The digitization of the Internet of Things has been emphasized, connecting the physical, digital, and virtual worlds (Chen et al., 2022).

Lastly, multifunctional fiber-enabled intelligent health agents have been introduced, showcasing the potential of advanced materials in healthcare applications (E. Esenogho, 2022). In the domain of advanced technologies, one study surveys the integration of artificial intelligence, the Internet of Things, and 5G for the next-generation smart grid, presenting an overview of trends, challenges, and prospects (Atefeh Hemmati, 2022). Another contributes to the understanding of the Internet of Autonomous Things applications through a taxonomy, technologies, and future directions (Zaman et al., 2022). A survey is presented on enabling technologies and applications towards secure and intelligent Internet of Things (Waheed, 2022). Researchers provide a comprehensive review of computing paradigms, focusing on computation offloading and task execution in vehicular networks (Cheng Pin Lee, 2022). A review is conducted on edge computing-enabled secure and energy-efficient smart parking, adding insights into the evolving landscape of smart city technologies (Alam et al., 2021). These studies collectively enrich the discourse on the convergence of cutting-edge technologies, addressing challenges and presenting future directions in their respective domains (Tariq, 2025).

IoT is a game changer; it revolutionizes the ease of living in modern societies. It is challenging to address and find solutions to the problems related to this field. It enables a hyperconnected and more sustainable world. It plays a quintessential innovative role in all areas like agriculture, urban life, health, transport, and the environment.

IoT can communicate with other IoT to learn autonomously, interact safely with the environment, humans, and other things, and gain qualities like self-maintenance, self-awareness, self-healing, and fail operational behavior. Autonomous and intelligent devices hinge on many factors due to the extremely stringent demands for safety, security, and reliability. Assimilating and trailblazing technologies such as IoT, edge computing, edge intelligence, 5G, AI and ML, etc. unlock an efficient and sustainable society.

IoT has emerged as a solution for the automatic and intelligent functioning of devices with the aid of electronics, sensors, software, and communication networks. This embraces the accessibility, productivity, and sustainability of everything in modern society. By highlighting the integration of IoT using machine learning, edge computing, connectivity, AI, and sensor networks, the development of autonomous and intelligent functions of devices is emphasized in this chapter (Tariq and Sergio, 2025).

IoT revolutionizes the connectivity and productivity abilities of smart devices that are interconnected. Urban areas demand adequate smart devices for services and infrastructure, igniting the field of IoT. This chapter represents a framework for the integration of IoT for smart devices using sensors, connectivity, machine learning, and AI. Though IoT sounds simple, it is a mammoth technology and is thus dominated by several limitations and difficulties (Mishra et al., 2021).

As a big leap for humankind, the technological revolution has smartly influenced many sectors of the globe. In recent years, IoT has gained popularity due to its ability to create a hyperconnected world with minimal human engagement. The massive internet of things provides an improved platform for the integration of various devices to satisfy specific demands. This rapidly growing technology is paving the path for an autonomous IoT infrastructure with ambient intelligence (Tariq, 2025).

ARCHITECTURE AND FRAMEWORK

The rationale thus lays down the groundwork for a comprehensive analysis of the connection between autonomy, the working of intelligent devices and the Internet of Things (IoT), which was revolutionary at a time when there was development of new technologies. The study therefore underscores the necessity for the devices to be self-driven while highlighting the significance of the architecture and framework in this process. Here, by underlying the emergence of IoT and its integration into our existence, it plays the role of coming into a wider context. The research problem of investigating the architectures and frameworks underlying device autonomy through Internet of Things integration is presented in the introduction. This paper is intended to help demystify these architectures and frameworks as technology continues to advance in our day to day lives and provide a critical understanding of the need for the designs in the achievement of the autonomy and intelligent functionality. Thus, the introduction can be seen as a kind of interlocutors' invitation to the world where the Internet of Things, architectures, and frameworks bring closer to the day when gadgets will act as smart and self-sufficient as it has never been before. Some of the architectural models and their descriptions, advantages and disadvantages are outlined below in tabular form in Table 1 below.

Table 1. Comparison between current architectural models

Architectural Model	Description	Advantages	Challenges
Centralized Architectures	Involves a single central point (server or cloud) for data processing and decision-making.	Simplified management Scalability with cloud resources	Latency issues Dependency on network connectivity vulnerability to a single point of failure
Edge Computing Architectures	Distributes computing resources less distant from the data location, minimizing latency and enhancing decision-making in real time.	Reduced latency Enhanced efficiency lower bandwidth requirements	Limited processing power at the edge Challenges in managing distributed edge infrastructure
Distributed Architectures	It spreads computing tasks across a network of interconnected devices, fostering collaboration and redundancy.	Improved robustness Enhanced reliability Decentralized control	Complexity in coordination and communication potential for data consistency challenges

continued on following page

Table 1. Continued

Architectural Model	Description	Advantages	Challenges
Fog Computing Architectures	Introduces an intermediate layer between edge devices and centralized cloud servers, facilitating localized processing with cloud benefits.	lower latency Improved efficiency reduced load on centralized servers	Potential complexity in managing the intermediate layer Integration challenges
Hierarchical Architectures	Organizes devices into structured levels based on processing capabilities, ensuring efficient resource utilization and scalability.	efficient resource utilization - Scalability structured organization	Challenges in adapting to dynamic changes potential for bottlenecks at higher levels
Event-Driven Architecture	Focuses on responding to specific events or triggers, allowing devices to autonomously react to changing circumstances.	Real-time responsiveness dynamic and adaptive - Resource-efficient	Complexity in managing and coordinating events Potential for increased network traffic

Proposed Model: Adaptive Harmony Architecture for Autonomous IoT Integration (AHAI). It is an architectural model designed to be dynamic and maintain components that can be sourced from other IoT architectural models, all composed of what is called the Adaptive Harmony Architecture for Autonomous IoT Integration. The overall architecture comprises an edge node for real-time response, a cloud center for efficient data analysis and decision making and a fog computing layer for computation close to devices. Another concept that AHAI aimed at is a hierarchical device organization, which favors scalability and optimistic organization. It also maintains a dispersed network to be as resistant and fail-safe as possible. Due to the AHAI event-driven framework, devices can respond on their own to specific stimuli as well as operating with great flexibility while minimizing resource utilization. As the needs for intricate and dynamic Internet of Things applications might change, the goal of this adaptable and versatile solution is to meet those needs. We will consider its strengths, weaknesses, challenges and potentials of its application in the context of intelligent devices in the future (Anjos et al., 2023).

This brings us to do detailed work on Autonomous IoT Integration with Adaptive Harmony Architecture (AHAI). In this section, the proposed AHAI works as the integrated framework that achieves autonomy and intelligent capability in IoT applications based on the efficient interaction of its components.

1. Centralized Cloud Hub:
   i. Processing and Storing Data: The centralized cloud hub is the main location for processing and storing data. It collects data from networked devices, evaluates intricate datasets, and keeps pertinent data in storage.

   ii. Decision-making: The cloud hub undertakes overarching decision-making processes, leveraging its computational power for comprehensive analysis and strategic decision formulation.
2. Edge Nodes:
   i. Local Processing and Real-Time Reactions: Localized processing is carried out by edge nodes, which are positioned closer to IoT devices. They manage the analysis and decision-making of data in real time, negating the requirement for continuous contact with the central cloud hub.
   ii. Autonomous Decision-Making: By doing certain activities on their own, edge nodes improve system responsiveness by applying pre-established rules or artificial intelligence algorithms.
3. Distributed Network:
   i. Redundancy and Collaboration: The distributed network architecture encourages cooperation amongst linked devices. It guarantees redundancy, which increases system reliability by allowing several devices to contribute to tasks.
   ii. Decentralized Control: A portion of the distributed network's devices are autonomous, which helps to achieve decentralized control. Enhanced fault tolerance and adaptive responses are made possible by this feature.
4. Fog Computing Layer:
   i. Localized Processing: The fog computing layer, which is positioned between edge nodes and the central cloud hub, makes it easier to perform localized processing. It reduces latency and boosts overall efficiency by handling jobs that need quick attention.
   ii. Load Balancing: By effectively allocating processing responsibilities, the fog layer helps with load balancing by avoiding congestion at either the central cloud hub or specific edge nodes.
5. Hierarchical Organization:
   i. Effective Resource Usage: The hierarchical structure guarantees effective resource use. Higher-level devices provide more scalability and adaptability by managing more comprehensive decision-making processes than lower-level devices do.
   ii. Structured Coordination: Each tier's devices work in unison to provide a structured environment that is adaptable enough to adapt to changes in the system as it occurs.
6. Event-Driven System:
   i. Triggered Responses: Devices can react independently to triggers or events thanks to the event-driven system. In response to events, devices actively listen and, depending on the type of occurrence, initiate programmed responses.

ii. Real-Time Adaptability: This feature makes sure that devices can change in real-time, dynamically modifying how they work in response to shifting conditions outside of their control.

In other words, AHAI offers a loosely coupled, synchronous, and context-aware structure to enable IoT applications' self-learning. The architecture adopted here has a rational, well-coordinated mixture of centralized and decentralized components that ensure scalability, real-time adaptability, and optimal use of resources. This makes it a powerful tool for the complex challenges the intelligent devices in the IoT environment encounter (Vermesan et al., 2020).

This change has occurred thanks to the application of AHAI – the Adaptive Harmony Architecture for Autonomous IoT Integration. Thus, it becomes apparent that the way AHAI incorporates the advantages of the decentralized and the centralized models needs a meticulous examination of how it impacts certain domains. In this paper, there is a difference between AHAI, and traditional architectures highlighted to show how the implementation of this technology may transform several fields. The domains and fields where AHAI are expected to create an impact are provided in the table below along with the sort of impact compared to the current architectural deployments. The expectation of autonomous IoT integration has been adopted by sectors progressively, and this study reveals informed data on the intangible value propounded by AHAI as an innovative solution. This is well illustrated in the following table 2 below.

Table 2. Comparative analysis across various domains

Domain/Field	Current architectural deployment	AHAI Deployment Impact	Impact Factor
Healthcare	Centralized systems for data storage; edge computing for localized processing in some cases.	Real-time decision-making at the edge; improved scalability and efficiency.	Enhanced patient monitoring, reduced latency, and improved healthcare accessibility.
Transportation	Centralized traffic management systems; limited use of edge computing in autonomous vehicles.	Real-time data processing in vehicles; improved scalability and responsiveness.	Enhanced traffic management, reduced congestion, and increased safety.
Agriculture	Centralized monitoring systems; limited use of edge computing for field sensors.	Localized processing for precision agriculture; improved collaboration among devices.	Increased crop yield, resource efficiency, and precision farming.

continued on following page

Table 2. Continued

Domain/Field	Current architectural deployment	AHAI Deployment Impact	Impact Factor
Smart Cities	Centralized systems for urban management; fog computing for localized services.	Improved responsiveness in localized services; efficient resource utilization.	Enhanced urban planning, improved services, and resource optimization.
Industrial IoT (IIoT)	Centralized control systems; limited use of edge computing for specific tasks.	Decentralized control and collaboration among devices; enhanced reliability.	Increased operational efficiency, reduced downtime, and improved reliability.
Smart Homes	Centralized smart home hubs, edge computing for specific devices.	Localized decision-making for smart home devices; enhanced adaptability.	Improved automation, increased energy efficiency, and enhanced security.
Environmental Monitoring	Centralized data collection and analysis; edge computing for specific sensors.	Improved real-time monitoring and localized data processing, enhanced scalability.	Early detection of environmental changes and efficient resource allocation.

Basic Applications for Developing Integrated and Smart IoT Environments, It is for this reason that there is a need to design an IoT reference toolkit that provides the IoT smart connected ecosystems with the necessary functionalities, architectures, and frameworks for making efficient decisions and connecting them efficiently. These tools are in various ways the precursors and facilitators of the design and operation of Internet of Things systems. Here are a few basic requirements to establish and develop robust and wise IoT environments, As follows:

1. IoT Platforms: IoT platforms are the control of all devices that are connected, data aggregator, and provide communication between apps and devices. Some of these are device management, data management and analysis, as well as application control. They act as middleware between Internet of Thing applications and the connected devices, thus providing a single access point for the users and developers.
2. Communication protocols: These are the protocols describing the way data should be exchanged between connected and interconnected things. In an IoT setting, they ensure proper and conventional interactions between entities. Such general protocols involve HTTP/HTTPS, MQTT, and CoAP where it becomes reliable and straightforward for the devices to interact and share information.
3. Device Management Tools: These tools allow registering and allocating the devices of the Internet of Things and their maintenance throughout their life cycle. The given instruments ensure proper control and monitoring of connected devices. These technologies enhance the stability and security of the IoT

ecosystem and some of the operations include remote configurations, security updates as well as device adoptions (Tariq, 2025).

4. Security Solutions: To ensure Internet of Things (IoT) devices, data, and communication, from unauthorized access or breach and cyber threats, security solutions entail several measures. AIDS such as intrusion detection systems, secure authentication techniques, and encryption procedures are examples of security tools. They are required to secure one's personal information and ensure the authenticity of the Internet of Things.

In conclusion, these fundamental instruments work together to build intelligent and networked IoT ecosystems. These technologies facilitate the creation of autonomous and intelligent Internet of Things solutions for a variety of industries by offering the required infrastructure, security precautions, and data processing and transmission capabilities.

INDUSTRY-SPECIFIC FRAMEWORKS

Internet of Things (IoT) technology is a promising phenomenon that has come to play an essential role in changes in the concept and the introduction of autonomy in various fields in an epoch characterized by the constant appearance of new advanced technologies. This paper aims to analyze dynamic interactions between frameworks and architectures that are supposed to provide support for IoT integration-based intelligent capabilities of devices. As society navigates through the uncertain journey of a world that is becoming increasingly complex it is crucial to fully appreciate the significance of these frameworks. The superposition of IoT technology and autonomy pushes us towards innovative thinking and redesigns the way the device's work, its decision-making, and interaction with the surrounding reality in various fields, including manufacturing, smart cities, healthcare, and agriculture. This study also helps to define the situation now and the opportunity to imagine the future when autonomy backed by robust foundations will become an important component of our smart and connected world (Vermesan, Bröring, et al., 2022).

1. Healthcare: FHIR (Fast Healthcare Interoperability Resources)

Today, HIR means a set of messages that is used for exchange between healthcare information systems. It fosters the development of healthcare applications and standardizes the sharing of the Electronic Health Record (EHR). FHIR helps to transfer patients' information easily across different healthcare systems in a convenient way, which helps care coordination and provides a guarantee that decisions are made based on, are well informed decisions. In that capacity, it promotes patient autonomy since the data must be protected to allow users to freely exchange their health information.

2. Agriculture: FarmBeats

FarmBeats is the IoT platform of agriculture that was developed by Microsoft. It also provides farmers with more practical and valuable information by utilizing artificial intelligence, machine learning as well as Internet of Things. Crop health, weather, and soil information is captured in the field for FarmBeats through sensors and cameras. Through this data processing in the cloud, the framework offers information to the farmer regarding plant management, pests, and watering at a real-time basis. By enhancing the making of informed decisions this in turn increases the autonomy of agriculture.

3. Smart Cities: CityIQ

CityIQ is an Internet of Things platform designed for smart city applications developed by Current, powered by GE. Originally it deals in real-time data analysis and intelligent infrastructure. Collecting information about air quality, parking space, traffic, and other topics, CityIQ is an integration of sensors in metropolitan regions. It is applied onto the data to enhance traffic conditions, increase safety and increase general efficiency within the municipal area. Thus, autonomy results from enabling cities to respond to new conditions in advance with the help of real-time data.

4. Manufacturing: Predix

General Electric created Predix, an industrial IoT platform intended for the manufacturing industry. The framework makes use of analytics to schedule maintenance activities, to predict equipment failure and to improve manufacturing operations. This can be attained through minimizing time when a machine or equipment is not producing through predictive maintenance of the same and the use of real-time data to reduce and make processes more efficient.

5. Retail: Intel Retail Sensor Platform

Intel has developed one sort of IoT solution that is aimed strictly at the retail market being Intel Retail Sensor Platform. There is significant emphasis on increasing customers' satisfaction and effectively managing the stock. The specified technology employs sensors to monitor the actual status of products in the stores in real-time. It assists the retailers to place products in a better way and to handle their stocks and inventory in the best possible manner. It is how the level of autonomy is achieved that, for example, will allow retailers to make decisions about when to restock goods and when to rotate the look of the shop.

6. Logistics and Supply Chain: IBM Watson Supply Chain

Since its introduction, IBM Watson Supply Chain has been equipped with Blockchain, AI, and IoT, finding ways to further improve and ensure the integrity of the supply chain. The framework is designed to provide real-time visibility through the supply chain from production all the way to delivery. Through sensors the location and condition of products and vehicles can be tracked, thus making it possible to engage in proactive problem-solving. This enhances the self-organizing capability of the supply chain by allowing a faster response to interferences as well as the optimization of the supply chain processes.

7. Education: Aruba Meridian

Aruba Meridian is a system of IoT that might be applied in educational facilities for indoor positioning. At college, the framework utilizes mobile applications and Bluetooth beacons for the provision of LBS. It can increase self-organization of the campus movement, asset or resource utilization and provide location-aware notices.

The following frameworks lay down how IoT applications can bring freedom and automation in several sectors: Channeling on specialized features and data analysis, they solve sectoral challenges and optimize work processes and decisions at the same time (Chen, 2018).

IOT TECHNOLOGIES DRIVING AUTONOMY

The introduction locates the discussion of IoT's role in autonomy against the backdrop of a Digital reformation. The new call for self-organizing systems in many fields pushes us to ask how Internet of Things (IoT) technologies – networks of smart sensors and actuators, of communication and computing – are taking us to a future where machinery and apparatus have unprecedented degrees of self-organization. Even though this study aims to determine the radical impact of IoT in various sectors such as smart cities, transportation, health and agriculture this study considers IoT in a broader perspective (Leiding, 2019). The targets of the research gradually evolve into a set of objectives that are designed to enlighten the readers on the relationship between autonomy and IoT. It is with these considerations in mind that this introduction aims to set up the basis for a more comprehensive examination as well as offering an understanding of what constitutes the state of the art, the challenges that are ahead, and some direction to the research domain of IoT-driven autonomy.

Key IoT Technologies Enabling Autonomy

This is made possible by a cluster of technologies that enable systems and objects to operate with ever-increasing levels of independence and self-awareness in the presence of the Internet of Things (IoT). This section presents a comprehensive review of the basic IoT technologies that support the autonomy of the system, and it discovers the Internet of Things communication protocols, Edge and cloud computing, and the Internet of Things sensors (Balsamo, 2017).

1. Sensors

Turning to IoT-based autonomy, it is essential to rely on the vast array of sensors that gather specific kinds of data from the outer environment. Such sensors make it possible for devices to feel their environment and respond thereto; hence, they act as the feelers of the IoT systems. In the health care field, wearable sensors are utilized for tracking the data of blood pressure, pulse, and other features of the human body constantly (Jyothi et al., 2023). Engines are monitored as well as changes on the road through vehicle-mounted sensors while ADAS in the transport industry also benefits from the sensors. Agriculture sensors make farmers More efficient in the ways they need to attend to their crops to ensure the right temperature, concentration of the fertilizer used and the appropriate moisture content of the soil. Sensors are used prevalently throughout smart cities to gauge energy consumption, traffic flow and air quality indices. All this data is used to make decisions as well as facilitate the autonomous management of cities.

2. Communication Protocols

Good communication protocols that surround the connectivity and network that allow for the flow of information between devices are important for an IoT system. Sensors, actuators and other components that constitute the Internet of Things are connected by weaves, like communication protocols. Communication represents a critical variable for decision-implementation-effort coherence, as well as for timely and synchronized action in the environment promoting autonomy.

MQTT is Message Queuing Telemetry Transport, CoAP is Constrained Application Protocol and HTTP is Hypertext Transfer Protocol as the IoT communication protocols. These protocols can make devices capable of transmitting data to cloud platforms, to centralized servers or other networked devices. For instance, in smart agriculture, some sensors present in the ground may transmit over MQTT the information about the moisture present in the ground, and then a server will process it and order an irrigation system present in the field to the water. Functions such as traffic management and prevention of collision in the L3 self-driving cars have provisions to ensure communication synchronization between the cars and the pavement.

3. Edge and Cloud Computing

The IoT technologies enable and manage the vast quantity of data produced through connected devices by consuming cloud and edge computing. Edge computing is the handling of data in another tier, most of the time as near to the source as a device or a local gateway. This method reduces the pressures from centralized Cloud servers and reduces the latency level. This is because the above applications for Cloud Computing place more emphasis on the use of distributed servers. While on the other side cloud computing provides the scalability and storage solutions needed for analyzing big datasets, and the ability to sort and give access to information from a distance.

A common characteristic of edge computing is that quick decisions at the local edge must often be made. For example, edge computing enables self-driving cars to decide various decisions within moments by processing data from the car's sensors locally and without the help of the cloud. Similarly, edge computing helps in processing sensitive data locally minimizing the transfer of huge amounts of data over the network, which enhances privacy and security (Balsamo, 2017).

CASE STUDY

In a time of rapid technological advancement, the Internet of Things (IoT) and artificial intelligence (AI) must work together urgently to address difficult problems. This case study delves into the field of self-governing pandemic management, focusing on the iResponse framework—a novel structure designed to enable and carry out intelligent actions in the event of global health emergencies

The fundamental need for streamlined and autonomous worldwide administration gives rise to the iResponse system. The virus that caused COVID-19, SARS-CoV-2, has severely impacted global supportability, necessitating creative solutions to lessen its effects. The iResponse system is a unique technological configuration that combines AI and IoT to promote effective and widespread responses (Kulkarni, 2019).

This case study aims to:

i. Investigate the IoT-enabled intelligence aspects within the iResponse framework.
ii. Examine the role of each iResponse module in autonomous pandemic management.
iii. Evaluate real-world case studies to understand the framework's impact on sustainability-related strategies.

As seen in Figure 1 below, the iResponse Framework is made up of five interconnected modules, each of which performs a distinct purpose.

Figure 1. One afflicted person can spread the infection rapidly.

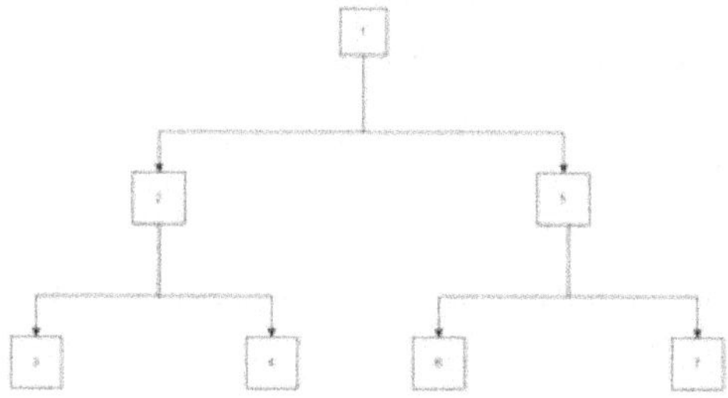

The Figure 2 shows how the transmission is broken by reducing the number of newly infected cases.

Figure 2. Chain of transmission

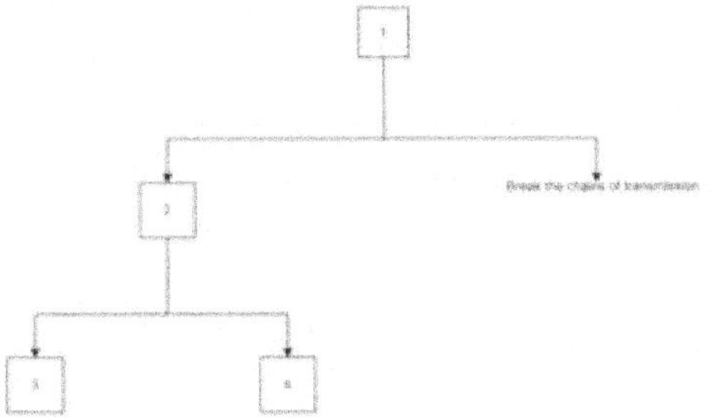

The Cure Development and Treatment (CDT) Module leverages IoT data in the development of pandemic cures, including drugs, vaccines, and natural remedies. Resource Planner (RP) Module uses Internet of Things data to inform resource delivery and planning includes human resources, medical supplies, testing equipment, and quarantine facilities. Data Analytics and Decision Making (DADM) Module utilizes AI and data analytics for informed decision-making in pandemic management. Data Storage and Management (DSM) Module manages big data, ensuring its storage, preparation, and provisioning within the iResponse framework. Monitoring and Break-the-Chain (MBC) Module deploys IoT sensors to monitor environmental conditions, detect potential pandemics, and initiate necessary actions. Actions include contact tracing, physical distancing enforcement, and other preventive measures. In Cure Development and Treatment (CDT) Module IoT data is instrumental in the development of cures, encompassing drugs, vaccines, and natural remedies. The CDT module facilitates a collaborative approach to finding effective treatments for pandemics. In Resource Planner (RP) Module IoT data informs resource planning, covering testing equipment, medical supplies, quarantine centers, and human resources. The RP module ensures efficient allocation and provisioning of resources based on real-time data.

Data Analytics and Decision Making (DADM) Module harnesses AI and data analytics for informed decision-making in pandemic management. Real-time data analysis contributes to agile and effective responses to emerging situations. Data Storage and Management (DSM) Module manages big data generated within the iResponse framework, whether located remotely (cloud), locally (edge), or in a hybrid setting (fog). Efficient data storage and retrieval are critical for the seamless functioning of the entire system.

There are difficulties in navigating the field of IoT-enabled intelligence in pandemic management. These include the requirement for established protocols, interoperability, and data security. The case study addresses these issues to present a thorough understanding of the potential roadblocks to implementing the iResponse architecture (Varadam, Shankar, Bharadwaj, et al., 2024). Using the iResponse paradigm, the case study concludes by highlighting the critical role that IoT-enabled intelligence plays in autonomous pandemic management. The combination of IoT and AI opens new possibilities for pandemic preparedness in the future by enabling flexible resource allocation, agile decision-making, and sustainability-focused tactics.

SECURITY AND ETHICAL CONSIDERATIONS

Security and Ethical Considerations in Enabling Autonomous and Intelligent Device Functionality through IoT Integration:

The Internet of Things (IoT) has changed inter-device relations and the ability of devices to perform operations on their own. It also comes along with several enormous threats such as data privacy issues, vulnerabilities, and moral dilemmas. Addressing these issues is important to utilize the full capability of IoT devices morally. The following opportunities are enabled by the integration of IoT: Health care, manufacturing, transportation and Smart cities. This makes gadgets to be able to work on their own, acquire data and even analyze such information, as well as make wise decisions at any given time. Some crucial security and ethical issues should be considered and resolved as more and more IoT application is integrated.

Security Considerations

IoT has completely changed the ways devices or things communicate and work in different contexts or settings and at the same time, has opened enormous possibilities in terms of optimization and new possibilities. However, as all IoT technologies involve questions of integrity, confidentiality, and resilience of IoT ecosystems, the integration of the IoT technologies discussed poses a difficult many-variate security problem. As IoT devices are proliferating, they are increasingly appealing to hackers who seek to use vulnerability and crack network security. In the deployment of IoT, there are different and diverse security risks which therefore require a systematic approach to handling security risks and threats. When it comes to Connected IoT devices for their autonomous and intelligent functions Privacy of Data, Identification of Devices, Network Security and firmware upgrades are the key security concerns to be considered.

Data Privacy

Since IoT devices often collect, transfer and store information about users, contexts, and processes, data protection is an integral element of IoT security. This data may contain personal identifiers, location information, health information and indeed any other data. It is evident that to prevent unauthorized access, misuse, and confidentiality breaches the privacy of this data needs to be protected.

It is recommended that IoT systems conform to privacy by design, which implies considering privacy at the time of designing and developing IoT systems. This incorporates placing higher and more effective methods of data encryption whenever the data is transferred and whenever it is stored, adequately minimizing the data that can be collected in the first place and allowing users to control their data through settings and consents.

Device Authentication

To ensure the IoT devices' authenticity, and to ensure that only the approved entities have the right to communicate with the devices, the authentication of the IoT devices is essential. IoT ecosystems contain insufficient or even inexistence of authentication protections making them vulnerable to hackers, device impersonation, and unauthorized admittance.

To establish a trust relationship between devices and backend systems, device authentication mechanisms are employed properly, which may include certificate-based authentication and mutual authentication. This requires ascribing to each device an authorized secured identification number and a crypt key to identify itself during communications.

Network Security

Data integrity and confidentiality demand the protection of all the networking pathways for IoT devices, gateways and cloud platforms. The IoT networks are exposed to various security threats such as replay attacks, packet sniffing and man-in-the-middle attacks that may potentially compromise the data integrity and confidentiality.

To mitigate these risks organizations should employ secure communication channels that employ Datagram Transport Layer Security (DTLS) or Transport Layer Security (TLS) and encrypt data payloads. Restricting IoT networks and the implementing of network access policies also help in avoiding unauthorized access and help to minimize the possible security breach's impact.

Ethical Considerations

Concerning the ethics of IoT, there are fundamental questions regarding privacy and autonomy as well as privacy's social implications in addition to the security issue. The following ethical issues need to be addressed when intelligent and autonomous technologies proliferate:

Among the important functions that ethics serve is that of guiding the development, deployment, and utilization of IoT systems. People use connected devices in their daily lives and that increases the development of IoT gadgets. However, the technology faces various ethical questions, which is why it is crucial to consider and discuss to which extent it should respect individuals' and societies' rights and values to provide benefit to the society. Let us examine the ethical issues raised in the study and discuss their importance in more detail:

Data Ownership and Consent

The populations may be assured that their privacy rights will always be protected in IoT systems and that they always retain control over their data by applying the concept of informed consent and ownership of data. It encourages accountability and openness among participants, strengthening the bond of trust between customers and technology suppliers.

i. Data Ownership: When discussing the Internet of Things, the term "data ownership" refers to the rights and obligations pertaining to the data that connected devices gather. Whether consumers, manufacturers, or service providers hold the data created by IoT devices, it is imperative to set clear norms in this regard.
ii. Informed Consent: Consumers ought to be free to choose how Internet of Things devices gather, use, and distribute their personal data. Transparent communication regarding data practices, any hazards, and the intended uses of data is necessary for obtaining informed permission. Users must have the authority to decide for themselves whether to consent to data collection and sharing.

Algorithmic Bias and Prejudice

For the likes of equity and justice to prevail in enduring deployments of the IoT, intervention on algorithmic bias and prejudice must embrace serious consideration. In a way, the developers may enhance the inclusion and protection of vulnerable groups by making sure intelligent IoT devices can make neutral and fair decisions by eradicating bias.

Bias in Algorithms: Issues of biases in algorithms that mean the intelligent Internet of Things devices give biased results due to algorithms learned from the training data. Bias as a concept can be racial, gender or even economic docket. They may result in prejudice or even exacerbate inequity within disabled people.

Mitigation methods: Therefore, algorithmic bias must be avoided, and justice achieved in the decision-making process, where these methods seem to be indispensable for developers. For this, it may require techniques for data collection such as when bias arises, algorithm review, bias detection algorithms and the constant assessment of the presence of bias in the system.

Accountability and Transparency

Being accountable and displaying transparency in IoT environments contributes to credibility and confidence. It provides technology providers with carrots to encourage ethical behavior, and it arms the consumers with the tools to protect themselves and make rational decisions; it provides outlets for people to seek justice when ethical behavior is compromised, or trust is violated.

i. Transparency: Being transparent means being open and honest about how Internet of Things systems internally operate, including data practices, algorithmic decision-making procedures, and possible hazards related to device functionality. Users need to have access to pertinent information so they can decide wisely and comprehend the effects of utilizing IoT technology.
ii. Accountability: Accountability entails making decision-making and action about IoT deployments accountable to stakeholders. Businesses have a responsibility to uphold moral principles, fix security flaws, and lessen the suffering caused by their goods and services (Varadam, Shankar, Shankar, et al., 2024).

Protecting security and maintaining ethical compliance is paramount in many industries in the current rapidly growing technological environment. The stakeholders can be involved in healthcare organizations, leaders at the financial level, education institutions or governing bodies, to name but a few and all of these find it hard to protect data from being leaked, respect the rights of privacy and be moral in this world that has gone global. New technologies, new laws and shifting societal values have raised the issues of security and ethics to the top of organizational agendas and affected strategies, guidelines and developments. This essay focuses on analyzing what is currently happening in the various industries in terms of security and ethics emphasizing on the issues, the new ideas, and the recommended processes. Our goal is to provide the analysis of new trends in digital governance, responsible innovation and societal impact under the circumstances of technological disruption and ethical concerns by investigating the sphere of security needs and ethical dilemmas in synergy. When organizations operate in the digital environment, security and ethical issues must be resolved to ensure Accountability, Create Trust, and enhance welfare for Communities.

i. Healthcare: The continued adoption of digital health technology in the delivery of care like wearable technology, telemedicine and electronic health records (EHRs) has changed the approach to clinical decision-making and patient care. But then there is the matter of ethics and security that comes with such development. One of the emerging trends that people have taken note of is the use of IoT in healthcare and connected medical devices, which throws up concerns regarding security, patient information as well as their privacy rights. High levels of security need to be implemented as the priority of healthcare to avoid cases of leakage of data and cyber threats. The other ethical challenges in health care include patient's self-determination and autonomy, their consent to treatment, diagnosing, the use of AI, and the use of decision-support technologies.
ii. Finance and Banking: The openness of users' personal data and the security of their monetary deals, security and ethical issues are vital in banking and finance. Ensuring the availability, confidentiality and integrity of financial systems and services is important in meeting the regulatory requirements as well as the customer expectations especially because of the emerging Fintech innovations, mobile payment solutions and online banking. Cryptocurrencies and blockchain industries are rapidly expanding and new threats related to digital assets, smart contracts and decentralized structures appear. Compared to non-financial risks, translating financial fairness, openness, and accountability into policy has been challenging mainly while dealing with cybersecurity threats like ransomware, phishing, and other internal threats.
iii. Retail and E-commerce: Security and moral concerns are significant in retail and e-commerce businesses since they enable buyers to trust them, defend individual data, and fight scams. As the population of outside actors in the retail industry expands so too do the threats of cybercrime threatening retailers' customers, payment cards, and supply chains. Encryption techniques, applications to detect fraud, and proper authentication measures must be adopted and implemented to cover the data of consumers and to ensure secure and protected online purchase transactions. Some of the ethical issues that are very crucial to retailers include the issue of fair and reasonable charges, issues of data integrity and the privacy of the consumer. Retailers must respect the rights and preferences of the cus-

tomers while at the same time adhering to ethical practices in data gathering, customer profiling, and personalized marketing.

iv. Education: Admission of inequity in the distribution of educational material, promoting acceptance of computer illiteracy, and safeguarding student information are all hinged on security and ethical problems in the education sector. As open learning environments, network-based collaboration and remote learning and teaching technology are highly employed, education institutional are vulnerable to cybersecurity threats such as virus attacks, data leakage and abuse, and cyberbullying and cyberstalking. In the context of education, the principals and educators have a singular emphasis on the protection of academic data, the privacy of learners and ensuring the overall safety of the online learning environment. Ethical issues in education involve eight concepts which are digital citizenship, student data privacy, and instructional use of technology.

v. Government and Public Sector: Security and moral issues remain essential, especially in the public sector and government to defend the electoral procedures of a nation, protect citizen's rights, and ensure transparency and accountability in the governmental procedures. Data protection for citizens, cybersecurity resilience, and privacy compliance are top concerns for public agencies and legislators as government services become more digitally integrated. To protect sensitive data, prevent cyberattacks, and maintain public confidence, government agencies need to put strong cybersecurity frameworks, incident response procedures, and risk management techniques into place. In the public sector, ethical considerations include things like citizen involvement, data responsibility, and government openness. Using technology to improve service delivery, encourage civic engagement, and address social issues, public authorities must uphold the moral principles of impartiality, integrity, and respect for civil freedoms (Friess, et al., 2022).

CONCLUSION

As IoT technology becomes increasingly prevalent, there are important questions regarding the security and ethics of enabling of autonomous and intelligent device functionality. To exploit as much as possible the benefits of the Internet of Things technology that has enormous values and at the same time minimize or avoid the threats involved and promote responsible use, those involved in the use of the technology have to prioritize data protection, implement stringent measures in security, and uphold ethical standards of use. Security, privacy and moral issues in the IoT environment can only be realized if the industry, government, academics and NGOs come up with frameworks, standards and best practices for the Internet of Things. In conclusion, it is possible to state that while technical solutions and software development are crucial for achieving the goal of connecting autonomous and intelligent devices through the Internet of Things, the security and ethical considerations intertwined with sociological ones are equally important if treated as a single complex. It is indeed possible to build on the positive impacts that the IoT promotes while ensuring the respect of rights and values of people and communities as the world becomes interconnected in more ways than one by focusing on the culture of innovation and collaboration.

REFERENCES

Agyemang, J. O., Yu, D., & Kponyo, J. J. (2021, September). Autonomic IoT: towards smart system components with cognitive IoT. In *Pan-African Artificial Intelligence and Smart Systems Conference* (pp. 248-265). Cham: Springer International Publishing.

Al Ridhawi, I., Aloqaily, M., & Karray, F. (2022). Intelligent blockchain-enabled communication and services: Solutions for moving internet of things devices. *IEEE Robotics & Automation Magazine, 29*(2), 10–20.

Alam, F., Almaghthawi, A., Katib, I., Albeshri, A., & Mehmood, R. (2021). iResponse: An AI and IoT-Enabled Framework for Autonomous COVID-19 Pandemic Management. *Sustainability (Basel), 13*(7), 3797. https://www.mdpi.com/2071-1050/13/7/3797. DOI: 10.3390/su13073797

Anjos, J. C. S. D., Matteussi, K. J., Orlandi, F. C., Barbosa, J. L. V., Silva, J. S., Bittencourt, L. F., & Geyer, C. F. R. (2023). A Survey on Collaborative Learning for Intelligent Autonomous Systems. *ACM Computing Surveys, 56*(4), 98. Advance online publication. DOI: 10.1145/3625544

Arisdakessian, S., Wahab, O. A., Mourad, A., Otrok, H., & Kara, N. (2020). FoGMatch: An intelligent multi-criteria IoT-fog scheduling approach using game theory. *IEEE/ACM Transactions on Networking, 28*(4), 1779–1789. DOI: 10.1109/TNET.2020.2994015

Atefeh Hemmati, A. M. R. (2022). *The Internet of Autonomous Things applications: A taxonomy, technologies, and future directions, Internet of Things* (Vol. 20) https://doi.org/https://doi.org/10.1016/j.iot.2022.100635

Bagdadee, A. H., Hoque, M. Z., & Zhang, L. (2020). IoT based wireless sensor network for power quality control in smart grid. *Procedia Computer Science, 167*, 1148–1160. DOI: 10.1016/j.procs.2020.03.417

Balsamo, D., Merrett, G. V., Zaghari, B., Wei, Y., Ramchurn, S., Stein, S., . . . Beeby, S. (2017, September). Wearable and autonomous computing for future smart cities: Open challenges. In 2017 25th International Conference on Software, Telecommunications and Computer Networks (SoftCOM) (pp. 1-5). IEEE.

Bu, L., Zhang, Y., Liu, H., Yuan, X., Guo, J., & Han, S. (2021). An IIoT-driven and AI-enabled framework for smart manufacturing system based on three-terminal collaborative platform. *Advanced Engineering Informatics, 50*, 101370. DOI: 10.1016/j.aei.2021.101370

Butt, F. A., Chattha, J. N., Ahmad, J., Zia, M. U., Rizwan, M., & Naqvi, I. H. (2022). On the integration of enabling wireless technologies and sensor fusion for next-generation connected and autonomous vehicles. *IEEE Access : Practical Innovations, Open Solutions*, *10*, 14643–14668. DOI: 10.1109/ACCESS.2022.3145972

Chen, B. W., Wan, J., Celesti, A., Li, D., Abbas, H., & Zhang, Q. (2018). Edge Computing in IoT-Based Manufacturing. *IEEE Communications Magazine*, *56*(9), 103–109. DOI: 10.1109/MCOM.2018.1701231

Chen, M., Li, P., Wang, R., Xiang, Y., Huang, Z., Yu, Q., He, M., Liu, J., Wang, J., Su, M., Zhang, M., Jian, A., Ouyang, J., Zhang, C., Li, J., Dong, M., Zeng, S., Wu, J., Hong, P., & Tao, G. (2022). Multifunctional Fiber-Enabled Intelligent Health Agents. *Advanced Materials*, *34*(52), 2200985. https://doi.org/https://doi.org/10.1002/adma.202200985. DOI: 10.1002/adma.202200985 PMID: 35820163

Dai, M., Su, Z., Li, R., & Yu, S. (2021). A software-defined-networking-enabled approach for edge-cloud computing in the internet of things. *IEEE Network*, *35*(5), 66–73. DOI: 10.1109/MNET.101.2100052

Dhar Dwivedi, A., Singh, R., Kaushik, K., Rao Mukkamala, R., & Alnumay, W. S. (2024). Blockchain and artificial intelligence for 5G-enabled Internet of Things: Challenges, opportunities, and solutions. *Transactions on Emerging Telecommunications Technologies*, *35*(4), e4329. DOI: 10.1002/ett.4329

Dushku, E., Rabbani, M. M., Conti, M., Mancini, L. V., & Ranise, S. (2020). SARA: Secure asynchronous remote attestation for IoT systems. *IEEE Transactions on Information Forensics and Security*, *15*, 3123–3136. DOI: 10.1109/TIFS.2020.2983282

Esenogho, E., Djouani, K., & Kurien, A. M. (2022). Integrating artificial intelligence Internet of Things and 5G for next-generation smartgrid: A survey of trends challenges and prospect. *IEEE Access : Practical Innovations, Open Solutions*, *10*, 4794–4831.

Feng, Z., Wei, Z., Chen, X., Yang, H., Zhang, Q., & Zhang, P. (2021). Joint communication, sensing, and computation enabled 6G intelligent machine system. *IEEE Network*, *35*(6), 34–42. DOI: 10.1109/MNET.121.2100320

Friess, P. (2016). *Digitising the industry-internet of things connecting the physical, digital and virtual worlds*. River Publishers.

Girma, A., Bahadori, N., Sarkar, M., Tadewos, T. G., Behnia, M. R., Mahmoud, M. N., Karimoddini, A., & Homaifar, A. (2020). IoT-enabled autonomous system collaboration for disaster-area management. *IEEE/CAA Journal of Automatica Sinica*, *7*(5), 1249-1262.

González, I., Calderón, A. J., & Portalo, J. M. (2021). Innovative multi-layered architecture for heterogeneous automation and monitoring systems: Application case of a photovoltaic smart microgrid. *Sustainability (Basel)*, *13*(4), 2234. DOI: 10.3390/su13042234

Guo, F., Yu, F. R., Zhang, H., Li, X., Ji, H., & Leung, V. C. (2021). Enabling massive IoT toward 6G: A comprehensive survey. *IEEE Internet of Things Journal*, *8*(15), 11891–11915. DOI: 10.1109/JIOT.2021.3063686

Hewa, T., Ylianttila, M., & Liyanage, M. (2021). Survey on blockchain based smart contracts: Applications, opportunities and challenges. *Journal of Network and Computer Applications*, *177*, 102857. DOI: 10.1016/j.jnca.2020.102857

Javaid, M., & Khan, I. H. (2021). Internet of Things (IoT) enabled healthcare helps to take the challenges of COVID-19 Pandemic. *Journal of Oral Biology and Craniofacial Research*, *11*(2), 209–214. https://pmc.ncbi.nlm.nih.gov/articles/PMC7897999/pdf/main.pdf. DOI: 10.1016/j.jobcr.2021.01.015 PMID: 33665069

Jyothi, A. P., Shankar, A., Narayan, J. R. A., Gaur, P., & Kumar, S. S. (2023, December). Smart Non-Invasive Real-Time Health Monitoring using Machine Learning and IoT. In 2023 IEEE International Symposium on Smart Electronic Systems (iSES) (pp. 437-440). IEEE.

Ke, R., Zhuang, Y., Pu, Z., & Wang, Y. (2020). A smart, efficient, and reliable parking surveillance system with edge artificial intelligence on IoT devices. *IEEE Transactions on Intelligent Transportation Systems*, *22*(8), 4962–4974.

Khan, M. A. (2021). Intelligent environment enabling autonomous driving. *IEEE Access : Practical Innovations, Open Solutions*, *9*, 32997–33017. DOI: 10.1109/ACCESS.2021.3059652

Khan, W. Z., Rehman, M., Zangoti, H. M., Afzal, M. K., Armi, N., & Salah, K. (2020). Industrial internet of things: Recent advances, enabling technologies and open challenges. *Computers & Electrical Engineering*, *81*, 106522. DOI: 10.1016/j.compeleceng.2019.106522

Kulkarni, S. G., Gu, Q., Myers, E., Polepeddi, L., Liptak, S., Beyah, R., & Divan, D. (2019). Enabling a Decentralized Smart Grid Using Autonomous Edge Control Devices. *IEEE Internet of Things Journal*, *6*(5), 7406–7419. DOI: 10.1109/JIOT.2019.2898837

Lee, C. P., Leng, F. T. J., Habeeb, R. A. A., & Amanullah, M. A., & ur Rehman, M. H. (2022). Edge computing-enabled secure and energy-efficient smart parking: A review. *Microprocessors and Microsystems*, *93*, 104612.

Leiding, B. (2019). The M2X Economy–Concepts for Business Interactions, Transactions and Collaborations Among Autonomous Smart Devices (Doctoral dissertation, Dissertation, Göttingen, Georg-August Universität, 2019).

Manoharan, H., Basha, A. R., Teekaraman, Y., & Manoharan, A. (2020). Smart home autonomous water management system for agricultural applications. *World Journal of Engineering*, *17*(3), 445–455.

Menon, V. G., Jacob, S., Joseph, S., Sehdev, P., Khosravi, M. R., & Al-Turjman, F. (2022). An IoT-enabled intelligent automobile system for smart cities. *Internet of Things : Engineering Cyber Physical Human Systems*, *18*, 100213. DOI: 10.1016/j.iot.2020.100213

Mishra, D., Zema, N. R., & Natalizio, E. (2021). A high-end IoT devices framework to foster beyond-connectivity capabilities in 5G/B5G architecture. *IEEE Communications Magazine*, *59*(1), 55–61. DOI: 10.1109/MCOM.001.2000504

Mishra, D. Z., Zema, N. R., & Natalizio, E. (2021). A High-End IoT Devices Framework to Foster Beyond-Connectivity Capabilities in 5G/B5G Architecture. *IEEE Communications Magazine*, *59*(1), 55–61. DOI: 10.1109/MCOM.001.2000504

Pajooh, H. H., Demidenko, S., Aslam, S., & Harris, M. (2022). Blockchain and 6G-enabled IoT. *Inventions (Basel, Switzerland)*, *7*(4), 109. DOI: 10.3390/inventions7040109

Qi, Q., Tao, F., Hu, T., Anwer, N., Liu, A., Wei, Y., Wang, L., & Nee, A. Y. (2021). Enabling technologies and tools for digital twin. *Journal of Manufacturing Systems*, *58*, 3–21. DOI: 10.1016/j.jmsy.2019.10.001

Quy, V. K., Hau, N. V., Anh, D. V., Quy, N. M., Ban, N. T., Lanza, S., Randazzo, G., & Muzirafuti, A. (2022). IoT-Enabled Smart Agriculture: Architecture, Applications, and Challenges. *Applied Sciences (Basel, Switzerland)*, *12*(7), 3396. https://www.mdpi.com/2076-3417/12/7/3396. DOI: 10.3390/app12073396

Ramu, S. P., Boopalan, P., Pham, Q. V., Maddikunta, P. K. R., Huynh-The, T., Alazab, M., & Gadekallu, T. R. (2022). Federated learning enabled digital twins for smart cities: Concepts, recent advances, and future directions. *Sustainable Cities and Society*, *79*, 103663.

Shahraki, A., Abbasi, M., Piran, M. J., & Taherkordi, A. (2021). A comprehensive survey on 6G networks: Applications, core services, enabling technologies, and future challenges. *arXiv preprint arXiv:2101.12475.*

Tan, L., Yu, K., Lin, L., Cheng, X., Srivastava, G., Lin, J. C.-W., & Wei, W. (2021). Speech emotion recognition enhanced traffic efficiency solution for autonomous vehicles in a 5G-enabled space–air–ground integrated intelligent transportation system. *IEEE Transactions on Intelligent Transportation Systems*, *23*(3), 2830–2842. DOI: 10.1109/TITS.2021.3119921

Tariq, M. U. (2025). *AI-Driven Learning and Engagement in Higher Education*. IGI Global., DOI: 10.4018/979-8-3693-4074-5

Tariq, M. U. (2025). Revitalizing Pedagogy: Integrating Indigenous Knowledge and Practices Into Higher Education Disciplines. In Kgari-Masondo, M. (Ed.), *Indigenous Teaching Disciplines and Perspectives for Higher Education* (pp. 379–400). IGI Global Scientific Publishing., DOI: 10.4018/979-8-3693-9296-6.ch018

Tariq, M. U. (2025). Revolutionizing Communication: EEG-Based Brain-Computer Interface for Speech and Mood Detection. In Prabhakar, A., Yu, P., & Erokhin, V. (Eds.), *Rural Social Entrepreneurship Development: Network-Based Manufacturing System Model* (pp. 237–264). IGI Global Scientific Publishing., DOI: 10.4018/979-8-3693-7515-0.ch008

Tariq, M. U. (2025). AI-Powered Business Process Automation: Revolutionizing Firm Management. In Derbali, A. (Ed.), *AI, Economic Perspectives, and Firm Business Management* (pp. 199–224). IGI Global Scientific Publishing., DOI: 10.4018/979-8-3693-7357-6.ch010

Tariq, M. U., & Sergio, R. (Eds.). (2025). *Convergence of AI, Education, and Business for Sustainability*. IGI Global., DOI: 10.4018/979-8-3373-1917-9

Tavakoli, M., Carriere, J., & Torabi, A. (2020). Robotics, smart wearable technologies, and autonomous intelligent systems for healthcare during the COVID-19 pandemic: An analysis of the state of the art and future vision. *Advanced Intelligent Systems*, *2*(7), 2000071. DOI: 10.1002/aisy.202000071

Varadam, D., Shankar, S. P., Bharadwaj, A., Saxena, T., Agrawal, S., & Dayananda, S. (2024). Enhancing Industrial Robotics Performance and Security With AI and Blockchain Technologies. In R. C. Biradar, G. D, N. Tabassum, N. Hegde, & M. Lazarescu (Eds.), *AI and Blockchain Applications in Industrial Robotics* (pp. 58-81). IGI Global. https://doi.org/DOI: 10.4018/979-8-3693-0659-8.ch003

Varadam, D., Shankar, S. P., Shankar, A., & Narayan, A. N. P, N., & Kumar, T. N. (2024). Application of AI and Blockchain Technologies in the Medical Domain. In R. C. Biradar, G. D, N. Tabassum, N. Hegde, & M. Lazarescu (Eds.), *AI and Blockchain Applications in Industrial Robotics* (pp. 200-225). IGI Global. https://doi.org/DOI: 10.4018/979-8-3693-0659-8.ch008

Venkataswamy, R., Paul, R., & Jogy, J. (2020). Internet of Things Based Autonomous Borewell Management System. In Cyber-physical Systems and Digital Twins: Proceedings of the 16th International Conference on Remote Engineering and Virtual Instrumentation 16 (pp. 229-243). Springer International Publishing.

Vermesan, O., Bahr, R., Ottella, M., Serrano, M., Karlsen, T., Wahlstrøm, T., & Gamba, M. T. (2020). Internet of robotic things intelligent connectivity and platforms. *Frontiers in Robotics and AI*, *7*, 104.

Vermesan, O., Bröring, A., Tragos, E., Serrano, M., Bacciu, D., Chessa, S., & Bahr, R. (2022). Internet of robotic things–converging sensing/actuating, hyperconnectivity, artificial intelligence and IoT platforms. In *Cognitive hyperconnected digital transformation* (pp. 97–155). River Publishers.

Vermesan, O., Bröring, A., Tragos, E., Serrano, M., Bacciu, D., Chessa, S., Gallicchio, C., Micheli, A., Dragone, M., & Saffiotti, A. (2022). Internet of robotic things–converging sensing/actuating, hyperconnectivity, artificial intelligence and IoT platforms. In *Cognitive hyperconnected digital transformation* (pp. 97–155). River Publishers. DOI: 10.1201/9781003337584-4

Vermesan, O., Friess, P., Guillemin, P., Gusmeroli, S., Sundmaeker, H., Bassi, A., Jubert, I. S., Mazura, M., Harrison, M., & Eisenhauer, M. (2022). Internet of things strategic research roadmap. In *Internet of things-global technological and societal trends from smart environments and spaces to green ICT* (pp. 9–52). River Publishers.

Waheed, A., Shah, M. A., Mohsin, S. M., Khan, A., Maple, C., Aslam, S., & Shamshirband, S. (2022). A comprehensive review of computing paradigms, enabling computation offloading and task execution in vehicular networks. *IEEE Access : Practical Innovations, Open Solutions*, *10*, 3580–3600.

Yang, T., & Han, J. (2021). Integrated management strategy with feasible smartness over heterogeneous IoT environments. *Electronics (Basel)*, *10*(2), 149. DOI: 10.3390/electronics10020149

Zaidan, A., & Zaidan, B. (2020). A review on intelligent process for smart home applications based on IoT: Coherent taxonomy, motivation, open challenges, and recommendations. *Artificial Intelligence Review*, *53*(1), 141–165. DOI: 10.1007/s10462-018-9648-9

Zaman, U. (1893). Imran, Mehmood, F., Iqbal, N., Kim, J., & Ibrahim, M. (2022). Towards secure and intelligent internet of health things: A survey of enabling technologies and applications. *Electronics (Basel)*, *11*(12).

KEY TERMS AND DEFINITIONS

Internet of Things (IoT): A network of interconnected devices that can communicate and exchange data with each other, enabling them to perform intelligent and autonomous operations.

Machine Learning (ML): A subset of artificial intelligence (AI) that enables devices or systems to learn from data and improve their performance over time without being explicitly programmed.

Artificial Intelligence (AI): The simulation of human intelligence in machines that are programmed to think, learn, and make decisions based on data, often applied in conjunction with IoT to enhance device autonomy.

Sensor Networks: A system of interconnected sensors that gather real-time data about the environment, which can then be processed and used by devices to make decisions or respond to changing conditions.

Edge Computing: A computing paradigm where data processing occurs closer to the source of data generation (at the "edge" of the network), reducing latency, improving response times, and decreasing the load on central servers.

Smart Devices: Devices embedded with sensors, processors, and connectivity that allow them to collect data, make intelligent decisions, and communicate with other devices in an IoT ecosystem.

Chapter 3
Role of AI in Revolutionizing Clinical Care

Oindrilla Ghosh
https://orcid.org/0000-0001-9609-0781
Christ University, India

Binod Kumar
https://orcid.org/0009-0002-9721-1367
Radha Govind University, India

Parimal Kumar Roy
https://orcid.org/0000-0002-0461-2587
Bangladesh Public Administration Training Centre, Bangladesh

ABSTRACT

Artificial Intelligence (AI) is significantly reshaping clinical care, driving advancements in diagnostics, treatment, and patient management. With its ability to analyze vast amounts of data rapidly, AI is transforming how healthcare professionals approach patient care, thereby improving outcomes and operational efficiency. One of the primary roles of AI in clinical care is enhancing diagnostic precision. Furthermore, AI-driven chatbots and virtual health assistants are revolutionizing patient engagement and support. These tools provide instant responses to health inquiries, manage appointment scheduling, and offer reminders for medication, which fosters adherence and ensures that patients remain informed about their health journey. Lastly, AI's role in streamlining administrative processes cannot be overstated. Automating routine tasks such as billing and data entry allows healthcare professionals to focus more on patient care rather than administrative burdens.

DOI: 10.4018/979-8-3373-4332-7.ch003

INTRODUCTION

Artificial Intelligence (AI) can be defined as the simulation of human intelligence processes by machines, particularly computer systems (Chiruvella & Guddati, 2021). It encompasses various sub-fields, including machine learning (ML), natural language processing (NLP), robotics, and computer vision (McKeown et al., 2021). The primary goal of AI is to enable machines to perform tasks that typically require human intelligence, such as understanding language, recognizing patterns, solving problems, and making decisions (Ukato et al., 2024). AI systems can either operate based on predefined rules (symbolic AI) or learn from data (sub-symbolic AI). Deep learning, a subset of machine learning (Oguanobi & Joel, 2024) that involves neural networks with many layers, has achieved notable success in tasks involving unstructured data, such as images and text. AI's application within the healthcare sector is extensive (Dash et al., 2019) and multifaceted, offering innovations that can potentially revolutionize patient care, streamline operations, and enhance research and development (Nakamoto, 2008). The definition of AI encompasses a broad range of technologies and capabilities aimed at simulating human intelligence (Kuo et al., 2017). Within the healthcare context, the scope of AI is vast and continuously evolving, promising to enhance clinical decision-making, improve diagnostic accuracy, streamline operations, and foster innovative treatments (Zhang et al., 2018). While the potential benefits are significant, addressing ethical and practical challenges is crucial for the successful and responsible integration of AI in healthcare (Sikdar & Guha, 2020). As technology advances, the collaboration between healthcare professionals and AI systems will likely lead to more effective and personalized patient care, ultimately transforming the healthcare landscape for the better (Rehman et al., 2022).

HISTORICAL PERSPECTIVE ON AI'S EVOLUTION IN CLINICAL CARE

The intersection of artificial intelligence (AI) and medicine represents one of the most transformative evolutions in healthcare over the past few decades (Zhang et al., 2022). AI's journey in medicine has been characterized by an interplay of technological advancements, research breakthroughs, and shifts in healthcare practices (Olatoye et al., 2024). This brief historical perspective delves into the significant milestones, challenges, and futuristic potential of AI in the medical field, offering insight into its evolution from its inception to the contemporary landscape (Awonuga et al., 2024).

Early Beginnings in AI

Artificial intelligence, as a formal concept, originated in the mid-20th century (Hassan et al., 2022). The term "artificial intelligence" was coined in 1956 at the Dartmouth Conference, where researchers gathered to explore ways machines could simulate human intelligence (Odeyemi et al., 2024). While the focus of early AI was often on solving problems through logic and reasoning, these foundational ideas paved the way for its application in various fields, including medicine (Arinze et al., 2024). One of the earliest examples of AI in medicine dates to the 1970s with the development of the MYCIN expert system at Stanford University (Oladipo et al., 2024). MYCIN was designed to diagnose bacterial infections and recommend antibiotics. It used a rule-based system to evaluate the symptoms and lab results, showcasing the potential of AI to aid clinical decision-making (Oriekhoe et al., 2024). Although MYCIN was never deployed in clinical settings due to regulatory and ethical concerns, it highlighted the practicality of computational models in improving diagnostic accuracy (Siddiq, 2022).

The Rise of Advanced Machine Learning

The 1980s and 1990s saw an evolution in AI techniques, notably the advent of machine learning a sub-field of AI that focused on creating algorithms that learn from data (Ofodile et al., 2024). With the gradual digitization of healthcare information and the introduction of electronic health records (EHRs), the potential for machine learning algorithms to analyze vast datasets became evident (Cozzoli et al., 2022). During this period, AI applications extended to medical imaging, with early systems designed to assist radiologists in interpreting X-rays and CT scans (Alotaibi et al., 2020). Researchers began experimenting with neural networks, forming the basis of deep learning. However, due to limited computational power, data access, and resources, the progress was slow, and the adoption of AI in clinical practice remained limited (Meriton et al., 2021).

The 2000s: Data and Computational Power

The new millennium marked a turning point in AI applications in medicine (Weerasinghe et al., 2022). The exponential increase in computational power and the widespread availability of large datasets provided fertile ground for more sophisticated AI algorithms (Chao et al., 2023). The field of genomics emerged as a prime example of how AI could enhance medical research and practice (Chatzopoulou et al., 2020). Researchers began using machine learning techniques to analyze genomic data, leading to discoveries in personalized medicine and targeted therapies (Perez-

Pozuelo et al., 2021). In 2004, the introduction of the "Data Mining in Healthcare" report highlighted the potential role of data-driven approaches in improving patient care (Williamson et al., 2024). By leveraging predictive analytics, healthcare providers could identify at-risk patients and improve outcomes through early intervention.

AI in Diagnostics and Imaging

As medical imaging technologies advanced, so did the application of AI algorithms in interpreting the resulting data (Padmanaban, 2024). Breakthroughs in convolutional neural networks (CNNs) demonstrated significant promise in image recognition tasks (Banerjee et al., 2020). AI systems began achieving or even surpassing human radiologists in detecting abnormalities in images (Nandi et al., 2024). The introduction of platforms like Google's DeepMind in the healthcare sector exemplified the potential to revolutionize diagnostics, becoming adept at identifying conditions like diabetic retinopathy and age-related macular degeneration (Bhatt and Chakraborty, 2023). These advances prompted regulatory bodies such as the U.S. Food and Drug Administration (FDA) to develop frameworks for evaluating and approving AI-based diagnostic tools (Gibney, 2019). In 2018, the FDA granted approval for the first AI system capable of detecting certain eye diseases from retinal photographs, marking a critical milestone for integrating AI into clinical practice (Arute et al., 2019).

THE SCOPE OF AI IN HEALTHCARE

AI's evolution in medicine reveals a journey filled with innovation, challenges, and transformative potential (Xu et al., 2018). From early expert systems to advanced machine learning applications and natural language processing capabilities, AI has dramatically reshaped how healthcare is delivered and understood (Schulte and Lee, 2019). Although navigating the complexities and ethical considerations remains crucial, ongoing advancements suggest that scope of AI will continue to play an instrumental role in shaping the future of medicine, ultimately enhancing patient care and healthcare outcomes worldwide (Secundo et al., 2019).

Clinical Decision Support

AI systems can analyze vast amounts of medical data to assist healthcare providers in making informed clinical decisions (Sestino et al., 2020). Utilizing algorithms trained on extensive datasets, AI can help predict patient outcomes, suggest treatment plans, and identify potential complications (Sousa et al., 2019). For instance,

AI models can analyze historical patient data and current clinical guidelines to recommend the most effective therapies for specific conditions, enabling personalized medicine (Yuan et al., 2020).

Diagnostic Tools

AI-powered diagnostic tools, especially in medical imaging, have shown remarkable accuracy in identifying diseases such as cancer, pneumonia, and retinal disorders (Coccia, 2020). AI algorithms can process and analyze images from CT scans, MRIs, and X-rays more rapidly than traditional methods (Coccia et al., 2018).. For example, Google's DeepMind demonstrated significant success in diagnosing eye diseases, achieving accuracy comparable to that of leading ophthalmologists (Cobianchi et al., 2020a).

Robotic Surgery

The integration of AI in robotic surgical systems has transformed surgical procedures, enhancing precision and minimizing invasiveness (Cerchione, Centobelli, Riccio, Abbate, & Oropallo, 2023). Robot-assisted surgeries, such as those performed using the da Vinci Surgical System, enable surgeons to execute highly intricate operations with enhanced dexterity and control (Colombo et al., 2020). AI algorithms can assist by analyzing real-time data during surgeries, providing surgeons with insights that can enhance decision-making and improve surgical outcomes (Mullins et al., 2020).

Virtual Health Assistants

AI-driven virtual health assistants, often implemented as chat-bots or mobile applications, offer patients immediate access to medical information and support (Tariq, 2024). These tools can answer health-related inquiries, provide medication reminders, and even conduct preliminary assessments based on symptoms described by the user (Fahy, Mauer, & Panteli, 2021). By offering 24/7 support, virtual health assistants alleviate the burden on healthcare providers and empower patients to take charge of their health (Melchiorre et al., 2020).

Telemedicine and Remote Monitoring

The rise of telemedicine, greatly accelerated by the COVID-19 pandemic, has seen the introduction of AI technologies to enhance patient monitoring and care (Tayefi et al., 2021). AI algorithms can analyze data from wearable devices, such as heart rate monitors and glucose level trackers, allowing for real-time health assessments (Duggineni, 2023). This enables healthcare providers to detect anomalies early and engage with patients proactively, often before more serious conditions arise.

Health Data Management

AI can enhance the management and utilization of healthcare data (Tan et al., 2020). The ever-growing volume of medical records, lab results, and research findings can overwhelm traditional data management systems (Nchinda et al., 2019). AI-driven data management solutions can help automate tasks such as data entry, cleaning, and analysis, enabling healthcare organizations to derive actionable insights quickly (Biancone et al., 2021). This not only improves efficiency but also enhances data accuracy and accessibility.

THE SHIFT TOWARDS PERSONALIZED MEDICINE ENHANCED BY AI ALGORITHMS ANALYZING INDIVIDUAL PATIENT DATA

In recent years, the medical field has experienced a remarkable transformation, largely driven by advancements in technology and a deeper understanding of the human genome (Malina & Woerner, 2019). The shift towards personalized medicine an approach that tailors medical treatment to the individual characteristics of each patient has gained momentum (Timpson, 2019). Central to this revolution is the integration of artificial intelligence (AI) algorithms that analyze vast amounts of patient data (Nchinda et al., 2019). By utilizing these technologies, healthcare providers can offer more precise and effective treatments, ultimately improving patient outcomes while potentially reducing costs (Urbinati et al., 2019).

Understanding Personalized Medicine

Personalized medicine, also known as precision medicine, is an innovative healthcare concept that takes into account an individual's genetic makeup, environment, and lifestyle when devising treatment plans (Wimelius et al., 2021). Unlike the traditional "one-size-fits-all" approach, which often relies on generalized treatment protocols, personalized medicine seeks to customize interventions for each patient (Drago et al., 2021). This approach is especially compelling in fields such as oncology, where genetic variations can significantly influence how patients respond to certain therapies (Biancone et al., 2021). For instance, the development of targeted therapies has improved the prognosis for patients with cancers that have specific genetic mutations (Abduljalil and Abduljalil, 2020). Drugs like trastuzumab (Herceptin) are designed to target the HER2 protein, effectively treating breast cancer in patients with HER2-positive tumors (Aftab et al., 2020). This level of specificity not only enhances therapeutic effectiveness but also minimizes adverse effects that can arise from more generalized treatments (Hashemzadeh et al., 2020).

THE ROLE OF AI IN PERSONALIZING CARE

While the concept of personalized medicine is not new, its execution has been revolutionized by the application of artificial intelligence (Loeppky, 1999). AI algorithms can analyze massive datasets that encompass genomic information, electronic health records (EHRs), clinical trial data, and even social determinants of health (Sneha and Priya Doss, 2016). This capability allows clinicians and researchers to draw insights that were previously unachievable, paving the way for more nuanced decision-making in patient care (Szefler, 2018).

Data Integration and Analysis

One of the primary challenges in personalized medicine is the sheer volume and complexity of data generated in healthcare (Fixson et al., 2017). AI algorithms can sift through this data at incredible speeds, identifying patterns and correlations that might escape human analysis (Mikalef & Pateli, 2017). Machine learning models can consider a multitude of variables such as genetic information, biomarkers, lifestyle choices, and socio-economic factors to generate predictions about disease progression and treatment response (Yang & Hsiao, 2009).

Predictive Analytics

AI enhances personalized medicine through predictive analytics, which can forecast health outcomes based on an individual's health data (Fairbank et al., 2006). For example, machine learning algorithms can analyze historical patient outcomes to identify which patients are at higher risk for certain diseases or complications (Cobianchi et al., 2020b). Programs that harness AI can thus preemptively alert healthcare providers to potential health issues, enabling earlier interventions (Secundo et al., 2019).

Identification of Biomarkers

Biomarker discovery is essential for the advancement of personalized medicine (Cobianchi et al., 2021). AI can streamline this process by evaluating data from various experiments and clinical studies to identify novel biomarkers that indicate how a patient might respond to treatment (McCracken, 1988). This not only facilitates the development of targeted therapies but also allows for earlier diagnosis and tailored screening strategies (Solenov et al., 2018).

Enhancing Clinical Trials

The integration of AI in the design and management of clinical trials offers immense potential for personalized medicine (Zinner et al., 2021). By utilizing algorithms to identify suitable candidates based on genetic and phenotypic data, researchers can enroll patients more likely to benefit from a treatment (Yuan et al., 2020). AI can also monitor the trial's progress and outcomes, allowing for real-time adjustments based on interim results, thereby accelerating the drug development process (Zheng et al., 2019).

Improving Decision Support

AI-powered clinical decision-support tools enable healthcare providers to make better-informed decisions based on individual patient data (Abduljalil & Abduljalil, 2020). By presenting relevant information tailored to the patient's condition, preferences, and history, these tools align treatment plans with personalized medicine principles (Pappalardo et al., 2019). This real-time support enhances the clinician's ability to choose the most effective treatment options (Flother et al., 2020).

AI-POWERED CHATBOTS AND VIRTUAL HEALTH ASSISTANTS REVOLUTIONIZE PATIENT INTERACTION AND ADHERENCE TO TREATMENT

In recent years, the integration of Artificial Intelligence (AI) in healthcare has ushered in a new era of patient interaction and treatment adherence (Schulte and Lee, 2019). Among the myriad applications of AI in healthcare, chatbots and virtual health assistants have emerged as game-changers (Secundo et al., 2019). These intelligent systems not only enhance patient-provider communication but also facilitate adherence to treatment protocols, leading to improved health outcomes (Sestino et al., 2020).

Understanding AI-Powered Chatbots and Virtual Health Assistants

AI-powered chatbots are software applications equipped with natural language processing capabilities that allow them to understand and respond to human queries in real time (Sousa et al., 2019). Virtual health assistants go a step further, often incorporating more advanced functionalities, including personalized health recommendations, reminders, and insights derived from patient data (Yuan et al., 2020). They emulate human interactions, making it easier for patients to seek information, schedule appointments, and manage their health effectively.

Accessibility and Availability

One of the most profound impacts of AI-powered chatbots is their ability to provide immediate support to patients (Yang and Hsiao, 2009). Unlike traditional healthcare settings, which may limit access to information during business hours, chatbots are available 24/7 (Wang et al., 2018a). This constant availability enables patients to ask questions, receive guidance, and obtain information about their conditions whenever they need it (Arthur, 2017). This seamless access reduces the time patients spend in uncertainty and empowers them to take charge of their health.

Reducing Barriers to Communication

Often, patients may feel intimidated or uncomfortable when engaging with healthcare professionals (Coccia, 2020). Chatbots create a more relaxed environment for patients to inquire about sensitive topics, which is particularly beneficial for discussions surrounding mental health, sexual health, and chronic illnesses (Coccia et al., 2018). By offering an anonymous platform, patients are more likely to seek help and receive the information they need without fear of judgment (Montes & Goertzel, 2019).

Personalized Engagement

AI chatbots can analyze patient data and interactions to provide tailored responses and recommendations (Cobianchi et al., 2020a). This personalization fosters a connection between patients and their care, ensuring that individuals feel understood and valued. For example, chatbots can track patients' concerns over time and adjust their responses (Ienca & Vayena, 2020) based on historical data, heightening the relevance of the information provided.

Medication Reminders and Education

One of the key challenges in healthcare is medication non-adherence (Rippa & Secundo, 2019). Reports indicate that approximately 50% of patients do not take their medications as prescribed. AI chatbots can play a pivotal role in combatting this issue by sending reminders at critical times (Blencowe, 2010). These reminders can include not only simple notifications but also educational content, helping patients understand the importance of their medications and the potential consequences of non-adherence (Nakamura et al., 1999).

Interactive Treatment Plans

Virtual health assistants can help create interactive treatment plans for patients, allowing them to track their progress in real time (Basile et al., 2022). For instance, if a patient is managing diabetes, a chatbot can facilitate daily check-ins, enabling patients to report their blood sugar levels, dietary habits, and exercise routines (Spanò et al., 2021). This close monitoring fosters accountability, as patients are more likely to adhere to their plans when continuously engaged.

Behavioral Nudges

AI chatbots employ behavioral psychology principles to provide nudges that promote positive health behaviors (Cerchione et al., 2023). For instance, if a patient misses a medication dose, a chatbot can use supportive language to encourage them to stay on track (Colombo et al., 2020). Messaging can include motivational quotes, success stories, or even empathetic responses that resonate with the patient's struggles, thus enhancing their adherence to treatment (Mullins et al., 2020; Tariq, 2025).

Continuous Engagement

AI-powered chatbots can maintain an ongoing dialogue with patients, building a sense of rapport over time (Tariq, 2024). Regular interactions can help establish trust, making it more likely that patients will seek advice and adhere to treatment plans (Fahy, Mauer, & Panteli, 2021). This continuous engagement is particularly important for patients with chronic conditions, as these individuals often require long-term support and monitoring (Melchiorre et al., 2020).

Feedback and Adaptation

Another advantage of AI chatbots is their ability to learn and adapt based on patient feedback (Tayefi et al., 2021). Through machine learning algorithms, these systems can refine their responses and better address common concerns and inquiries (Gamal et al., 2021). This adaptability not only enhances the quality of patient interaction but also ensures that the information being communicated is the most relevant and accurate, further solidifying trust between patients and their virtual assistants (Jambol et al., 2024).

Human-Like Interaction

Advancements in natural language processing enable chatbots to engage in conversations that feel increasingly human-like (Ochulor et al., 2024). This humanization of technology can make patients feel heard and understood, mirroring the empathetic presence of healthcare professionals. The better the interaction feels, the more likely patients are to return for ongoing support, ultimately improving adherence to their treatment regimens (Duggineni, 2023; Tariq and Sergio, 2025).

THE USE OF AI IN MONITORING CHRONIC DISEASES AND ALERTING HEALTHCARE PROVIDERS TO POTENTIAL HEALTH DETERIORATIONS

The proliferation of chronic diseases, such as diabetes, heart disease, asthma, and hypertension, places a substantial burden on healthcare systems worldwide (Uzougbo et al., 2024). According to the World Health Organization, chronic diseases account for 60% of all deaths and 43% of the global disease burden (Yang & Gu, 2021). This alarming trend has prompted researchers, healthcare professionals, and innovators to seek advanced methodologies for managing these conditions effectively (Joel & Oguanobi, 2024a). One promising avenue is the application of Artificial Intelligence (AI) in the realm of chronic disease management, particularly in monitoring patients and alerting healthcare providers to potential health deteriorations (Towett et al., 2023).

Understanding Chronic Disease Management

Chronic diseases often require continuous monitoring to manage symptoms and prevent complications (Balch et al., 2023). Traditional methods of monitoring, such as periodic check-ups and outdated manual tracking systems, may not suffice, especially for patients living with multiple conditions or those who are geographically isolated (Benson & Grieve, 2021). The advent of AI technologies has revolutionized how healthcare providers approach chronic disease management, offering timely insights and intervention strategies to improve patient outcomes (de Villiers, 2021).

Wearable Technology and IoT Devices

AI-enabled wearables, such as smartwatches, glucose monitors, heart rate sensors, and blood pressure cuffs, have become integral tools in continuously monitoring vital health metrics (Jacks, Ajala, Lottu, & Okafor, 2024). These devices collect real-time data, which can then be analyzed using AI algorithms for patterns that may indicate potential health deteriorations.

Data Integration and Analysis

AI systems can aggregate data from multiple sources, including electronic health records (EHRs), lab results, and patient self-reports (Colombo et al., 2020). This holistic approach provides a comprehensive view of a patient's health, allowing the AI to identify trends or anomalies that may require attention (Shrivastava, Song, Han, & Dietzman, 2021). For instance, sudden weight gain in a patient with heart disease could trigger alerts to healthcare providers for an evaluation.

Real-Time Alerts

AI systems equipped with real-time analytics can send immediate alerts to healthcare providers when they detect deviations from a patient's baseline health status (Joel & Oguanobi, 2024d). For instance, if a diabetic patient's blood sugar levels fluctuate significantly beyond their typical range, an alert can be dispatched to the healthcare provider, enabling timely intervention (Chiruvella & Guddati, 2021).

Automated Communication

AI-driven communication platforms can automate regular check-ins with patients, reminding them about medication adherence and lifestyle modifications (Joel & Oguanobi, 2024c). Such proactive engagement not only reinforces the importance of self-management but can also provide critical data points for healthcare providers, helping them adjust treatment plans as necessary (McKeown et al., 2021).

Telehealth Integration

With the growing acceptance of telehealth, AI can further enhance remote monitoring capabilities (Jacks, Ajala, Lottu, & Okafor, 2024). Remote consultations facilitated by AI can provide healthcare professionals with insights derived from collected data, making virtual appointments more productive (Tamiminia et al., 2020). This integration can streamline the process of assessing chronic disease states, allowing healthcare providers to make informed decisions without requiring in-person visits (McGhin et al., 2019).

Patient Apps

Mobile applications infused with AI capabilities can empower patients to take an active role in their health management (Aceto et al., 2020). These apps can not only track symptoms and medication schedules but can also provide reminders and educational resources (Abouelmehdi et al., 2018). By fostering a sense of autonomy, patients may feel more motivated to stay compliant with their treatment regimens, leading to better health outcomes (Firouzi et al., 2018).

ANALYSIS OF AI-DRIVEN TELEHEALTH SOLUTIONS AND THEIR IMPACT DURING THE COVID-19 PANDEMIC

The COVID-19 pandemic has significantly reshaped healthcare delivery worldwide, compelling rapid adaptations and innovations to ensure patient care continuity while minimizing the risk of virus transmission (McGhin et al., 2019). Among the transformative developments, telehealth solutions have emerged as pivotal in maintaining healthcare access (Manogaran et al., 2018). Notably, artificial intelligence (AI)-driven technologies have accelerated the effectiveness of these telehealth solutions (Dash et al., 2019).

Rapid Expansion of AI-Driven Telehealth During COVID-19

The onset of the COVID-19 pandemic acted as a catalyst, propelling the adoption and innovation of telehealth solutions (Aceto et al., 2020). Healthcare systems quickly pivoted towards remote care, leading to a surge in telemedicine visits (Hariri et al., 2019). In the United States, for example, telehealth visits increased from approximately 10,000 per week in 2019 to over 1 million in April 2020. AI-driven solutions became instrumental in handling this unprecedented influx (Saggi & Jain, 2018). Chatbots implemented via online platforms screened individuals for COVID-19 symptoms, offering guidance on testing and treatment (Choi et al., 2018). Moreover, AI systems facilitated patient triage, allowing healthcare providers to prioritize virtual consultations based on urgency, which was vital in managing limited resources amidst rising cases (Wang et al., 2018; Tariq, and Sergio, 2025).

REGULATORY AND GOVERNANCE FRAMEWORKS

Given the complexities surrounding AI in healthcare, regulatory frameworks must be established to ensure responsible deployment (Sestino et al., 2020). Collaborative engagement between policymakers, healthcare professionals, technologists, and ethicists are essential:

Transparency and Explainability: AI systems should provide clear explanations for their recommendations (Elhoseny et al., 2018). This empowers healthcare providers and patients to understand the rationale behind AI-driven decisions, fostering trust and facilitating informed decision-making (Hariri et al., 2019).
Accountability: Clear lines of accountability should be established regarding AI usage in clinical environments (Mehta & Pandit, 2018). Whether it is the healthcare provider, the AI developer, or the institution deploying the technology, clear responsibilities must be defined to address potential errors or adverse outcomes stemming from AI decisions (Ivanov et al., 2019).
AI Model Validation and Clinical Trials: AI-driven healthcare solutions, like diagnostic algorithms and robotic-assisted surgeries, require rigorous validation through clinical trials (Zhang et al., 2018). Regulatory agencies such as the FDA (USA), EMA (Europe), and MHRA (UK) mandate that AI models undergo similar testing as pharmaceutical drugs before approval (Ukato, Sofoluwe, Jambol, & Ochulor, 2024).
Compliance with Medical Device Regulations: Many AI-based solutions, such as AI-powered radiology tools or robotic surgeries, are classified as medical devices (Sikdar and Guha, 2020). Thus, they must comply with medical device regulations like MDR (EU Medical Device Regulation) and FDA's Software as a Medical Device (SaMD) framework to ensure safety and effectiveness (Rehman et al., 2022).
Integration with Existing Healthcare Systems: AI governance must ensure seamless integration with electronic health records (EHRs) and telemedicine while maintaining interoperability standards like FHIR (Fast Healthcare Interoperability Resources) (Zhang et al., 2022). This prevents fragmentation and ensures AI enhances rather than disrupts healthcare workflows.

Global AI Governance and Cross-Border Collaboration: Harmonization of AI healthcare regulations across countries is essential (Hassan et al., 2022) for cross-border research, telemedicine, and AI-powered drug trials. Global bodies like the WHO, OECD, and UN AI for Good Initiative push for uniform AI governance frameworks, ensuring ethical and safe international AI deployment in healthcare (Arinze et al., 2024).
Continuous Monitoring and Feedback: AI systems should be continuously monitored post-deployment to assess their performance and impact on patient care (Jabbour et al., 2018). Mechanisms for feedback must be integrated, allowing healthcare providers to report issues and for developers to iterate on algorithms based on real-world outcomes (Radanliev et al., 2019).

CONCLUSION

The integration of Artificial Intelligence (AI) into healthcare represents one of the most transformative advancements in recent history (Xu et al., 2018). As we have explored throughout this discourse, AI's potential to revolutionize clinical care is profound, affecting diverse areas from diagnostics to treatment planning, patient management, and personalized medicine (Azadegan et al., 2020). The evolution in this sector is not merely a trend but a paradigm shift, fundamentally altering how healthcare providers deliver services and how patients experience care (Gong and Ribiere, 2020). This conclusion aims to encapsulate the highlights of AI's role in clinical care, emphasizing its challenges, ethical considerations, and the future trajectory (Tan et al., 2020).

At the heart of AI's impact is its ability to process and analyze vast amounts of data far beyond human capability (Chiruvella & Guddati, 2021). Machine learning algorithms can sift through millions of medical records, identifying patterns and correlations that could lead to better diagnostics and treatment options (McKeown et al., 2021). This analytical power enhances clinical decision-making, enabling healthcare professionals to deliver more accurate and evidence-based care (Ukato, Sofoluwe, Jambol, & Ochulor, 2024). For instance, AI applications in radiology have shown a remarkable ability to detect anomalies in imaging studies, often with greater accuracy than seasoned radiologists (Oguanobi & Joel, 2024). This not only improves patient outcomes but also alleviates the burden on healthcare systems plagued by staff shortages and increasing patient loads (Dash et al., 2019).

Moreover, AI fosters personalized medicine by tailoring treatment plans to the individual characteristics of each patient (Nakamoto, 2008). By analyzing factors such as genetics, lifestyle, and environment, AI can help clinicians devise treatments that are more effective and have fewer side effects (Kuo et al., 2017). The development of AI algorithms that predict the likelihood of disease progression allows physicians to intervene earlier, potentially changing the trajectory of chronic conditions like diabetes or heart disease (Zhang et al., 2018). This shift from a one-size-fits-all approach to more personalized care represents a significant advancement in how clinicians interact with patients and manage their health (Sikdar & Guha, 2020).

Furthermore, AI plays a crucial role in enhancing operational efficiency in clinical settings (Rehman et al., 2022). Administrative burdens, from scheduling appointments to managing billing, consume valuable time that healthcare professionals could otherwise spend with patients (Zhang et al., 2022). AI-driven solutions, such as natural language processing, can automate these tasks, allowing clinicians to focus on providing care rather than bureaucratic processes (Olatoye et al., 2024). The reduction of administrative workload also has the potential to increase job satisfaction among healthcare workers, ultimately benefiting patient care by creating a more engaged workforce (Awonuga et al., 2024).

However, the integration of AI into clinical practice is not without its challenges (Hassan et al., 2022). One of the primary concerns centers around the quality and representativeness of the data used to train AI models (Odeyemi et al., 2024). Algorithms developed on biased datasets can yield skewed results, potentially leading to disparate healthcare outcomes among various demographic groups (Arinze et al., 2024). It is crucial that AI systems undergo rigorous testing and validation across diverse populations to ensure equitable access to advanced care (Oladipo et al., 2024).

Ethical considerations surrounding patient privacy and data security cannot be overlooked (Cozzoli et al., 2022). The use of AI often necessitates access to sensitive patient information, raising concerns about confidentiality and consent (Alotaibi et al., 2020). Healthcare organizations must navigate the delicate balance between harnessing AI capabilities and safeguarding patient rights. Transparency in how data is used and robust cybersecurity measures are crucial (Meriton et al., 2021) to maintaining patient trust and ensuring compliance with legal and ethical standards.

Moreover, the implementation of AI in clinical practice raises questions about the future role of healthcare professionals (Martínez-Vergara and Valls-Pasola, 2020). While AI can assist in diagnosis and decision-making, it does not replace the human element crucial to patient care (Craighead et al., 2007). Compassion, empathy, and ethical judgement are intrinsic qualities that AI cannot replicate (Kull and Closs, 2008). Healthcare professionals must be trained not only to utilize AI tools but also to complement these technologies with the interpersonal skills that define quality care (Mehring, 2007). As AI continues to evolve, it is essential for the healthcare workforce to embrace lifelong learning and adapt to changing roles where technology becomes an integral part of clinical practice (Cuijpers et al., 2011).

Ultimately, the revolutionization of clinical care through AI is not just about technology; it is about enhancing the quality of care and improving health outcomes for all individuals (Alrahbi et al., 2021). As artificial intelligence becomes more entrenched in healthcare systems, it is imperative to keep patient welfare at the forefront (Hansen et al., 2019). The promise of AI lies in its potential to enable more informed decision-making, enhance the precision of treatments, improve accessibility, and, most importantly, foster a more compassionate and effective healthcare environment (Hewitt-Horsman, 2009).

Thus, AI's role in revolutionizing clinical care is multifaceted, presenting both extraordinary opportunities and significant challenges (Georgescu, 2014). As we stand on the precipice of this transformation, the journey ahead necessitates careful navigation of ethical landscapes, investment in training and education, and a resolute commitment to equitable patient care (Bub, 2010). The collaboration between technology and healthcare professionals must be guided by the core values of compassion, empathy, and a dedication to improving the human experience through better health care (Preskill, 2012). By prioritizing these principles, AI has the potential to truly revolutionize clinical care, paving the way for a healthcare future that is not just more efficient, but also more humane (Basile et al., 2022).

FUTURE RESEARCH DIRECTIONS

Artificial Intelligence (AI) is transforming healthcare by enhancing diagnostics, personalized medicine, and operational efficiency (Massaro, 2021). However, to maximize its potential, further research and strategic implementations are required.

Data Quality and Standardization: AI algorithms rely on high-quality data, but healthcare data is often fragmented and inconsistent (Ozercan et al., 2018). Future efforts should focus on developing standardized data formats, interoperability between systems, and improved data-sharing policies while maintaining patient privacy (Wang et al., 2018b).

Explainable AI (XAI): Many AI models, especially deep learning algorithms, operate as "black boxes," making it difficult for healthcare professionals to understand their decision-making processes (Plachkinova et al., 2018). Research should focus on developing transparent AI models that provide interpretable insights to build trust and improve adoption (Rajabion et al., 2019).

Ethical AI and Bias Mitigation: AI models can inherit biases from training data, leading to disparities in healthcare outcomes (Dai and Tayur, 2020). Future research should explore techniques for bias detection, mitigation strategies, and diverse datasets that ensure fairness across different demographic groups (Martínez-Román et al., 2020).

AI-Driven Drug Discovery: AI can accelerate drug development by identifying potential compounds, predicting drug interactions, and optimizing clinical trials (Martínez-Vergara and Valls-Pasola, 2020). Research should explore AI's role in repurposing existing drugs for new diseases.

Robotics and AI in Surgery: AI-powered robotic systems can enhance precision in surgical procedures (Shaygan and Daim, 2021). Future research should refine robotic-assisted surgeries, ensuring safety, adaptability, and cost-effectiveness.

AI for Mental Health: AI chatbots and virtual therapists are emerging tools for mental health support Ng and Bezak, 2020). Studies should examine their effectiveness, user engagement, and long-term impacts on patient well-being.

AI in Remote and Personalized Care: Telemedicine, wearable health monitors, and AI-driven diagnostics are expanding healthcare access (Ienca and Vayena, 2020). Research should focus on optimizing these technologies for rural and underserved populations.

REFERENCES

Abduljalil, J. M., & Abduljalil, B. M. (2020). *Epidemiology, genome, and clinical features of the pandemic SARS-CoV-2: a recent view*. New Microb. New Infect., DOI: 10.1016/j.nmni.2020.100672

Abouelmehdi, K., Beni-Hessane, A., & Khaloufi, H. (2018). Big healthcare data: Preserving security and privacy. *Journal of Big Data, 5*(1), 1–18. DOI: 10.1186/s40537-017-0110-7

Aceto, G., Persico, V., & Pescapé, A. (2020). Industry 4.0 and health: Internet of things, big data, and cloud computing for Healthcare 4.0. *Journal of Industrial Information Integration, 18*, 100129. DOI: 10.1016/j.jii.2020.100129

Alotaibi, S., Mehmood, R., & Katib, I. (2020). The role of big data and twitter data analytics in healthcare supply chain management. *Smart Infrastructure and Applications: Foundations for Smarter Cities and Societies*, 267-279.

Arifiani, L., & Arifiani, L. (2019). The effect of disruption technology, opportunities and challenges of telecommunication industry 4.0 in Indonesia. *Int. J. Recent Technol. Eng, 7*(6), 808–819.

Arinze, C. A., Ajala, O. A., Okoye, C. C., Ofodile, O. C., & Daraojimba, A. I. (2024). *Evaluating the integration of advanced IT solutions for emission reduction in the oil and gas sector*. Engineering Science and Technology Journal., DOI: 10.51594/estj/v5i3.862

Arowoogun, J. O., Babawarun, O., Chidi, R., Adeniyi, A. O., & Okolo, C. A. (2024). A comprehensive review of data analytics in healthcare management: Leveraging big data for decision-making. *World Journal of Advanced Research and Reviews, 21*(2), 1810–1821. DOI: 10.30574/wjarr.2024.21.2.0590

Arute, F., Arya, K., Babbush, R., Bacon, D., Bardin, J. C., Barends, R., Biswas, R., Boixo, S., Brandao, F. G. S. L., Buell, D. A., Burkett, B., Chen, Y., Chen, Z., Chiaro, B., Collins, R., Courtney, W., Dunsworth, A., Farhi, E., Foxen, B., & Martinis, J. M. (2019). Quantum supremacy using a programmable superconducting processor. *Nature, 574*(7779), 505–510. DOI: 10.1038/s41586-019-1666-5 PMID: 31645734

Awonuga, K. F., Nwankwo, E. E., Oladapo, J. O., Okoye, C. C., Odunaiya, O. G., & Uzondu, C. S. (2024). *Driving sustainable growth in SME manufacturing: The role of digital transformation, project, and capture management. International Journal of Science and Research Archives*. IJSRA., DOI: 10.30574/ijsra.2024.11.1.0270

Balch, J. A., Ruppert, M. M., Loftus, T. J., Guan, Z., Ren, Y., Upchurch, G. R., Ozrazgat-Baslanti, T., Rashidi, P., & Bihorac, A. (2023). Machine learning–enabled clinical information systems using fast healthcare interoperability resources data standards: Scoping review. *JMIR Medical Informatics*, *11*, e48297. DOI: 10.2196/48297 PMID: 37646309

Banerjee, A., Chakraborty, C., Kumar, A., & Biswas, D. (2020). Emerging trends in IoT and big data analytics for biomedical and health care technologies. *Handbook of data science approaches for biomedical engineering*, 121-152.

Benson, T., & Grieve, G. (2021). *Principles of health interoperability*. Springer International. DOI: 10.1007/978-3-030-56883-2

Bhatt, V., & Chakraborty, S. (2023). Improving service engagement in healthcare through internet of things based healthcare systems. *Journal of Science and Technology Policy Management*, *14*(1), 53–73. DOI: 10.1108/JSTPM-03-2021-0040

Biancone, P., Secinaro, S., Marseglia, R., & Calandra, D. (2021). E-health for the future. Managerial perspectives using a multiple case study approach. *Technovation*. Advance online publication. DOI: 10.1016/j.technovation.2021.102406

Blencowe, M. (2010). Quantum RAM. *Nature*, *468*(7320), 44–45. DOI: 10.1038/468044a PMID: 21048756

Bub J. (2010) Quantum computation: where does the speed-up come from? *Phil. Quantum Info. Entang*. 231–246.

Chakraborty, D., & Paul, J. (2022). Healthcare apps' purchase intention: A consumption values perspective. *Technovation*. Advance online publication. DOI: 10.1016/j.technovation.2022.102481

Chao, K., Sarker, M. N. I., Ali, I., Firdaus, R. R., Azman, A., & Shaed, M. M. (2023). Big datadriven public health policy making: Potential for the healthcare industry. *Heliyon*, *9*(9), e19681. DOI: 10.1016/j.heliyon.2023.e19681 PMID: 37809720

Chiruvella, V., & Guddati, A. K. (2021). Ethical issues in patient data ownership. *Interactive Journal of Medical Research*, *10*(2), e22269. DOI: 10.2196/22269 PMID: 34018968

Choi, T. M., Wallace, S. W., & Wang, Y. (2018). Big data analytics in operations management. *Production and Operations Management*, *27*(10), 1868–1883. DOI: 10.1111/poms.12838

Chua, H. N., Teh, J. S., & Herbland, A. (2021). Identifying the effect of data breach publicity on information security awareness using hierarchical regression. *IEEE Access : Practical Innovations, Open Solutions*, *9*, 121759–121770. DOI: 10.1109/ACCESS.2021.3107426

Cobianchi, L., Dal Mas, F., & Angelos, P. (2021). One size does not fit all–translating knowledge to bridge the gaps to diversity and inclusion of surgical teams. *Annals of Surgery*, *273*(2), e34–e36. DOI: 10.1097/SLA.0000000000004604 PMID: 33201134

Coccia, M. (2020) Global Encyclopedia of Public Administration, Public Policy, and Governance. *Springer Nature Switzerland AG*; Disruptive technologies for industrial and corporate change; pp. 1–7.

Colombo, F., Oderkirk, J., & Slawomirski, L. (2020). Health information systems, electronic medical records, and big data in global healthcare: Progress and challenges in oecd countries. *Handbook of Global Health*, 1-31

Cozzoli, N., Salvatore, F. P., Faccilongo, N., & Milone, M. (2022). How can big data analytics be used for healthcare organization management? Literary framework and future research from a systematic review. *BMC Health Services Research*, *22*(1), 809. DOI: 10.1186/s12913-022-08167-z PMID: 35733192

Cuijpers, M., Guenter, H., & Hussinger, K. (2011). Costs and benefits of inter-departmental innovation collaboration. *Research Policy*, *40*(4), 565–575. DOI: 10.1016/j.respol.2010.12.004

Dai, T., & Tayur, S. (2020). Healthcare operations management: A snapshot of emerging research. *Manufacturing & Service Operations Management*, *22*(5), 869–887. DOI: 10.1287/msom.2019.0778

Dash, S., Shakyawar, S. K., Sharma, M., & Kaushik, S. (2019). Big data in healthcare: Management, analysis and prospects. *Journal of Big Data*, *6*(1), 1–25. DOI: 10.1186/s40537-019-0217-0

de Villiers, K. (2021). Bridging the health inequality gap: An examination of South Africa's social innovation in health landscape. *Infectious Diseases of Poverty*, *10*(1), 1–7. DOI: 10.1186/s40249-021-00804-9 PMID: 33648585

Duggineni, S. S. (2023). *Data integrity as a code*. DIAC.

Elhoseny, M., Abdelaziz, A., Salama, A. S., Riad, A. M., Muhammad, K., & Sangaiah, A. K. (2018). A hybrid model of the Internet of Things and cloud computing to manage big data in health services applications. *Future Generation Computer Systems*, *86*, 1383–1394. DOI: 10.1016/j.future.2018.03.005

Fahy, N., Mauer, N., & Panteli, D. (2021). *European support for improving health and care systems.*

Fixson, S. K., Khachatryan, D., & Lee, W. (2017). Technological uncertainty and firm boundaries: The moderating effect of knowledge modularity. *IEEE Transactions on Engineering Management, 64*(1), 16–28. DOI: 10.1109/TEM.2016.2638847

Flother, F., Moose, C., & Tavernelli, I. (2020) IBM Institute for Business Value Institute for Business Value. *Expert Insights Exploring Quantum Computing Use Cases for Life Sciences: Decoding Secrets of Genomes, Drugs, and Proteins*.https://www.ibm.com/downloads/cas/EVBKAZGJ

Gamal, A., Barakat, S., & Rezk, A. (2021). Standardised electronic health record data modeling and persistence: A comparative review. *Journal of Biomedical Informatics, 114*, 103670. DOI: 10.1016/j.jbi.2020.103670 PMID: 33359548

Hansen, J., Wilson, P., Verhoeven, E., Kroneman, M., Kirwan, M., Verheij, R., & van Veen, E.-B. (2021). *Assessment of the EU Member States' rules on health data in the light of GDPR*

Hariri, R. H., Fredericks, E. M., & Bowers, K. M. (2019). Uncertainty in big data analytics: Survey, opportunities, and challenges. *Journal of Big Data, 6*(1), 1–16. DOI: 10.1186/s40537-019-0206-3

Hartzog, W., & Richards, N. (2020). *Privacy's constitutional moment and the limits of data protection.*

Hassan, M., Awan, F. M., Naz, A., deAndrés-Galiana, E. J., Alvarez, O., Cernea, A., Fernández-Brillet, L., Fernández-Martínez, J. L., & Kloczkowski, A. (2022). Innovations in genomics and big data analytics for personalized medicine and health care: A review. *International Journal of Molecular Sciences, 23*(9), 4645. DOI: 10.3390/ijms23094645 PMID: 35563034

Hassan, S., Dhali, M., Zaman, F., & Tanveer, M. (2021). Big data and predictive analytics in healthcare in Bangladesh: Regulatory challenges. *Heliyon, 7*(6), e07179. DOI: 10.1016/j.heliyon.2021.e07179 PMID: 34141936

Hutchings, E., Loomes, M., Butow, P., & Boyle, F. M. (2020). A systematic literature review of health consumer attitudes towards secondary use and sharing of health administrative and clinical trial data: A focus on privacy, trust, and transparency. *Systematic Reviews, 9*(1), 235. DOI: 10.1186/s13643-020-01481-9 PMID: 33036664

Ienca, M., & Vayena, E. (2020). On the responsible use of digital data to tackle the COVID-19 pandemic. *Nature Medicine, 26*(4), 463–464. DOI: 10.1038/s41591-020-0832-5 PMID: 32284619

Ivanov, D., Dolgui, A., & Sokolov, B. (2019). The impact of digital technology and Industry 4.0 on the ripple effect and supply chain risk analytics. *International Journal of Production Research, 57*(3), 829–846. DOI: 10.1080/00207543.2018.1488086

Iyamu, I., Gómez-Ramírez, O., Xu, A. X., Chang, H. J., Watt, S., Mckee, G., & Gilbert, M. (2022). Challenges in the development of digital public health interventions and mapped solutions: Findings from a scoping review. *Digital Health, 8*, 20552076221102255. DOI: 10.1177/20552076221102255 PMID: 35656283

Lopes de Sousa Jabbour, A. B., Jabbour, C. J. C., Godinho Filho, M., & Roubaud, D. (2018). Industry 4.0 and the circular economy: A proposed research agenda and original roadmap for sustainable operations. *Annals of Operations Research, 270*, 273–286.

Jacks, B. S., Ajala, O. A., Lottu, O. A., & Okafor, E. S.Boma Sonimiteim JacksOlakunle Abayomi AjalaOluwaseun Augustine LottuEnyinaya Stefano Okafor. (2024). Theoretical frameworks for ICT for development: Impact assessment of telecommunication infrastructure projects in Africa and the US. *World Journal of Advanced Research and Reviews, 21*(3), 394–400. DOI: 10.30574/wjarr.2024.21.3.0721

Jambol, D. D., Sofoluwe, O. O., Ukato, A., & Ochulor, O. J. (2024). Enhancing oil and gas production through advanced instrumentation and control systems. *GSC Advanced Research and Reviews*, 19(3), 043-056

Joel, O. T., & Oguanobi, V. U.Oloruntosin Tolulope JoelVincent Ugochukwu Oguanobi. (2024a). Entrepreneurial leadership in startups and SMEs: Critical lessons from building and sustaining growth. *International Journal of Management & Entrepreneurship Research, 6*(5), 1441–1456. DOI: 10.51594/ijmer.v6i5.1093

Joel, O. T., & Oguanobi, V. U.Oloruntosin Tolulope JoelVincent Ugochukwu Oguanobi. (2024b). Geotechnical assessments for renewable energy infrastructure: Ensuring stability in wind and solar projects. *Engineering Science & Technology Journal, 5*(5), 1588–1605. DOI: 10.51594/estj.v5i5.1110

Joel, O. T., & Oguanobi, V. U.Oloruntosin Tolulope JoelVincent Ugochukwu Oguanobi. (2024c). Leadership and management in high-growth environments: Effective strategies for the clean energy sector. *International Journal of Management & Entrepreneurship Research, 6*(5), 1423–1440. DOI: 10.51594/ijmer.v6i5.1092

Joel, O. T., & Oguanobi, V. U.Oloruntosin Tolulope JoelVincent Ugochukwu Oguanobi. (2024d). Navigating business transformation and strategic decision-making in multinational energy corporations with geodata. *International Journal of Applied Research in Social Sciences, 6*(5), 801–818. DOI: 10.51594/ijarss.v6i5.1103

Kazançoğlu, Y., Sağnak, M., Lafcı, Ç., Luthra, S., Kumar, A., & Taçoğlu, C. (2021). Big dataenabled solutions framework to overcoming the barriers to circular economy initiatives in healthcare sector. *International Journal of Environmental Research and Public Health, 18*(14), 7513. DOI: 10.3390/ijerph18147513 PMID: 34299964

Kumar, B. K., Srinivas, U. M., Kumar, M. B. B., Rao, N. S., Mishra, M. K., Srinivas, P., & Ramanaiah, M. P. (2023). Health informatics and social determinants utilizing big data to address health disparities. *Journal of Namibian Studies: History Politics Culture, 35*, 2398–2414.

Kuo, T. T., & Liu, J. (2017). Blockchain in healthcare applications: A survey. *Health Care Management Review, 42*(4), 357–366.

Lancieri, F. (2022). *Narrowing data protection's enforcement gap.*

Lapalme, A. N., Corbin, D., Tastet, O., Avram, R., & Hussin, J. G. (2024). Advancing Fairness in Cardiac Care: Strategies for Mitigating Bias in Artificial Intelligence Models within Cardiology. *The Canadian Journal of Cardiology*. PMID: 38735528

Manogaran, G., Varatharajan, R., Lopez, D., Kumar, P. M., Sundarasekar, R., & Thota, C. (2018). A new architecture of the Internet of Things and big data ecosystem for secured smart healthcare monitoring and alerting system. *Future Generation Computer Systems, 82*, 375–387. DOI: 10.1016/j.future.2017.10.045

McCoach, D.B., Dineen, J.N., Chafouleas, S.M., & Briesch, A. (2020). Reproducibility in the era of big data: Lessons for developing robust data management and data analysis procedures. *Big Data Meets Survey Science: A Collection of Innovative Methods*, 625- 655.

McGhin, T., Choo, K. K. R., Liu, C. Z., & He, D. (2019). Blockchain in healthcare applications: Research challenges and opportunities. *Journal of Network and Computer Applications, 135*, 62–75. DOI: 10.1016/j.jnca.2019.02.027

McKeown, A., Mourby, M., Harrison, P., Walker, S., Sheehan, M., & Singh, I. (2021). Ethical issues in consent for the reuse of data in health data platforms. *Science and Engineering Ethics, 27*(1), 1–21. DOI: 10.1007/s11948-021-00282-0 PMID: 33538942

Mehta, N., & Pandit, A. (2018). Concurrence of big data analytics and healthcare: A systematic review. *International Journal of Medical Informatics, 114*, 57–65. DOI: 10.1016/j.ijmedinf.2018.03.013 PMID: 29673604

Melchiorre, M. G., Papa, R., Quattrini, S., Lamura, G., Barbabella, F., & Consortium, I. E. (2020). Integrated care programs for people with multimorbidity in European countries: eHealth adoption in health systems. *BioMed Research International, 2020*(1), 9025326. DOI: 10.1155/2020/9025326 PMID: 32337283

Mensah, B. A. (2021). *Implementing blockchain technology to develop a national electronic data exchange system for medical records* (Doctoral dissertation, Colorado Technical University).

Meriton, R., Bhandal, R., Graham, G., & Brown, A. (2021). An examination of the generative mechanisms of value in big data-enabled supply chain management research. *International Journal of Production Research, 59*(23), 7283–7310. DOI: 10.1080/00207543.2020.1832273

Mora, R., & Maze, M. (2024). The role of cultural competency training to address health disparities in surgical settings. *British Medical Bulletin, 150*(1), ldae005. DOI: 10.1093/bmb/ldae005 PMID: 38465857

Mullins, A., O'Donnell, R., Mousa, M., Rankin, D., Ben-Meir, M., Boyd-Skinner, C., & Skouteris, H. (2020). Health outcomes and healthcare efficiencies associated with the use of electronic health records in hospital emergency departments: A systematic review. *Journal of Medical Systems, 44*(12), 200. DOI: 10.1007/s10916-020-01660-0 PMID: 33078276

Nakamoto, S. (2008). *Bitcoin: A peer-to-peer electronic cash system.* Bitcoin.org.

Nakamura, Y., Pashkin, Y. A., & Tsai, J. S. (1999). Coherent control of macroscopic quantum states in a single-Cooper-pair box. *Nature, 398*(6730), 786–788. DOI: 10.1038/19718

Nandi, S., Bhaduri, S., Das, D., Ghosh, P., Mandal, M., & Mitra, P. (2024). Deciphering the lexicon of protein targets: A review on multifaceted drug discovery in the era of artificial intelligence. *Molecular Pharmaceutics, 21*(4), 1563–1590. DOI: 10.1021/acs.molpharmaceut.3c01161 PMID: 38466810

Nchinda, N., Cameron, A., Retzepi, K., & Lippman, A. (2019) International Conference on Computing, Networking and Communications, ICNC 2019. 2019. *MedRec: a network for personal information distribution*; pp. 637–641.

Ng, K. H., & Bezak, E. (2020). Prejudice in science – lessons from the coronavirus story. *Physica Medica, 75*, 83–84. DOI: 10.1016/j.ejmp.2020.06.011 PMID: 32559649

Ochulor, O. J., Sofoluwe, O. O., Ukato, A., & Jambol, D. D. (2024). Technological innovations and optimised work methods in subsea maintenance and production.

Odeyemi, O., Mhlongo, N. Z., Nwankwo, E. E., Uzondu, C. S., & Okoye, C. C. (2024). *Big data applications in portfolio management: A review of techniques and strategies. International Journal of Science and Research Archives.* IJSRA., DOI: 10.30574/ijsra.2024.11.1.0268

Ofodile, O. C., Oyewole, A. T., Ugochukwu, C. E., Addy, W. A., Adeoye, O. B., & Okoye, C. C. (2024). *Predictive analytics in climate finance: Assessing risks and opportunities for investors.* GSC Advanced Research and Reviews., DOI: 10.30574/gscarr.2024.18.2.0076

Ofori-Asenso, R., Hallgreen, C. E., & De Bruin, M. L. (2020). Improving interactions between health technology assessment bodies and regulatory agencies: A systematic review and cross-sectional survey on processes, progress, outcomes, and challenges. *Frontiers in Medicine, 7*, 582634. DOI: 10.3389/fmed.2020.582634 PMID: 33178721

Oguanobi, V. U., & Joel, O. T.Vincent Ugochukwu OguanobiOloruntosin Tolulope Joel. (2024). Scalable business models for startups in renewable energy: Strategies for using GIS technology to enhance SME scaling. *Engineering Science & Technology Journal, 5*(5), 1571–1587. DOI: 10.51594/estj.v5i5.1109

Oladipo, J. O., Okoye, C. C., Elufioye, O. A., Falaiye, T., & Nwankwo, E. E. (2024). *Human factors in cybersecurity: Navigating the fintech landscape. International Journal of Science and Research Archives.* IJSRA., DOI: 10.30574/ijsra.2024.11.1.0258

Olatoye F. O., Elufioye O. A., Okoye C. C., Nwankwo E. E., & Oladapo O. O. (2024). Leadership styles and their impact on healthcare management effectiveness: A review. *International Journal of Science and Research Archives (IJSRA).* DOI: DOI: 10.30574/ijsra.2024.11.1.0271

Omotunde, H., & Ahmed, M. (2023). A comprehensive review of security measures in database systems: Assessing authentication, access control, and beyond. *Mesopotamian Journal of CyberSecurity, 2023*, 115–133. DOI: 10.58496/MJCSC/2023/016

Onunka, T., Raji, A., Osafiele, A. N., Daraojimba, C., Egbokhaebho, B. A., & Okoye, C. C. (2023). *Banking: A comprehensive review of the evolution and impact of innovative banking services on entrepreneurial growth. Economic Growth and Environment Sustainability.* EGNES., DOI: 10.26480/egnes.02.2023.50.62

Oriekhoe O. I., Addy W. A., Okoye C. C., Oyewole A. T., Ofodile O. C. & Ugochukwu C. E. (2024). The role of accounting in mitigating food supply chain risks and food price volatility. *International Journal of Science and Research Archives (IJSRA).* DOI: DOI: 10.30574/ijsra.2024.11.1.0340

Oshioste, E. E., Okoye, C. C., & Udokwu, S. T. C. (2023). *The effectiveness of CSR in sustainable development: A case-study of Total's oil exploration activities in the Niger-Delta region. Economic Growth and Environment Sustainability*. EGNES., DOI: 10.26480/egnes.02.2023.41.49

Ozercan, H. I., Ileri, A. M., Ayday, E., & Alkan, C. (2018). Realizing the potential of blockchain technologies in genomics. *Genome Research*, *28*(9), 1255–1263. DOI: 10.1101/gr.207464.116 PMID: 30076130

Padmanaban, H. (2024). Navigating the Complexity of Regulations: Harnessing AI/ML for Precise Reporting. *Journal of Artificial Intelligence General science (JAIGS)*, 3(1), 49-61.

Pappalardo, F., Russo, G., Tshinanu, F. M., & Viceconti, M. (2019). In silico clinical trials: Concepts and early adoptions. *Briefings in Bioinformatics*, *20*(5), 1699–1708. DOI: 10.1093/bib/bby043 PMID: 29868882

Peng, C., Goswami, P., & Bai, G. (2020). A literature review of current technologies on health data integration for patient-centered health management. *Health Informatics Journal*, *26*(3), 1926–1951. DOI: 10.1177/1460458219892387 PMID: 31884843

Perez-Pozuelo, I., Spathis, D., Clifton, E. A., & Mascolo, C. (2021). Wearables, smartphones, and artificial intelligence for digital phenotyping and health. In *Digital health* (pp. 33–54). Elsevier. DOI: 10.1016/B978-0-12-820077-3.00003-1

Perez-Pozuelo, I., Spathis, D., Gifford-Moore, J., Morley, J., & Cowls, J. (2021). Digital phenotyping and sensitive health data: Implications for data governance. *Journal of the American Medical Informatics Association : JAMIA*, *28*(9), 2002–2008. DOI: 10.1093/jamia/ocab012 PMID: 33647989

Pesqueira, A., Sousa, M. J., & Rocha, Á. (2020). Big data skills sustainable development in healthcare and pharmaceuticals. *Journal of Medical Systems*, *44*(11), 197. DOI: 10.1007/s10916-020-01665-9 PMID: 33034743

Petersen, E., Potdevin, Y., Mohammadi, E., Zidowitz, S., Breyer, S., Nowotka, D., Henn, S., Pechmann, L., Leucker, M., Rostalski, P., & Herzog, C. (2022). Responsible and regulatory conform machine learning for medicine: A survey of challenges and solutions. *IEEE Access : Practical Innovations, Open Solutions*, *10*, 58375–58418. DOI: 10.1109/ACCESS.2022.3178382

Plachkinova, M., Vo, A., Bhaskar, R., & Hilton, B. (2018). A conceptual framework for quality healthcare accessibility: A scalable approach for big data technologies. *Information Systems Frontiers*, *20*(2), 289–302. DOI: 10.1007/s10796-016-9726-y

Preskill, J. (2012). Quantum computing and the entanglement frontier. arXiv preprint arXiv:1203.5813.

Radanliev, P., De Roure, D., Nicolescu, R., & Huth, M. (2019). A reference architecture for integrating the Industrial Internet of Things in the Industry 4.0. University of Oxford combined working papers and project reports prepared for the PETRAS National Centre of Excellence and the Cisco Research Centre. https://doi. org/, 2(26854.47686).DOI: 10.13140/RG

Rajabion, L., Shaltooki, A. A., Taghikhah, M., Ghasemi, A., & Badfar, A. (2019). Healthcare big data processing mechanisms: The role of cloud computing. *International Journal of Information Management*, *49*, 271–289. DOI: 10.1016/j.ijinfomgt.2019.05.017

Rehman, A., Naz, S., & Razzak, I. (2022). Leveraging big data analytics in healthcare enhancement: Trends, challenges and opportunities. *Multimedia Systems*, *28*(4), 1339–1371. DOI: 10.1007/s00530-020-00736-8

Saggi, M. K., & Jain, S. (2018). A survey towards integrating big data analytics to big insights for value-creation. *Information Processing & Management*, *54*(5), 758–790. DOI: 10.1016/j.ipm.2018.01.010

Schulte, P., & Lee, D. K. C. (2019). *AI & Quantum Computing for Finance & Insurance: Fortunes and Challenges for China and America* (Vol. 1). World Scientific.

Secundo, G., Toma, A., Schiuma, G., & Passiante, G. (2019). Knowledge transfer in open innovation: A classification framework for healthcare ecosystems. *Business Process Management Journal*, *25*(1), 144–163. DOI: 10.1108/BPMJ-06-2017-0173

Sestino, A., Prete, M. I., Piper, L., & Guido, G. (2020). Internet of Things and Big Data as enablers for business digitalization strategies. *Technovation*, *98*, 102173. Advance online publication. DOI: 10.1016/j.technovation.2020.102173

Shahid, J., Ahmad, R., Kiani, A. K., Ahmad, T., Saeed, S., & Almuhaideb, A. M. (2022). Data protection and privacy of the internet of healthcare things (IoHTs). *Applied Sciences (Basel, Switzerland)*, *12*(4), 1927. DOI: 10.3390/app12041927

Shaygan, A., & Daim, T. (2021). Technology management maturity assessment model in healthcare research centers. *Technovation*. Advance online publication. DOI: 10.1016/j.technovation.2021.102444

Shrivastava, U., Song, J., Han, B. T., & Dietzman, D. (2021). Do data security measures, privacy regulations, and communication standards impact the interoperability of patient health information? A cross-country investigation. *International Journal of Medical Informatics*, *148*, 104401. DOI: 10.1016/j.ijmedinf.2021.104401 PMID: 33571743

Siddiq, M. (2022). Use of machine learning to predict patient developing a disease or condition for early diagnose. *International Journal of Multidisciplinary Sciences and Arts*, *1*(1). Advance online publication. DOI: 10.47709/ijmdsa.v1i1.2271

Sikdar, S., & Guha, S. (2020). Advancements of healthcare technologies: Paradigm towards smart healthcare systems. *Recent trends in image and signal processing in computer vision*, 113-132.

Singhal, S., Kayyali, B., Levin, R., & Greenberg, Z. (2020). *The next wave of healthcare innovation: The evolution of ecosystems.* McKinsey & co. 2020a. Disponível em: https://www.mckinsey.com/industries/healthcare-systems-and-services/our-nsights/the-next-wave-of-healthcare-innovation-the-evolution-of-ecosystems

Singhania, K., & Reddy, A. (2024). Improving preventative care and health outcomes for patients with chronic diseases using big data-driven insights and predictive modeling. *International Journal of Applied Health Care Analytics*, *9*(2), 1–14.

Sousa, M. J., Pesqueira, A. M., Lemos, C., Sousa, M., & Rocha, Á. (2019). Decision-making based on big data analytics for people management in healthcare organizations. *Journal of Medical Systems*, *43*(9), 1–10. DOI: 10.1007/s10916-019-1419-x PMID: 31332535

Sriram, R. D., & Subrahmanian, E. (2020). Transforming health care through digital revolutions. *Journal of the Indian Institute of Science*, *100*(4), 753–772. DOI: 10.1007/s41745-020-00195-0 PMID: 33132546

Tamiminia, H., Salehi, B., Mahdianpari, M., Quackenbush, L., Adeli, S., & Brisco, B. (2020). Google Earth Engine for geo-big data applications: A meta-analysis and systematic review. *ISPRS Journal of Photogrammetry and Remote Sensing*, *164*, 152–170. DOI: 10.1016/j.isprsjprs.2020.04.001

Tan, F. T. C., Ondrus, J., Tan, B., & Oh, J. (2020). Digital transformation of business ecosystems: Evidence from the Korean pop industry. *Information Systems Journal*, *30*(5), 866–898. DOI: 10.1111/isj.12285

Tariq, M. U. (2024). Enhancing cybersecurity protocols in modern healthcare systems: Strategies and best practices. In *Transformative Approaches to Patient Literacy and Healthcare Innovation* (pp. 223–241). IGI Global. DOI: 10.4018/979-8-3693-3661-8.ch011

Tariq, M. U. (2025). *AI-Driven Learning and Engagement in Higher Education*. IGI Global., DOI: 10.4018/979-8-3693-4074-5

Tariq, M. U. (2025). Innovative Business Models: Integrating Social, Economic, and Environmental Goals in Rural Manufacturing. In Prabhakar, A., Yu, P., & Erokhin, V. (Eds.), *Rural Social Entrepreneurship Development: Network-Based Manufacturing System Model* (pp. 147–174). IGI Global Scientific Publishing., DOI: 10.4018/979-8-3693-7515-0.ch005

Tariq, M. U., & Sergio, R. (Eds.). (2025). *Cases on AI-Driven Solutions to Environmental Challenges*. IGI Global, 2025. https://doi.org/DOI: 10.4018/979-8-3693-7483-2

Tariq, M. U., & Sergio, R. (Eds.). (2025). *Convergence of AI, Education, and Business for Sustainability*. IGI Global., DOI: 10.4018/979-8-3373-1917-9

Tayefi, M., Ngo, P., Chomutare, T., Dalianis, H., Salvi, E., Budrionis, A., & Godtliebsen, F. (2021). Challenges and opportunities beyond structured data in analysis of electronic health records. *Wiley Interdisciplinary Reviews: Computational Statistics*, *13*(6), e1549. DOI: 10.1002/wics.1549

Timpson C.G. (2019). *Old and New in Quantum Information*.

Towett, G., Snead, R. S., Grigoryan, K., & Marczika, J. (2023). Geographical and practical challenges in the implementation of digital health passports for cross-border COVID-19 pandemic management: A narrative review and framework for solutions. *Globalization and Health*, *19*(1), 98. DOI: 10.1186/s12992-023-00998-7 PMID: 38066568

Udokwu, S. T. C., Okoye, C. C., Oshioste, E. E., Nwankwo, E. F., & Azubuike, N. U. (2023). *The role of leadership on organization management: A case study of First Bank in Abuja Metropolis. Business and Economics in Developing Countries*. BEDC., DOI: 10.26480/bedc.02.2023.46.51

Ukato, A., Sofoluwe, O. O., Jambol, D. D., & Ochulor, O. J. (2024). *Optimising maintenance logistics on offshore platforms with AI: Current strategies and future innovations*.

Urbinati, A., Bogers, M., Chiesa, V., & Frattini, F. (2019). Creating and capturing value from Big Data: A multiple-case study analysis of provider companies. *Technovation*, *84*, 21–36. DOI: 10.1016/j.technovation.2018.07.004

Uzougbo, N. S., Ikegwu, C. G., & Adewusi, A. O. (2024). International enforcement of cryptocurrency laws: Jurisdictional challenges and collaborative solutions. *Magna Scientia Advanced Research and Reviews,* 11(1), 068-083

Vernacchia, S. (2019). *Quantum leap: advancing a strategy for quantum computing that will inspire, support and safeguard economic growth in the Middle East.* https://www.pwc.com/m1/en/world-government-summit/documents/wgs-quantum-leap.pdf World Government Summit.

Wang, Y., Kung, L., & Byrd, T. A. (2018). Big data analytics: Understanding its capabilities and potential benefits for healthcare organizations. *Technological Forecasting and Social Change, 126*, 3–13. DOI: 10.1016/j.techfore.2015.12.019

Weerasinghe, K., Scahill, S. L., Pauleen, D. J., & Taskin, N. (2022). Big data analytics for clinical decision-making: Understanding health sector perceptions of policy and practice. *Technological Forecasting and Social Change, 174*, 121222. DOI: 10.1016/j.techfore.2021.121222

Wesson, P., Hswen, Y., Valdes, G., Stojanovski, K., & Handley, M. A. (2022). Risks and opportunities to ensure equity in the application of big data research in public health. *Annual Review of Public Health, 43*(1), 59–78. DOI: 10.1146/annurev-publhealth-051920-110928 PMID: 34871504

Williamson, S. M., & Prybutok, V. (2024). Balancing privacy and progress: A review of privacy challenges, systemic oversight, and patient perceptions in AI-Driven healthcare. *Applied Sciences (Basel, Switzerland), 14*(2), 675. DOI: 10.3390/app14020675

Yang, F., & Gu, S. (2021). Industry 4.0, a revolution that requires technology and national strategies. *Complex & Intelligent Systems, 7*(3), 1311–1325. DOI: 10.1007/s40747-020-00267-9

Yao, R., Zhang, W., Evans, R., Cao, G., Rui, T., & Shen, L. (2022). Inequities in health care services caused by the adoption of digital health technologies: Scoping review. *Journal of Medical Internet Research, 24*(3), e34144. DOI: 10.2196/34144 PMID: 35311682

Yuan, X., Yang, C., He, Q., Chen, J., Yu, D., Li, J., Zhai, S., Qin, Z., Du, K., Chu, Z., & Qin, P. (2020). Current and perspective diagnostic techniques for COVID-19. *ACS Infectious Diseases, 6*(8), 1998–2016. DOI: 10.1021/acsinfecdis.0c00365 PMID: 32677821

Zafar, I., Anwar, S., Yousaf, W., Nisa, F. U., & Kausar, T. (2023). Reviewing methods of deep learning for intelligent healthcare systems in genomics and biomedicine. *Biomedical Signal Processing and Control*, *86*, 105263. DOI: 10.1016/j.bspc.2023.105263

Zeadally, S., Siddiqui, F., Baig, Z., & Ibrahim, A. (2020). Smart healthcare: Challenges and potential solutions using internet of things (IoT) and big data analytics. *PSU Research Review*, *4*(2), 149–168. DOI: 10.1108/PRR-08-2019-0027

Zhang, A., Xing, L., Zou, J., & Wu, J. C. (2022). Shifting machine learning for healthcare from development to deployment and from models to data. *Nature Biomedical Engineering*, *6*(12), 1330–1345. DOI: 10.1038/s41551-022-00898-y PMID: 35788685

Zinner, M., Dahlhausen, F., Boehme, P., Ehlers, J., Bieske, L., & Fehring, L. (2021). Quantum computing's potential for drug discovery: Early stage industry dynamics. *Drug Discovery Today*, *26*(7), 1680–1688. DOI: 10.1016/j.drudis.2021.06.003 PMID: 34119668

ADDITIONAL READINGS

Alshareef, I.M.A., & Al Shaman, H.M.H., & hadi Al Mansour, I. (2023). The role of data analytics in medical administration: Leveraging information for decision-making. *Journal of Namibian Studies: History Politics Culture*, *36*, 12–23.

Bincoletto, G. (2020). Data protection issues in cross-border interoperability of Electronic Health Record systems within the European Union. *Data & Policy*, *2*, e3. DOI: 10.1017/dap.2020.2

Freitas, A.T. (2023). Data-Driven approaches in healthcare: challenges and emerging trends. multidisciplinary perspectives on artificial intelligence and the law, 65-80.

Mbunge, E., Muchemwa, B., & Batani, J. (2021). Sensors and healthcare 5.0: Transformative shift in virtual care through emerging digital health technologies. *Global Health Journal (Amsterdam, Netherlands)*, *5*(4), 169–177. DOI: 10.1016/j.glohj.2021.11.008

Miranda, J. P. (2024). Mobile Applications in Advancing Health Literacy: A Scoping Review of Methodologies, Participants, and Challenges. In *Transformative Approaches to Patient Literacy and Healthcare Innovation*. IGI Global., DOI: 10.4018/979-8-3693-3661-8.ch003

KEY TERMS AND DEFINITIONS

Machine Learning (ML): A subset of AI that enables computers to learn from data patterns and make predictions or decisions without explicit programming. In healthcare, ML is used for disease prediction, medical image analysis, and personalized treatment recommendations.

Natural Language Processing (NLP): AI technology that enables machines to understand, interpret, and process human language. NLP is widely used in healthcare for analyzing clinical notes, automating medical documentation, and extracting insights from unstructured patient records.

Predictive Analytics: The use of AI algorithms to analyze historical healthcare data and predict future outcomes, such as disease progression, patient readmission risks, or potential outbreaks. This helps in early diagnosis, personalized treatment planning, and proactive healthcare management.

Computer Vision: A branch of AI that enables machines to interpret and analyze visual data, such as medical images (X-rays, MRIs, CT scans). AI-driven computer vision assists radiologists in detecting abnormalities like tumors and fractures with high accuracy.

Explainable AI (XAI): A field of AI focused on making AI-driven decisions more transparent and interpretable for healthcare professionals. Explainable AI ensures trust, accountability, and ethical implementation of AI models in clinical settings.

Chapter 4
Transforming Healthcare by Integrating IoT for Enhanced Patient Care and Innovation

Cheng Hang
https://orcid.org/0009-0001-2031-5910
Management and Science University, Malaysia

Safaa Najah Saud Al-Humairi
https://orcid.org/0000-0001-5978-7651
Management and Science University, Malaysia

Ahgalya Subbiah
https://orcid.org/0009-0006-4427-0872
Management and Science University, Malaysia

ABSTRACT

The healthcare landscape is undergoing a profound transformation, driven by technological innovation and evolving patient needs. As the global population ages, the sector faces unprecedented challenges in delivering high-quality, patient-centered care. Integrating Artificial Intelligence (AI), robotics, and the Internet of Things (IoT) has emerged as a key strategy to revolutionize healthcare and improve patient outcomes. AI enhances diagnostics, treatment, and management through machine learning algorithms that analyze data, predict outcomes, and enable early interventions. Robotics has improved surgical precision, rehabilitation, and elder care, elevating patients' quality of life. Meanwhile, IoT has facilitated remote patient monitoring, supported telehealth services, and integrated smart devices for contin-

DOI: 10.4018/979-8-3373-4332-7.ch004

uous health management. This chapter examines these technologies' applications, potential, benefits, and challenges, aiming to contribute to the ongoing discourse on the future of healthcare.

INTRODUCTION

The Internet of Things (IoT) has represented a disruption in healthcare, transforming patient care and encouraging innovation across all sectors within the practice of medicine. The healthcare industry uses IoT, which provides remote monitoring, data gathering, and patient-centric services, leading to superior health and operational efficiency (Furtado et al., 2023). In healthcare, a crucial advantage afforded by the IoT is that patients can be monitored 24/7 through wearables. They capture real-time health parameters like body temperature, blood pressure, or glucose level and send this data as messages for doctors to analyze. In addition to supporting early interventions, this capability also empowers the patient to manage their health states as they are actively partaking in it. For example, the systematic review supports the utility of internet-based wearables associated with IOTs for individual tracking to manage public health, such as, in this instance — transmittable diseases (Sivaraman, 2019). Moreover, when combined with cloud services, these devices improve data access and analytics, enabling better clinical decisions. The ease of implementation for healthcare IoT orders — in other terms, user experience must be best. (Fehringer & Stary, 2023) highlights the importance of user-informed adaptation in medical IoT applications and support for ground-up, evidence-based designs to deliver personalized patient-centered care. Not only does this make users healthier and happier, but it also improves the chances of following their health care plans. These bespoke IoT solutions for each user help healthcare providers offer more personalized healthcare essentials concerning recuperation and taking check of chronic diseases. While IoT in healthcare has promising advantages, some challenges must be addressed before its standardized implementation (Tariq, 2024).

The primary reason for this is security — health data will always be seen as a high-value target, and it could have profound consequences in the wrong hands. (Abdulmalek et al., 2022) address the need for reliable, secure IoT healthcare monitoring systems to increase patient quality of life. Furthermore, the use of blockchain technology to increase the security and integrity of data in IoT healthcare applications has been suggested (Ratta et al., 2021). With IoT and blockchain, these dual methods lead to a very safe keeping of patient data, establishing faith in digital health solutions. In addition, just as the pandemic has reduced deployment cycles in IoT technology for healthcare with clear examples of success and challenges. Al-Atawi et al. overview the wide range of applications and challenges that IoT healthcare systems face today during this pandemic times, stressing the growing need for a more reliable infrastructure capable of monitoring remote patients and telemedicine (Al-Atawi et al., 2022). During the pandemic, the need for collecting and monitoring real-time data was also highlighted, something IoT allows to react quickly when new health crises emerge. The idea of Colleges emphasizing innovative healthcare facilities is also becoming popular, especially in areas like Gauteng, South Africa, where the effects and impact of IoT are toward the goal of building a Smart primary healthcare building facility (Gambo & Musonda, 2022). IoT technologies in healthcare infrastructure provide better resource management, patient flow, and overall service delivery. The growth of intelligent healthcare environments indicates a more significant trend toward digitization within the health industry. It can even be described as a requirement to meet current patient care expectations. IoT technologies have transformed patient care and added to healthcare operations. In addition to patient care, IoT has applications in healthcare for managing the inventory and allocation of staff based on real-time data analytics (Tariq, 2025).

This effective operation helps save money and makes Khan's research illustrate the vast use cases of IoT in Healthcare, from device enablement to remote medical care, showcasing how these technologies are poised to revolutionize all aspects of health delivery (Rashid, 2024). Personalized treatment plans could be developed when AI models can predict a person's health risks using the data gathered through sensors. These insights could be used to combine digital services with additional clinical markers. As reported (Zhou, 2024), the combination of AI and IoT in Healthcare will contribute to a diagnosis technique that is faster and more accurate; it has the potential to improve treatment effectiveness for certain diseases. Enter the marriage of IoT with AI — a frontier in healthcare innovation with immense potential to deliver quality human medical care and reshape patient-care paradigms altogether. There are also some hurdles in the case of IoT implementation, such as interoperability and standardization-related problems. Standardized protocols are necessary to facilitate the communication between devices and systems. These interoperability issues must materialize the full benefits of the IoT space within Healthcare. Also, healthcare considerations regarding IoT in Healthcare must not be neglected at any cost. Taking and using personal health data is one area where people can be easily freaked out thinking about privacy issues around consent. The ubiquity of the IoT means organizations in this particular industry must build a secure yet open model that allows patient records to be breached but does not quell innovation. IoT and blockchain technology enhance data security and patient privacy; hence, this research also supports how integration between them shall come to maximize those things (Kamangar et al., 2023) and improve patient care. In this chapter, we will cover different implementations of IoT in healthcare, the additional challenges it poses, and the solutions being provided to tackle these.

ROLE OF THE INTERNET OF THINGS IN THE HEALTHCARE INDUSTRY

The Internet of Things (IoT) has become a game-changer in many fields, and healthcare is no different (Akkaş et al., 2020; Yang & Xu, 2018). Patient care, medical research, and the healthcare industry as a whole have all been profoundly impacted by the IoT, sometimes known as Healthcare IoT or IoMT (Internet of Medical Things) (Al-Humairi & Hajamydeen, 2022; Safaa N Saud Al-Humairi et al., 2022). Patient's vital signs, chronic diseases, and postoperative recoveries can all be tracked in real-time with the help of IoT devices. Fitness trackers and smartwatches are only two examples of wearable sensors that gather data and send it to doctors in real time. Telemedicine and online medical consultations are made more accessible by the IoT. Videoconferencing, wearable health monitoring, and the secure transmission of medical data over the IoT allow doctors to diagnose and treat patients remotely (Kadhim et al., 2020). Intelligent medical devices such as insulin pumps, pacemakers, and medicine delivery systems have emerged and are recognized by the IoT. Remotely monitoring and changing these devices improves patient security and allows for more individualized treatment (Zeadally & Bello, 2021). Medical equipment theft, inefficient inventory management, and subpar patient care can all be mitigated with the help of IoT solutions implemented in hospitals. Medication adherence trackers and smart pill dispensers assist patients in taking their medications as prescribed. An example of the telemedicine service is illustrated in **Figure 1.**

Figure 1. Telemedicine service

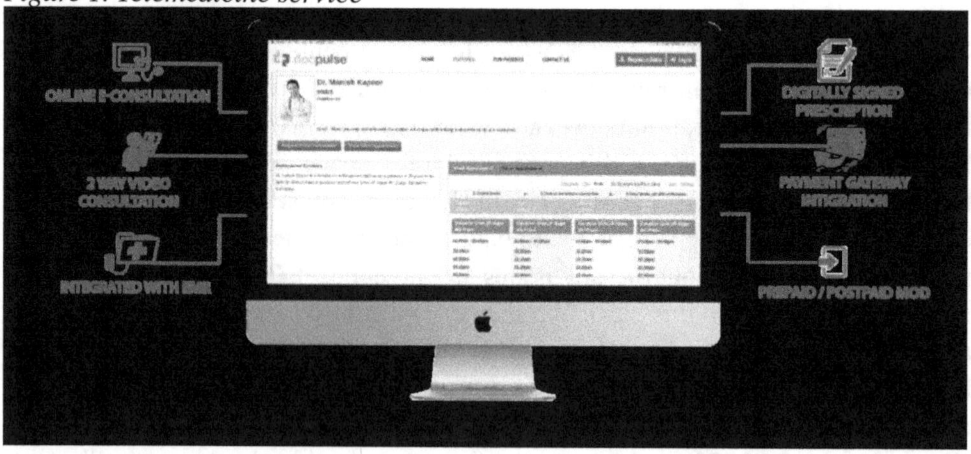

In particular, implementing Internet of Things (IoT)-based fall detection devices is crucial for senior patients. This technology alerts caregivers or medical professionals if a patient falls (Karar et al., 2022), ensuring swift response and intervention. Safekeeping vaccinations, drugs, and laboratory samples rely on the data gathered by IoT sensors installed across healthcare institutions (Safaa Najah Saud Al-Humairi et al., 2022). Early diagnosis of health problems by real-time monitoring and data collecting allows for prompt interventions and improved patient outcomes. With the help of IoT-enabled apps and devices, patients may take an active role in their healthcare by viewing their health records and receiving individualized suggestions. IoT reduces healthcare spending and optimizes overall efficiency by preventing unnecessary hospital visits. Asset tracking and predictive maintenance powered by the IoT can help hospitals and other healthcare institutions increase productivity, streamline processes, and utilize their available resources more effectively (Teoh et al., 2021). By producing massive datasets, IoT speeds up scientific findings in medicine, clinical trials, and epidemiological studies. With IoT, telehealth services have grown, bringing medical care to previously unreached areas and people without insurance (Tariq, 2024).

Compliance with healthcare privacy legislation and protecting sensitive patient data from cyber threats is paramount. To achieve success in healthcare IoT, it is imperative to surmount the challenge of ensuring interoperability and seamless connectivity among various IoT devices and systems. Furthermore, ethical and regulatory considerations, such as safeguarding patients' privacy and obtaining their informed consent before collecting and distributing data, play a pivotal role in shaping the responsible implementation of IoT in healthcare. The Healthcare Insurance Portability and Accountability Act (HIPAA) and the Food and Drug Administration's (FDA) licensing of medical devices must be followed (Kirchner, Kim, Kwart, et al., 2023). The potential for IoT to improve healthcare in the future is enormous. Obligations to developments in edge computing, AI-driven analytics, and 5G networks, as well as data processing speed and accuracy improvements, are on the horizon. Predictive analytics and machine learning techniques will significantly benefit disease prognosis and treatment planning. Regarding global healthcare issues like pandemics and an ageing population, IoT will continue to play a critical role in telehealth expansion and management. The IoT has ushered in a new era of patient-centered care, increased efficiency, and groundbreaking medical discoveries in healthcare. Better healthcare outcomes and experiences for patients and providers are on the horizon as an advantage of the IoT, which is poised to change the healthcare landscape as technology advances and healthcare stakeholders handle the accompanying problems.

IOT DEVICES AND APPLICATIONS IN HEALTHCARE

In healthcare, the proliferation of IoT-enabled connected devices and software has been nothing short of revolutionary. These IoT devices are revolutionizing the healthcare industry by allowing remote monitoring, data-driven decision-making, and individualized patient care. Devices such as Fitbit and Garmin enable tracking activity, heart rate, sleep, and calorie burn, fostering healthier choices and aiding in chronic disease management. Apple Watches and the Samsung Galaxy Watch have advanced features like ECG monitors, fall detectors, and irregular heart rhythm alerts. Patients with diabetes can now track their glucose levels and instantly share that information with their doctors (Montenegro & Dori-Hacohen, 2020). The user can track Connected blood pressure monitors and send them to their doctor for remote monitoring. These gadgets aid patients in taking their medications as prescribed by sending alerts and dosing them automatically. They can also notify doctors or nurses if a medication is forgotten. Utilizing tiny sensors that can be integrated into blister packs or medication bottles, medication adherence can be actively monitored. This technology proves particularly advantageous for patients with conditions like asthma and chronic obstructive pulmonary disease, as it enables the tracking of inhaler usage and provides valuable feedback (Tariq, 2025).

Patients with diabetes can better control their insulin regimen and monitor their injections using IoT pens. Cameras and sensors enabled by the IoT can be used in telehealth devices to examine and monitor patients remotely. IoT cameras allow dermatologists to monitor patient skin problems remotely. Complete IoT telehealth platforms allow for two-way video calls, encrypted data transfers, and immediate doctor-patient interaction. IoT sensors can monitor seniors' activity and alert care-takers or doctors if they detect falls. Sensors in these systems track the whereabouts of the elderly, how long they sleep, and the temperature and humidity of their house to ensure their health and safety (Del-Valle-Soto et al., 2023). Figure 2 illustrates photos of the devices installed for the monitoring system.

Figure 2. Real implementation of the sensor network

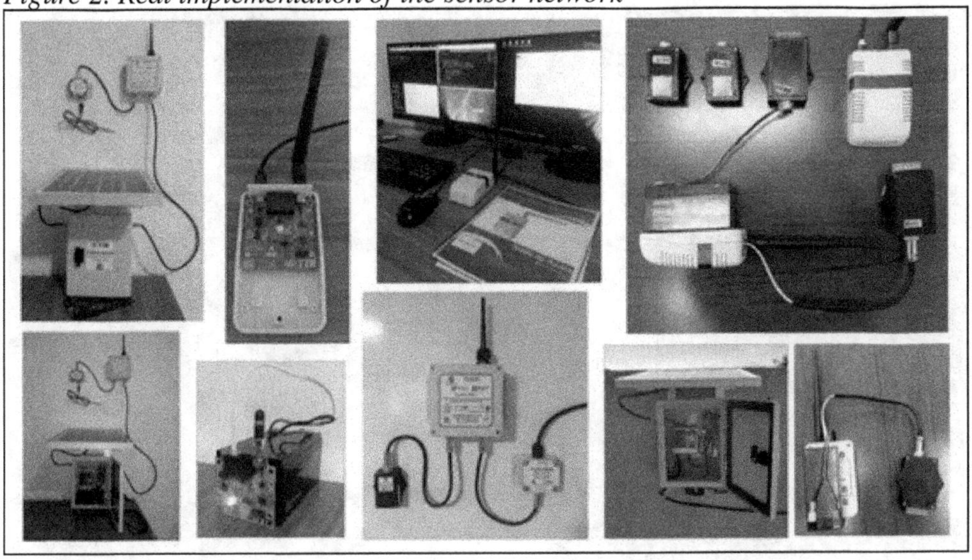

(Del-Valle-Soto et al., 2023)

Radio-frequency identification (RFID) tags are attached to hospital assets like medical supplies and devices. The IoT allows for real-time monitoring of the whereabouts and condition of these assets, which helps cut down on waste and maximize productivity. **Figure 3** mentions the advantages of IoT in healthcare. Medications, vaccinations, and laboratory samples may all be safely stored with the help of Internet of Things sensors that track environmental conditions.

Figure 3. IoT in healthcare

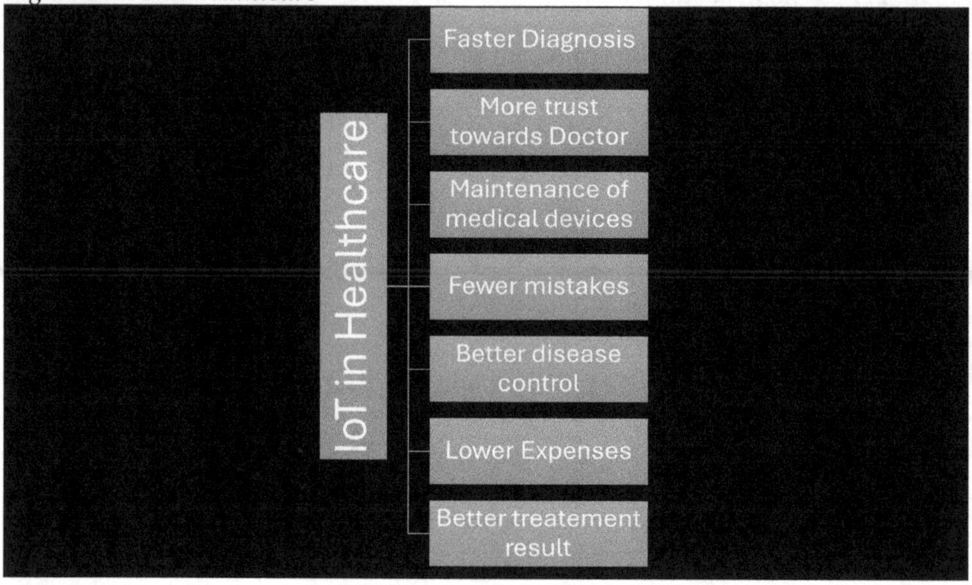

These sensors help hospitals with infection control by detecting airborne contaminants and bacteria. IoT devices are increasingly being employed in clinical trials because of their ability to collect real-time data from participants. Wearable biosensors continuously record physiological data for research and drug development in the medical and pharmaceutical industries (Pollard et al., 2021). In an emergency, assisted living or senior care facility residents can use IoT panic buttons to summon help immediately. Patient care, clinical efficiency, and the ability to make data-driven decisions are all improved using IoT devices and applications in the medical field. The healthcare IoT ecosystem is primed for growth and evolution as technology progresses, providing even more novel solutions to enhance the standard of healthcare delivery. Integrating the Internet of Things (IoT) into healthcare has ushered in a transformative era characterized by enhanced patient monitoring capabilities. This evolution is primarily driven by the convergence of advanced technologies such as 5G networks, artificial intelligence (AI), and cloud computing, which collectively facilitate real-time data acquisition and analysis, thereby improving patient outcomes and healthcare delivery systems. One of the most significant advancements in this domain is the emergence of the Internet of Medical Things (IoMT), which encompasses a wide array of connected medical devices that monitor, collect, and transmit health data. The IoMT is pivotal in enabling continuous patient monitoring, which is essential for managing chronic diseases and ensuring timely interventions. For instance, wearable devices equipped with sensors can track vital signs such as heart rate, blood pressure, and glucose levels, providing healthcare professionals with critical information that can be analyzed in real-time to make informed decisions regarding patient care (Mishra & Singh, 2023). The role of 5G technology in enhancing IoT applications in healthcare cannot be overstated. The high-speed, low-latency communication capabilities of 5G networks facilitate the seamless transmission of large volumes of health data from IoMT devices to healthcare providers. This capability is particularly crucial in emergency situations where timely access to patient data can significantly impact treatment outcomes. Moreover, 5G enables the deployment of telemedicine solutions, allowing healthcare providers to conduct remote consultations and monitor patients from afar, thus reducing the need for in-person visits and minimizing the risk of infection, especially in the context of the COVID-19 pandemic (Liu et al., 2023; Taiwo & Ezugwu, 2020; Tariq, 2024). Furthermore, integrating AI and machine learning algorithms into IoT healthcare systems enhances the analytical capabilities of these technologies. By leveraging vast amounts of data collected from various sources, AI can identify patterns and predict potential health issues before they escalate. This predictive analytics approach not only improves patient monitoring but also supports personalized treatment plans tailored to individual patient needs (Ragab & Binyamin, 2023). For example, AI-driven algorithms can analyze data from wearable devices to detect anomalies in a patient's health metrics, prompting immediate medical attention if necessary. The Cloud-of-Things (CoT) paradigm further complements the IoT in healthcare by providing a robust infrastructure for data storage and processing. Utilizing cloud computing allows healthcare providers to aggregate and analyze data from multiple IoMT devices, enabling a comprehensive view of a patient's health status. This centralized approach not only enhances data accessibility for healthcare professionals but also supports collaborative care models where multiple specialists can contribute to a patient's treatment plan (Alabdulatif et al., 2019; Chopade et al., 2023). Security and privacy concerns are paramount in the context of IoT healthcare applications, given the sensitive nature of health data. As such, developing secure frameworks for data transmission and storage is critical. Recent advancements in encryption and authentication protocols aim to safeguard patient information from unauthorized access and cyber threats. For instance, privacy-preserving frameworks have been proposed to ensure that patient data remains confidential while still allowing for effective monitoring and analysis (Azad et al., 2019; Shahid et al., 2022). Moreover, the application of IoT in healthcare extends beyond individual patient monitoring to encompass broader public health initiatives. For instance, smart healthcare systems can aggregate data from various sources to monitor population health trends, enabling public health officials to respond proactively to emerging health threats. This capability is particularly relevant in the context of infectious disease outbreaks, where real-time data can inform containment strategies and resource allocation (Meinert et al., 2018).

INTEGRATION OF IOT IN ROBOTIC COMMUNICATION

The IoT and robotic communication constitute a potent convergence of technologies dramatically altering the medical industry. Healthcare professionals may deliver cutting-edge, efficient care for patients by integrating Internet of Things devices and systems with robotics. Vital signs, activity levels, and other health parameters are continuously collected via data from IoT-enabled patient devices, such as fitness trackers or medical sensors. IoT networks enable real-time data transmission to a centralized healthcare system or cloud platform. This guarantees that doctors always have the most recent data on their patients. Integrating the Internet of Things (IoT) in robotic communication represents a transformative shift in how robots interact with their environments and human operators. This integration facilitates enhanced communication, data sharing, and operational efficiency, which is critical for developing autonomous systems. The Internet of Robotic Things (IoT) concept encapsulates this synergy, highlighting the convergence of robotics and IoT technologies to create intelligent, interconnected systems capable of performing complex tasks in various domains. One of the primary advantages of integrating IoT with robotic systems is enhancing communication capabilities. For instance, discuss how integrating IoT devices can facilitate the control of robots and data collection tailored to specific applications, thereby improving the robot's operational efficiency and adaptability to different tasks (Michalík et al., 2021). This is particularly relevant in scenarios where robots must respond to dynamic environments or user commands in real time. Moreover, using platforms like Blynk to control surveillance robots remotely exemplifies how IoT can streamline communication between users and robotic systems. Incorporating Wi-Fi modules allows for seamless interaction, enabling users to monitor and control robotic movements from there. This capability is essential for applications in security and surveillance, where timely responses to detected anomalies are crucial. The challenges and synergies of integrating robots within the IoT ecosystem have been explored in various studies. Highlighting the complexities of establishing reliable wireless connectivity for connected agricultural vehicles is a critical aspect of IoRT (Grétarsson et al., 2023). The need for robust communication protocols and the ability to predict connectivity issues are vital for ensuring that robotic systems can operate effectively in the field. In addition to communication, integrating IoT in robotics enhances data processing capabilities. emphasize the importance of understanding the characteristics and architectures of IoT systems as they relate to robotic applications. Their research indicates that incorporating IoT technologies allows robots to utilize a variety of sensors and actuators, thereby expanding their functional capabilities across different application areas (Kamilaris & Botteghi, 2020). This adaptability is crucial for deploying robots in diverse environments, from industrial settings to smart homes. The concept of

IoRT is further supported by the work of those who discuss energy-efficient computation offloading in IoRT environments. Their findings suggest that integrating IoT technologies can significantly improve robotic systems' energy management and operational efficiency (Menbawy et al., 2023). This is particularly important in scenarios where robots operate in remote or resource-constrained environments, as efficient energy use can extend operational lifetimes and reduce the need for frequent recharging. The role of cloud computing and edge computing in enhancing robotic communication is also noteworthy. argue that cloud and IoT technologies are essential for creating distributed robotic systems leveraging vast amounts of data for improved decision-making and operational efficiency (Simoens et al., 2018). The ability to process data at the edge, as discussed, allows for low-latency responses and high-bandwidth communication, which are critical for real-time robotic applications (Queralta et al., 2020). Furthermore, integrating artificial intelligence (AI) with IoT in robotics is a significant area of research. Tychola explores how AI can enhance the capabilities of IoT-enabled robotic systems, enabling them to learn from their environments and improve their performance over time (Tzafestas, 2018). This synergy between AI and IoT not only enhances the functionality of robots but also allows for more sophisticated interactions with human users and other devices. The potential applications of IoRT are vast, ranging from industrial automation to smart city initiatives. The work illustrates how the convergence of sensing, actuating, and hyperconnectivity can lead to the development of intelligent robotic systems capable of operating autonomously in complex environments (Vermesan et al., 2022). This is particularly relevant in smart cities, where robots can assist in various tasks such as waste management, traffic monitoring, and public safety. In conclusion, integrating IoT in robotic communication is a multifaceted endeavor that enhances the capabilities of robotic systems across various domains. By leveraging advanced communication technologies, data processing capabilities, and AI, IoRT represents a significant advancement in robotics. As research continues to evolve, the potential for innovative applications and improved operational efficiencies will likely expand, paving the way for a new era of intelligent, interconnected robotic systems. The IoT technology allows robots to interact with patients and gather information (Mukati et al., 2023). These robots can be placed in any healthcare setting, from hospitals and clinics to patients' homes. The IoT implementation process chart in the medical area is illustrated in **Figure 4**.

Figure 4. IoT implementation process chart in the medical area

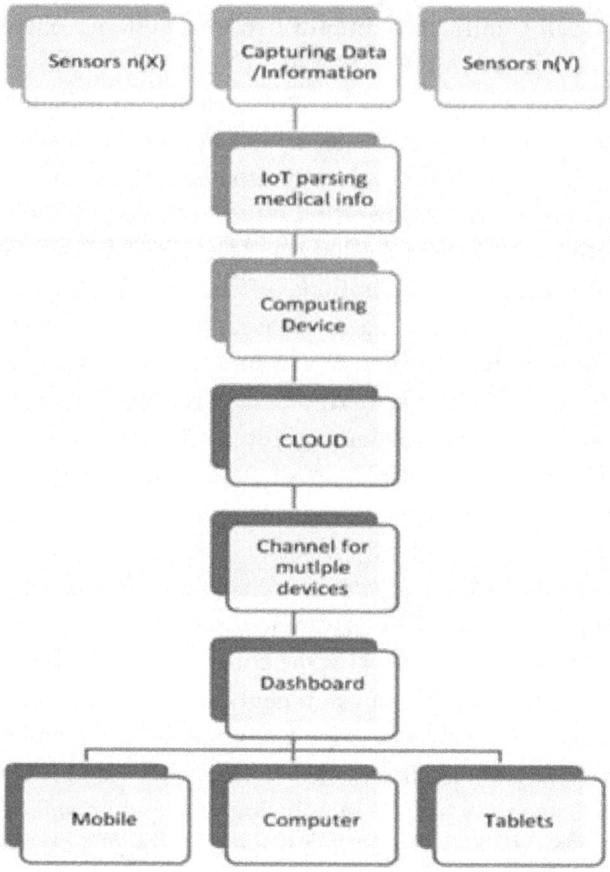

(Mukati et al., 2023)

Connected to the Internet of Things (IoT), robots can communicate with patients through their cameras and microphones. This enables them to inquire, instruct, and remind patients regarding medication schedules or therapy sessions. Moreover, leveraging IoT technology, these robots can read facial expressions and tones of voice, providing a comprehensive understanding of the patient's emotional state. Robots' IoT data can be evaluated with the help of artificial intelligence and machine learning programs. This approach enables the identification of patterns, outliers, and potential health issues, allowing for proactive and informed healthcare interventions. The process of IOET is shown in **Figure 5**.

Figure 5. Process of IOET

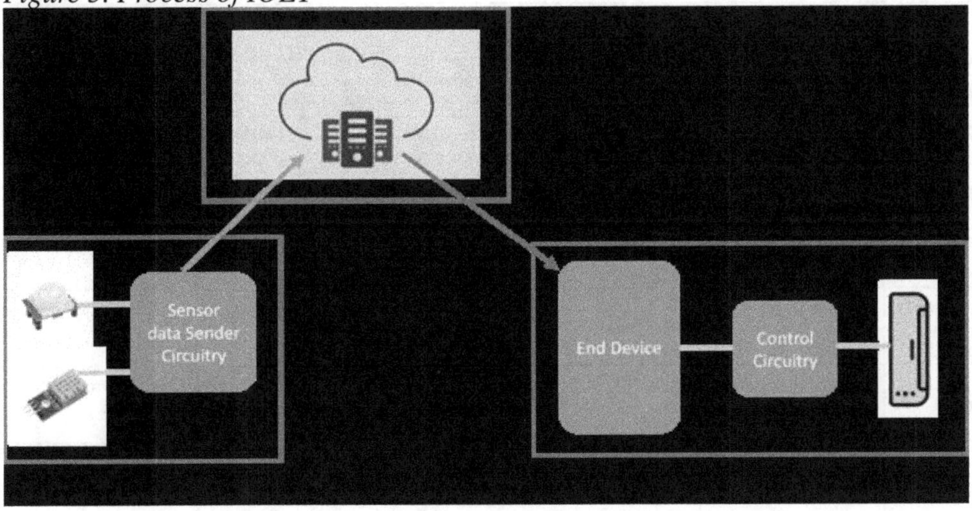

Robots can use IoT data to deliver instantaneous feedback to patients. For instance, if a patient's wearable device detects abnormal heart rhythms, the robot can notify them and suggest they see a doctor. Medical experts may conduct in-depth remote consultations with the robot's built-in IoT sensors and cameras. Sensors like these can record vital signs, take pictures, and gather other information about a patient. IoT networks safeguard patients' personal information during confidential telehealth consultations. Robots can distribute medications by prescription regimens kept in the Internet of Things databases (Zeadally & Bello, 2021). In addition, they can track drug compliance and give out reminders. The robot can inform doctors how well patients follow their prescribed drug schedules. Robots with IoT sensors can promptly notify medical staff or emergency personnel when patients fall. In order to assess the patient's condition and guide them until help arrives, the robot can establish two-way communication with them. Healthcare robots in the home can help patients stay on track with their treatment programs.

Table 1 presents the IoT technology built into these robots, enabling them to monitor and transmit data to healthcare providers in real-time continuously. Robotic communication systems offered by the IoT are flexible enough to meet the needs of hospitals of varying sizes. The accessibility and integrity of IoT data saved in the cloud are guaranteed for healthcare practitioners (Saha et al., 2019). There will be a dramatic change in how patients are cared for due to the IoT being applied to robotic communication in healthcare. Better patient outcomes and greater efficiency in healthcare settings result from expanded communication, real-time data analysis, and individualized treatment plans.

Table 1. IoT in healthcare

Author	Method Used	Setting/ Industry	Findings	Relevance to Study
	Data analysis and statistics	Spine surgery	Spine robots show low adverse events, enhancing robot adoption in surgery.	Robot adoption
	Qualitative study and discourse analysis	Diabetes care	Blood glucose discussions imply a moral judgment on patient self-care in diabetes treatment.	Insights into robotic adoption
(Del-Valle-Soto et al., 2023)	System design, implementation, and evaluation	Elderly care	Non-invasive vital sign monitoring improves elderly care and home security.	Virtual communication insights
(Pollard et al., 2021)	Literature review and analysis	Precision medicine	Electrochemical biosensors' role in precision medicine for enhanced diagnosis and treatment.	Impact on virtual healthcare
(Mukati et al., 2023)	System design, implementation, and evaluation	COVID-19 care	IoT healthcare assistance for COVID-19 enhances patient care reliability and usability.	Impact on virtual healthcare
(Zeadally & Bello, 2021)	Literature review and analysis	Healthcare	IoT improves healthcare through better remote monitoring and emergency response solutions.	IoT integration

BENEFITS AND CHALLENGES

With the use of IoT, healthcare providers may monitor their patient's vital signs and chronic illnesses from a distance in real time. By granting patients access to their health data, personalized advice, and self-monitoring tools, IoT devices and applications encourage patients to participate actively in their treatment. With IoT's ability to continuously gather and analyze data, medical practitioners can intervene before serious health problems escalate (Park & Han, 2023). By reducing unnecessary hospitalizations and enhancing resource allocation and healthcare delivery efficiency, the IoT helps bring down healthcare expenses. Predictive maintenance, inventory management, and asset tracking made possible by the IoT help healthcare facilities run more efficiently (Daily & Peterson, 2017). Data collected by the IoT helps enhance healthcare through clinical trials and the creation of new treatments and therapies.

The IoT allows for the growth of telehealth services, increasing access to healthcare for underserved and rural communities while relieving pressure on traditional medical institutions. A significant difficulty in using IoT is protecting sensitive patient data from cyber threats while maintaining compliance with healthcare privacy rules. For healthcare IoT to be effective, the difficulty of ensuring interoperability and seamless communication between different IoT devices and systems must be overcome. **Figure 6** illustrates the famous applications that could change healthcare.

Figure 6. AI applications

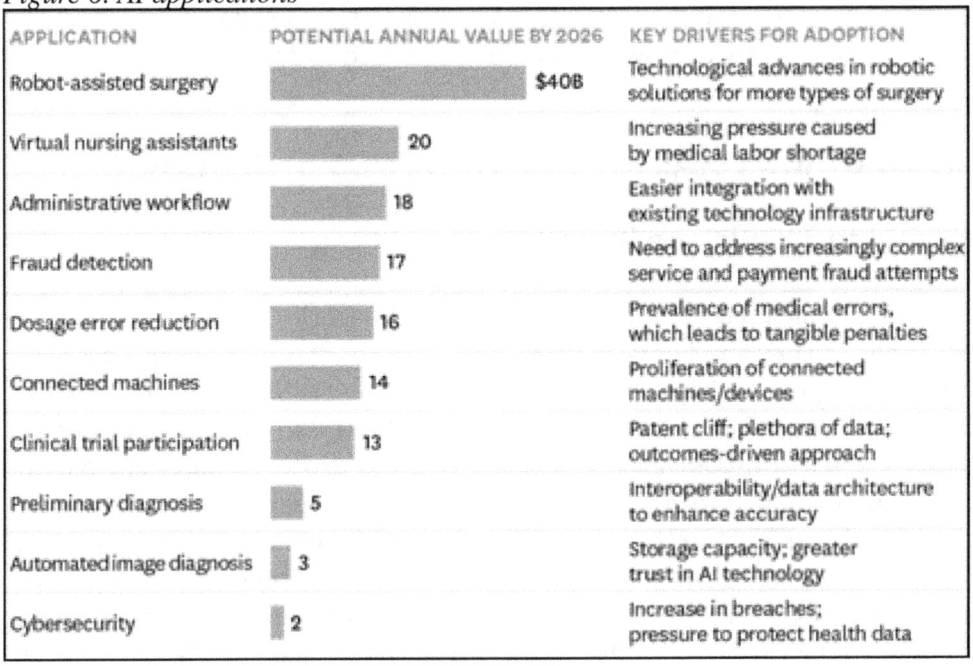

Managing ethical and regulatory considerations is crucial, as well as focusing on safeguarding patients' privacy and obtaining informed consent before collecting or disclosing any data. HIPAA (the Health Insurance Portability and Accountability Act) and the Food and Drug Administration (FDA) both have standards that healthcare providers must follow (Hayes & Vance, 2020). The success of IoT gadgets depends on stable network connections and supporting infrastructure. Data transmission and patient monitoring can be disrupted by lousy connectivity. In some cases, extensive IT resources and experience may be needed to successfully integrate IoT devices with preexisting healthcare IT systems.

Concerns about bias in AI-driven healthcare decisions and data ownership and access are just two examples of the ethical issues that arise from the usage of IoT in healthcare. IoT devices can improve healthcare management, but only if patients are aware of and open to their potential benefits and are willing to adopt and use them. In healthcare settings, the sheer volume of data created by IoT devices can be overwhelming, making it difficult to extract relevant insights. The safety of patients and the efficacy of healthcare decision-making depend on the precision and consistency of data collected by IoT devices (Aminizadeh et al., 2023). While acknowledging the potential benefits of Internet of Things (IoT) technologies in healthcare, it is essential to recognize that overemphasizing these technologies might compromise the doctor-patient connection and individualized treatment. To truly transform the healthcare industry, overcoming these challenges is paramount as we simultaneously leverage the advantages offered by IoT.

INTEGRATION OF AI, IOT, AND ROBOTICS IN HEALTHCARE

Emerging Trends and Innovations

Artificial intelligence (AI), the Internet of Things (IoT), and robotics are all advancing at a breakneck pace in the healthcare sector. Artificial intelligence algorithms are improving diagnostic imaging techniques, including CT scans and MRIs. Radiologists and other medical professionals use AI tools to spot abnormalities and diagnose diseases in patient images. AI-driven improvements to surgical robots increase their accuracy and make them less invasive. Robotic devices enhance surgeons' precision during delicate surgeries. The potential for remote patient monitoring is growing thanks to the proliferation of IoT-connected devices and AI-enabled platforms. Patients can monitor their vitals, chronic diseases, and overall health in real-time, while doctors and other medical professionals gain insight to act quickly when necessary. The AI systems sift through mountains of data to find promising new drugs and estimate how well they will do. New medicines and treatments can rapidly develop when drug discovery processes are hastened (Berdigaliyev & Aljofan, 2020).

Furthermore, smartwatches and fitness trackers are only two examples of wearable technologies that are progressively being included in medical infrastructure. These gadgets track exercise, sleep, and physiological data to promote preventative healthcare. Using factors including genetics, medical history, and lifestyle choices, AI can provide individualized treatment recommendations for patients. Because of precision medicine, therapeutic efficacy is increased while side effects are minimized. Rehabilitation robots aid recovery and physical therapy (Ju et al., 2023). The rehabilitation effects of different types of robots are shown in **Figure 7.**

Figure 7. Rehabilitation effects of different types of robots

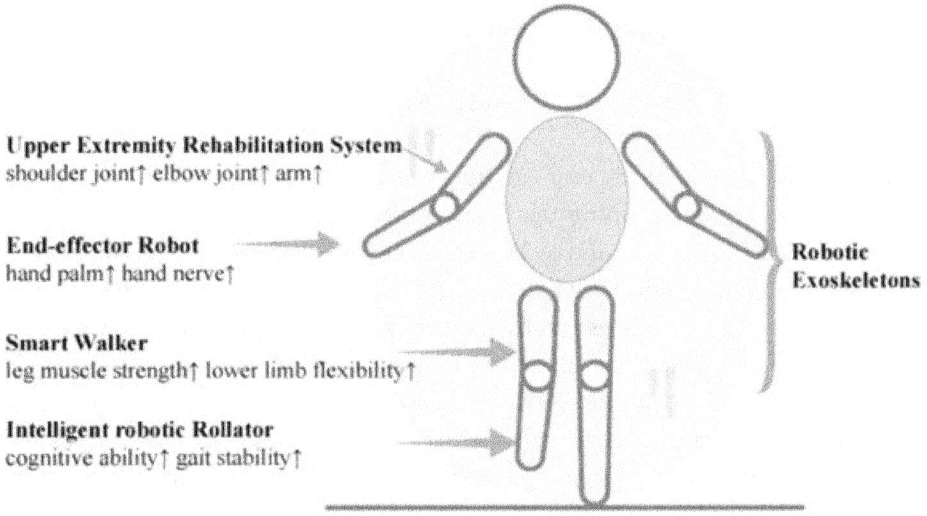

(Ju et al., 2023)

Artificial intelligence algorithms modify treatment programs in response to changing patient conditions. Chatbots and virtual assistants powered by artificial intelligence can help with medical questions and diagnoses around the clock. They improve communication with patients and lessen the workload of healthcare workers. Internet of Things devices track patient compliance with medication schedules and notify them when doses are due. AI-recommended medication regimes can help achieve better health outcomes.

Some features of AI-driven mental health services are support, counseling, and CBT. The increasing need for care in the mental health field is being met with these new approaches. Companionship and assistance with daily activities are two ways robots aid in caring for the elderly. The Internet of Things sensors that check on seniors regularly make ageing in place possible. Artificial intelligence examines healthcare data to spot patterns, forecast illness outbreaks, and optimize available resources. Healthcare firms can make better decisions with the help of predictive analytics. Connected medical devices safely send patient information for off-site analysis and treatment. This shift increases the availability of healthcare in underprivileged areas and facilitates better diagnosis of diseases at an early stage. Robots using artificial intelligence and Internet of Things technologies are helpful in search and rescue operations and managing potentially dangerous substances. First responders' protection is thereby improved. Cybersecurity systems driven by AI safeguard sensitive healthcare information. They monitor for and remediate security incidents, protecting sensitive patient information (Ju et al., 2023). These new developments show how AI, IoT, and robotics can be used to revolutionize the healthcare system. These developments have the potential to enhance patient care, increase healthcare efficiency, and meet the needs of a changing healthcare system.

Case Studies of Successful Integration

IBM Watson for Oncology is an artificial intelligence-driven platform designed to aid physicians in determining the best course of treatment for cancer patients. Integrating with EHRs, it analyzes large volumes of data from medical journals, clinical trials, and individual patients to develop tailored treatment plans. IBM Watson for Oncology was used to assess breast cancer cases and suggest therapies in a trial done in India. Treatment plans suggested by human oncologists were compared to those suggested by the AI system. According to the study's findings, Watson's advice agreed with that of human physicians in 96% of breast cancer cases. As a result of AI's incorporation into cancer care, doctors' workloads were lightened while patients gained access to cutting-edge research and treatment alternatives. Medtronic, a medical equipment manufacturer, has created the Internet of Things (IoT), which enables insulin pumps for people with diabetes. These pumps check glucose levels constantly and make real-time adjustments to insulin delivery. Patients' blood sugar levels can be tracked and treated accordingly because of the pumps' connectivity with a cloud-based platform.

A person with type 1 diabetes was using the Internet of Things-enabled Medtronic insulin pump. When the patient's glucose levels dropped significantly over the night, the pump adjusted insulin delivery to prevent hypoglycemia. At the same time, the pump notified the patient's doctor, who checked the data remotely to ensure the pump's response was suitable. In addition to averting a potentially disastrous incident, this Internet of Things connection showed how diabetes treatment may be enhanced by remote patient monitoring and intervention. Minimally invasive surgeries are among those that benefit from the da Vinci Surgical System, which is a robotic-assisted surgical platform. Robotic arms equipped with precise equipment and a 3D high-definition camera are controlled by surgeons, who, therefore, gain better control and precision during surgery. The da Vinci Surgical System was used to perform a robot-assisted prostatectomy on a patient who had been diagnosed with prostate cancer. Compared to open surgery, the minimally invasive treatment resulted in a shorter hospital stay, less discomfort, and faster recovery. Due to the precision of the robotic technology, the surgeon was able to avoid damaging any vital nerves, reducing the likelihood of unpleasant after-effects like incontinence and erectile problems. The incorporation of robotics not only led to better surgical outcomes but also increased the patient's postoperative quality of life.

Many mental health services are using chatbots powered by artificial intelligence. These chatbots use NLP to have meaningful conversations with their users, suggesting techniques for dealing with difficult situations and providing comfort and empathy. An anxious and depressed patient reached out for help from a mental health chatbot driven by artificial intelligence. The chatbot initiated discussion, suggested calming methods, and suggested the individual seek medical attention if necessary. The chatbot's suggestion of a mental health professional was followed up on and eventually contacted by the patient. This application of AI not only helped the patient at the right moment, but it also helped them get the right kind of help, demonstrating AI's potential to enhance mental health outcomes. These examples show how AI, IoT, and robotics have been successfully implemented in healthcare settings, leading to better patient outcomes, more efficient care delivery, and broader availability of cutting-edge medical technologies. There is much hope for the future of medical care in these developing technologies (Tariq, 2025).

ETHICAL AND PRIVACY CONSIDERATIONS

Gathering and sharing patient data via IoT devices and AI systems can pose serious privacy hazards without proper safeguards. In order to protect their patients' personal information, healthcare providers should use strong data encryption, access controls, and secure storage. All patients should be aware of their data's potential uses in AI and IoT projects. Patients can then weigh the risks and advantages of their data decisions with full knowledge of the merits of informed consent. Healthcare providers using AI algorithms should be able to explain the reasoning behind any recommendations they make to patients. The process through which medical decisions are made should be made transparent to the patient. The data used to train an AI system may contain inherent biases that could unfairly impact some demographics. Equal treatment requires that healthcare facilities actively confront and reduce bias. Determining the fault is crucial when an AI, IoT, or robotic system causes medical mistakes or complications. There should be well-defined procedures for responding to and resolving such crises.

The ownership of patient data produced by Internet of Things gadgets and artificial intelligence systems raises ethical questions. Patients should be able to access and make choices about using their data. While AI can help prescribe a course of treatment, the patient should always have the final say. Patients should be free to make decisions that respect their priorities and ethics. Hacking and cyberattacks on Internet of Things devices threaten patient safety. Both patient information and the physical safety of medical equipment require stringent cybersecurity safeguards. Professionals in the healthcare industry need to find a middle ground between relying on technology to aid decision-making and letting their clinical judgment take precedence. A lack of critical thinking or medical mistakes could result from an excessive dependence on technology. The increased use of robotics and artificial intelligence could affect the healthcare labour market. Ethical issues include helping people who have lost their jobs find new positions or retrain for new ones (Tariq, 2024). Access to Internet of Things devices and cutting-edge AI-powered healthcare is standard. When using these tools, healthcare institutions must consider the need for universal access. The potential use of AI algorithms in making decisions regarding end-of-life care raises moral concerns about the utility of technology in such intimate settings. Artificial intelligence (AI) in terminal care should be guided by ethical considerations and the patient's wishes. Suppose we want to maximize the advantages while avoiding the hazards and protecting patients' rights and well-being by integrating AI, IoT, and robotics in healthcare. In that case, we must address these ethical and privacy considerations. Regulatory and ethical frameworks for the healthcare industry should develop with innovations in the field.

FUTURE PROSPECTS AND POSSIBILITIES

Artificial intelligence (AI), the Internet of Things (IoT), and robotics all have promising applications and future potential in healthcare. The use of AI will further progress personalized medicine. By analyzing a patient's genetic, clinical, and lifestyle data, AI-driven algorithms will provide individualized treatment regimens and medicines that are both effective and safe. The prevalence of IoT-enabled gadgets will increase, allowing for continuous tracking of those with long-term illnesses. It is expected that telehealth and other forms of virtual care will continue to grow, making medical attention more widely available even in underserved and rural locations. Diagnostic tools powered by AI will advance, allowing for quicker and more precise disease diagnosis. Consumers and medical professionals will have access to these instruments via wearable and mobile applications. More and more surgeries will be performed with the help of robots because of the benefits they provide to surgeons in terms of increased precision, more dexterity, and better visualization. As less intrusive techniques become the norm, patients will spend less time in the hospital and return to normal activities sooner.

AI in the pharmaceutical industry will hasten the process of finding new medicines and cut down on the money and time needed to bring them to market. Artificial intelligence-powered simulations and predictive models will help find promising drug candidates. The ability of chatbots and other AI-powered virtual assistants to provide healthcare information, support, and even preliminary medical assessments will improve over time. They will be available to answer questions and provide help related to healthcare whenever you need them. Innovative healthcare facilities will be built with the help of IoT adoption in hospitals. These settings will improve patient safety, streamline operations, and maximize resource utilization.

The use of AI-powered mental health platforms to provide treatment, counselling, and emotional support to those who seek it will grow in popularity. As mental health care increases, they can help meet that need. Predictive analytics will help find disease outbreaks, distribute resources, and enhance healthcare delivery with the help of artificial intelligence and the Internet of Things. This preventative measure will strengthen public health readiness. With the help of IoT-connected gadgets and electronic health records, patients will have easier access to their medical records. They will actively participate in their healthcare, using AI-powered health apps and wearables to encourage prevention and early intervention. Clinical decision assistance for healthcare practitioners, including AI-assisted diagnostic and therapy suggestions, will be available in real-time. The safety of patients will be increased while medical mistakes are decreased. Emerging difficulties in the context of artificial intelligence (AI), the Internet of Things (IoT), and robotics in healthcare will necessitate new ethical and regulatory frameworks to solve them. To ensure that AI, IoT, and robotics can be easily integrated into the current healthcare infrastructure, efforts will continue to increase interoperability of healthcare systems and data exchange.

Improved cybersecurity safeguards will shield medical records from future cyberattacks, maintaining their authenticity and privacy. In the future, healthcare professionals and AI-driven systems will work more closely, with humans and machines working together to give optimal treatment. There is great potential for healthcare delivery, patient outcomes, and medical research to be improved through artificial intelligence (AI), the Internet of Things (IoT), and robotics. With the development and refinement of these technologies, the healthcare system will undergo a revolutionary change, resulting in healthcare that is more readily available, more efficient, and more focused on the individual patient. **Table 2** lists the most recent algorithms utilized for the image process embedded in the robot applications.

Table 2. Algorithms and the architecture used for image processing

Reference	Model used
(Zhang et al., 2023)	Various image processing techniques for tasks such as object detection, tracking, and human-robot gaze interaction.
(Abedin et al., 2021)	From image data, a deep learning model is used to learn the relationship between robots and humans.
(Fu et al., 2022)	Deep learning model to estimate the gaze of humans from image data.
(Kolling et al., 2016)	Various image processing techniques include face detection and recognition, emotion recognition, gesture recognition, and activity recognition.

CONSIDERATIONS

While reviewing robotic facilitation in healthcare communication, several significant problems need to be considered to ensure that the findings are accurate and reliable. This is the first thing that needs to be done: the scope of the review must be explicitly established, including explicit criteria for the included studies and those that need to be included. To do this, it is vital to determine the numerous types of robotic technologies that should be considered, the areas of healthcare communication that should be stressed, and the time range for the literature search.

Another significant aspect that must be considered is assessing the quality of the studies incorporated into the analysis. Due to the rapid advancement of technology in the medical field, it is of the utmost importance to evaluate the studies not only for the methodological rigor they possess but also for their relevance to the technological standards and practices currently in place in the medical field. This is because the technologies used in the medical field are constantly evolving. As a consequence of this, it is ensured that the evaluation is reflective of the most recent and pertinent knowledge that is currently available in the sector.

In addition, the evaluation needs to consider the different patient groups and the many surroundings associated with healthcare. Robotic facilitation in healthcare communication may have various implications depending on the context. For example, the use of robotic facilitation may have different implications depending on whether it is used in an inpatient or outpatient setting or whether it is used for pediatric or geriatric care services. It is essential to be aware of these complexities in order to have a comprehensive understanding of the subject matter.

In conclusion, the review should emphasize ethical considerations, particularly preserving patient privacy and data. Because robotic technologies usually include the collection and processing of sensitive patient information, it is of the utmost importance to consider how these concerns are addressed in the body of literature that is now available. As part of this process, a study into the ethical frameworks and privacy precautions utilized in the research currently being examined is being carried out. As a result of the study of these characteristics, the systematic review can provide a thorough and nuanced understanding of robotic facilitation's function in the healthcare industry's communication process.

CONCLUSION

The convergence of advanced Artificial Intelligence (AI), the Internet of Things (IoT), and telemedicine can revolutionize healthcare. By harnessing the power of these technologies, healthcare providers can create a more efficient, accessible, and personalized healthcare system that prioritizes patient needs and preferences. This integration is not merely a trend but a transformative leap toward addressing some of our time's most complex medical challenges. These technologies collectively promise to revolutionize healthcare delivery, improve patient outcomes, and create a more efficient and personalized healthcare system.

As the global population ages and grows, the healthcare sector faces unprecedented challenges in delivering high-quality, patient-centered care. Telemedicine, supported by AI and IoT, can reach new heights of efficiency and effectiveness and offers a beacon of hope in addressing these challenges, enabling the co-creation of value services that prioritize patient needs and preferences. This synergy can ease complex medical propositions, such as managing chronic diseases, aging populations, and pandemics, by fostering innovation and enhancing patient-centered care. Patients can be more involved in their care, accessing their health data and receiving personalized health advice. This participatory approach empowers patients and leads to better health outcomes as patients are more likely to adhere to treatment plans they understand and are involved in creating.

From a future perspective, the continued convergence of Artificial Intelligence, robotics, and the Internet of Things (IoT) in healthcare holds immense potential for transformation. AI-driven personalized medicine will become increasingly sophisticated, offering highly tailored treatments based on comprehensive genetic, lifestyle, and medical data analysis. Robotics will continue to evolve, providing even more precise surgical capabilities and expanding their role in patient care and rehabilitation. IoT will enable seamless and continuous health monitoring, generating vast data that can be leveraged to improve health outcomes and operational efficiency. The transformative potential of these technologies is vast and far-reaching. As they become more advanced and integrated, they will enhance the quality and efficiency of healthcare services and unlock new opportunities for innovation and improvement.

By harnessing the power of these cutting-edge technologies to address the complex challenges of healthcare, we can create a more responsive, personalized, and effective healthcare system that meets the unique needs of every patient. In conclusion, integrating AI, robotics, and IoT in healthcare represents a groundbreaking medical science and technology milestone. These advancements promise to revolutionize healthcare by easing complex medical propositions, improving service delivery, and unlocking new prospects of personalization and efficiency. As nations strive towards a safer, more harmonious, and healthier world, these technologies will play a pivotal role in achieving these goals, ensuring better health outcomes and quality of life for people everywhere.

REFERENCES

Abdulmalek, S., Nasir, A., Jabbar, W. A., Almuhaya, M. A., Bairagi, A. K., Khan, M. A.-M., & Kee, S.-H. (2022). IoT-based healthcare-monitoring system towards improving quality of life: A review. *Health Care*. PMID: 36292441

Akkaş, M. A., Sokullu, R., & Çetin, H. E. (2020). Healthcare and patient monitoring using IoT. *Internet of Things : Engineering Cyber Physical Human Systems, 11*, 100173. DOI: 10.1016/j.iot.2020.100173

Al-Atawi, A. A., Khan, F., & Kim, C. G. (2022). Application and Challenges of IoT Healthcare System in COVID-19. *Sensors (Basel), 22*(19), 7304. DOI: 10.3390/s22197304 PMID: 36236404

Al-Humairi, S. N. S., & Hajamydeen, A. I. (2022). IoT-based healthcare monitoring practices during Covid-19: Prospects and approaches. In *Healthcare Systems and Health Informatics* (pp. 163–185). CRC Press.

Al-Humairi, S. N. S., Hajamydeen, A. I., & Razalli, H. (2022). Internet of Medical Things: Smart Healthcare Monitoring Systems and Their Potential Implementations. In *Healthcare Systems and Health Informatics* (pp. 127–150). CRC Press.

Al-Humairi, S. N. S., Yih, A. E., & Daud, R. J. (2022). Design and Development of an IoT Virtual Doctor Robot (IVDR) for the Hospital Station. In *2022 IEEE 10th Conference on Systems, Process & Control (ICSPC)*

Alabdulatif, A., Khalil, I., Yi, X., & Guizani, M. (2019). Secure Edge of Things for Smart Healthcare Surveillance Framework. *IEEE Access : Practical Innovations, Open Solutions, 7*, 31010–31021. DOI: 10.1109/ACCESS.2019.2899323

Aminizadeh, S., Heidari, A., Toumaj, S., Darbandi, M., Navimipour, N. J., Rezaei, M., Talebi, S., Azad, P., & Unal, M. (2023). The applications of machine learning techniques in medical data processing based on distributed computing and the Internet of Things. *Computer Methods and Programs in Biomedicine, 241*, 107745. DOI: 10.1016/j.cmpb.2023.107745 PMID: 37579550

Azad, M. A., Arshad, J., Mahmoud, S., & Imran, M. (2019). A Privacy-preserving Framework for Smart Context-aware Healthcare Applications. *Transactions on Emerging Telecommunications Technologies, 33*(8), e3634. Advance online publication. DOI: 10.1002/ett.3634

Berdigaliyev, N., & Aljofan, M. (2020). An overview of drug discovery and development. *Future Medicinal Chemistry, 12*(10), 939–947. DOI: 10.4155/fmc-2019-0307 PMID: 32270704

Chopade, S., Gupta, H. P., & Dutta, T. (2023). Survey on Sensors and Smart Devices for IoT Enabled Intelligent Healthcare System. *Wireless Personal Communications, 131*(3), 1957–1995. DOI: 10.1007/s11277-023-10528-8 PMID: 37360143

Daily, J., & Peterson, J. (2017). Predictive maintenance: How big data analysis can improve maintenance. In *Supply chain integration challenges in commercial aerospace: A comprehensive perspective on the aviation value chain* (pp. 267–278).

Del-Valle-Soto, C., Briseño, R. A., Valdivia, L. J., Velázquez, R., & Nolazco-Flores, J. A. (2023). Non-Invasive Monitoring of Vital Signs for the Elderly Using Low-Cost Wireless Sensor Networks: Exploring the Impact on Sleep and Home Security. *Future Internet, 15*(9), 287. DOI: 10.3390/fi15090287

Fehringer, H., & Stary, C. (2023). User-Informed Adaptation in IoT Home Healthcare: Grounding Development in Empirical Evidence. *Journal of Theoretical and Applied Electronic Commerce Research, 18*(4), 1901–1925. DOI: 10.3390/jtaer18040096

Furtado, D., Gygax, A. F., Chan, C. A., & Bush, A. I. (2023). Time to forge ahead: The Internet of Things for healthcare. *Digital Communications and Networks, 9*(1), 223–235. DOI: 10.1016/j.dcan.2022.06.007

Gambo, N., & Musonda, I. (2022). Impact of IoT on Achieving Smart Primary Healthcare Building Facilities in Gauteng, South Africa. *International Journal of Environmental Research and Public Health, 19*(18), 11147. DOI: 10.3390/ijerph191811147 PMID: 36141421

Grétarsson, B. Ö., Orfanidis, C., Marchegiani, L., & Fafoutis, X. (2023). Short-Term Wireless Connectivity Prediction for Connected Agricultural Vehicles. https://doi.org/DOI: 10.1109/ITSC57777.2023.10421944

Hayes, E. L., & Vance, K. A. (2020). Health insurance portability and accountability act of 1996: Health & Public Welfare. *Georgia State University Law Review, 37*(1), 153.

Ju, F., Wang, Y., Xie, B., Mi, Y., Zhao, M., & Cao, J. (2023). The Use of Sports Rehabilitation Robotics to Assist in the Recovery of Physical Abilities in Elderly Patients with Degenerative Diseases: A Literature Review. *Health Care*. PMID: 36766901

Kadhim, K. T., Alsahlany, A. M., Wadi, S. M., & Kadhum, H. T. (2020). An overview of patient's health status monitoring system based on internet of things (IoT). *Wireless Personal Communications, 114*(3), 2235–2262. DOI: 10.1007/s11277-020-07474-0

Kamangar, Z. U., Memon, R. A., Murtaza Memon, G., & Kamangar, U. A. (2023). Integration of internet of things and blockchain technology in healthcare domain: A systematic literature review. *International Journal of Communication Systems*, *36*(16), e5582. DOI: 10.1002/dac.5582

Kamilaris, A., & Botteghi, N. (2020). The Penetration of Internet of Things in Robotics: Towards a Web of Robotic Things. *Journal of Ambient Intelligence and Smart Environments*, *12*(6), 491–512. DOI: 10.3233/AIS-200582

Karar, M. E., Shehata, H. I., & Reyad, O. (2022). A survey of IoT-based fall detection for aiding elderly care: Sensors, methods, challenges and future trends. *Applied Sciences (Basel, Switzerland)*, *12*(7), 3276. DOI: 10.3390/app12073276

Kirchner, G. J., Kim, A. H., Kwart, A. H., Weddle, J. B., & Bible, J. E. (2023). Reported Events Associated With Spine Robots: An Analysis of the Food and Drug Administration's Manufacturer and User Facility Device Experience Database. *Global Spine Journal*, *13*(3), 21925682221126116. DOI: 10.1177/21925682221126116 PMID: 36074993

Kirchner, G. J., Kim, A. H., Smith, N. P., Martinazzi, B. J., Hines, S. M., Weddle, J. B., & Bible, J. E. (2023). Few randomized controlled trials in spine surgery in the United States include sociodemographic patient data: A systematic review. *JAAOS—. The Journal of the American Academy of Orthopaedic Surgeons*, *31*(8), 421–427. DOI: 10.5435/JAAOS-D-22-00838 PMID: 36735417

Liu, I. H., Lee, M.-H., Huang, H.-C., & Li, J.-S. (2023). 5g-Based Smart Healthcare and Mobile Network Security: Combating Fake Base Stations. *Applied Sciences (Basel, Switzerland)*, *13*(20), 11565. DOI: 10.3390/app132011565

Meinert, E., Velthoven, M. H., Brindley, D., Alturkistani, A., Foley, K., Rees, S., Wells, G., & Pennington, N. (2018). The Internet of Things in Health Care in Oxford: Protocol for Proof-of-Concept Projects. *JMIR Research Protocols*, *7*(12), e12077. DOI: 10.2196/12077 PMID: 30514695

Menbawy, N. E., Ali, H. A., Saraya, M. S., Ali-Eldin, A., & Abdelsalam, M. M. (2023). Energy-Efficient Computation Offloading Using Hybrid GA With PSO in Internet of Robotic Things Environment. *The Journal of Supercomputing*, *79*(17), 20076–20115. DOI: 10.1007/s11227-023-05387-w

Michalík, R., Janota, A., Gregor, M., & Hruboš, M. (2021). Human-Robot Motion Control Application With Artificial Intelligence for a Cooperating YuMi Robot. *Electronics (Basel)*, *10*(16), 1976. DOI: 10.3390/electronics10161976

Mishra, P., & Singh, G. (2023). Internet of Medical Things Healthcare for Sustainable Smart Cities: Current Status and Future Prospects. *Applied Sciences (Basel, Switzerland), 13*(15), 8869. DOI: 10.3390/app13158869

Montenegro, R. E., & Dori-Hacohen, G. (2020). Morality in sugar talk: Presenting blood glucose levels in routine diabetes medical visits. *Social Science & Medicine, 253*, 112925. DOI: 10.1016/j.socscimed.2020.112925 PMID: 32244153

Mukati, N., Namdev, N., Dilip, R., Hemalatha, N., Dhiman, V., & Sahu, B. (2023). Healthcare assistance to COVID-19 patient using internet of things (IoT) enabled technologies. *Materials Today: Proceedings, 80*, 3777–3781. DOI: 10.1016/j.matpr.2021.07.379 PMID: 34336599

Park, J.-H., & Han, M.-H. (2023). Enhancing Livestock Management with IoT-based Wireless Sensor Networks: A Comprehensive Approach for Health Monitoring, Location Tracking, Behavior Analysis, and Environmental Optimization. *Journal of Sustainable Urban Futures, 13*(6), 34–46.

Pollard, T. D., Ong, J. J., Goyanes, A., Orlu, M., Gaisford, S., Elbadawi, M., & Basit, A. W. (2021). Electrochemical biosensors: A nexus for precision medicine. *Drug Discovery Today, 26*(1), 69–79. DOI: 10.1016/j.drudis.2020.10.021 PMID: 33137482

Queralta, J. P., Li, Q., Zou, Z., & Westerlund, T. (2020). Enhancing Autonomy With Blockchain and Multi-Access Edge Computing in Distributed Robotic Systems. https://doi.org/DOI: 10.1109/FMEC49853.2020.9144809

Ragab, M., & Binyamin, S. S. (2023). Intelligent Energy-Aware Thermal Exchange Optimization With Deep Learning Model for IoT-Enabled Smart Healthcare. *Journal of Healthcare Engineering, 2023*(1), 1–13. DOI: 10.1155/2023/3830857 PMID: 37483302

Rashid, A. (2024). Internet of Things (IoT) Its Applications and Enabling Technologies in Healthcare: A Review. *Educational Administration: Theory and Practice, 30*(4), 9699–9714.

Ratta, P., Kaur, A., Sharma, S., Shabaz, M., & Dhiman, G. (2021). Application of blockchain and internet of things in healthcare and medical sector: Applications, challenges, and future perspectives. *Journal of Food Quality, 2021*(1), 7608296. DOI: 10.1155/2021/7608296

Saha, R., Kumar, G., Rai, M. K., Thomas, R., & Lim, S.-J. (2019). Privacy Ensured e{e}e-healthcare for fog-enhanced IoT based applications. *IEEE Access : Practical Innovations, Open Solutions, 7*, 44536–44543. DOI: 10.1109/ACCESS.2019.2908664

Shahid, J., Ahmad, R., Kiani, A. K., Ahmad, T., Saeed, S., & Almuhaideb, A. M. (2022). Data Protection and Privacy of the Internet of Healthcare Things (IoHTs). *Applied Sciences (Basel, Switzerland), 12*(4), 1927. DOI: 10.3390/app12041927

Simoens, P., Dragone, M., & Saffiotti, A. (2018). The Internet of Robotic Things. *International Journal of Advanced Robotic Systems, 15*(1), 172988141875942. DOI: 10.1177/1729881418759424

Sivaraman, H. (2019). IoT-Enabled Healthcare Monitoring: A Systematic Review of Wearable Devices. *Information Technology In Industry, 7*(3), 78–86. DOI: 10.17762/itii.v7i3.815

Taiwo, O., & Ezugwu, A. E. (2020). Smart Healthcare Support for Remote Patient Monitoring During Covid-19 Quarantine. *Informatics in Medicine Unlocked, 20*, 100428. DOI: 10.1016/j.imu.2020.100428 PMID: 32953970

Tariq, M. U. (2024). Application of blockchain and Internet of Things (IoT) in modern business. In Sinha, M., Bhandari, A., Priya, S., & Kabiraj, S. (Eds.), *Future of customer engagement through marketing intelligence* (pp. 66–94). IGI Global., DOI: 10.4018/979-8-3693-2367-0.ch004

Tariq, M. U. (2024). AI and IoT in flood forecasting and mitigation: A comprehensive approach. In Ouaissa, M., Ouaissa, M., Boulouard, Z., Iwendi, C., & Krichen, M. (Eds.), *AI and IoT for proactive disaster management* (pp. 26–60). IGI Global., DOI: 10.4018/979-8-3693-3896-4.ch003

Tariq, M. U. (2024). The transformation of healthcare through AI-driven diagnostics. In Sharma, A., Chanderwal, N., Tyagi, S., Upadhyay, P., & Tyagi, A. (Eds.), *Enhancing medical imaging with emerging technologies* (pp. 250–264). IGI Global., DOI: 10.4018/979-8-3693-5261-8.ch015

Tariq, M. U. (2024). Emerging trends and innovations in blockchain-digital twin integration for green investments: A case study perspective. In Jafar, S., Rodriguez, R., Kannan, H., Akhtar, S., & Plugmann, P. (Eds.), *Harnessing blockchain-digital twin fusion for sustainable investments* (pp. 148–175). IGI Global., DOI: 10.4018/979-8-3693-1878-2.ch007

Tariq, M. U. (2025). Empowering Learning Through Networked and Connected Education: Case Studies in Digital Engagement. In Gierhart, A. (Ed.), *Cases on Enhancing P-16 Student Engagement With Digital Technologies* (pp. 169–198). IGI Global Scientific Publishing., DOI: 10.4018/979-8-3693-5633-3.ch007

Tariq, M. U. (2025). Quantum Algorithms and Predictive Analytics: Revolutionizing Consumer Insight. In Gajbhiye, P., Kannan, H., Rodriguez, R., & Rojas-Méndez, J. (Eds.), *The Quantum AI Era of Neuromarketing* (pp. 173–202). IGI Global Scientific Publishing., DOI: 10.4018/979-8-3693-7673-7.ch008

Tariq, M. U., & Sergio, R. P. (2025). Ethical Considerations and Social Impact of AI and Computational Intelligence in Society 5.0. In Pandey, R., Srivastava, N., Prasad, R., Prasad, J., & Garcia, M. (Eds.), *Open AI and Computational Intelligence for Society 5.0* (pp. 225–248). IGI Global Scientific Publishing., DOI: 10.4018/979-8-3693-4326-5.ch010

Teoh, Y. K., Gill, S. S., & Parlikad, A. K. (2021). IoT and fog computing based predictive maintenance model for effective asset management in industry 4.0 using machine learning. *IEEE Internet of Things Journal*.

Tzafestas, S. G. (2018). Synergy of IoT and AI in Modern Society: The Robotics and Automation Case. *Robotics & Automation Engineering Journal*, *3*(5). Advance online publication. DOI: 10.19080/RAEJ.2018.03.555621

Vermesan, O., Bröring, A., Τράγος, H., Serrano, M. M., Bacciu, D., Chessa, S., Gallicchio, C., Micheli, A., Dragone, M., Saffiotti, A., Simoens, P., Crea, F., & Bahr, R. (2022). *Internet of Robotic Things – Converging Sensing/Actuating*. Hyperconnectivity, Artificial Intelligence and IoT Platforms., DOI: 10.1201/9781003337584-4

Zeadally, S., & Bello, O. (2021). Harnessing the power of Internet of Things based connectivity to improve healthcare. *Internet of Things : Engineering Cyber Physical Human Systems*, *14*, 100074. DOI: 10.1016/j.iot.2019.100074

Zhou, X. (2024). IoT Sensor-Based Big Data and Information Systems in Smart Healthcare. *Highlights in Science. Engineering and Technology*, *92*, 397–402.

KEY TERMS AND DEFINITIONS

Internet of Things (IoT): A network of interconnected devices that collect and exchange data to enhance remote healthcare monitoring and service delivery.

Remote Patient Monitoring: The continuous tracking of patients' health metrics via connected devices, enabling timely interventions and personalised care.

Blockchain Technology: A decentralised digital ledger that ensures data security, integrity, and privacy in healthcare applications.

Artificial Intelligence (AI): Advanced algorithms used to analyse sensor data, predict health risks, and support the development of personalised treatment plans.

Interoperability: The capability of diverse healthcare systems and devices to communicate effectively, facilitating seamless data exchange and integration.

Chapter 5
Qualitative Analysis of Authentication Mechanisms for Secure Communication in the Internet of Medical Things

Shivam K. Prajapati
https://orcid.org/0009-0002-6318-9958
Symbiosis Institute of Digital and Telecom Management, India

Prasanna Kulkarni
https://orcid.org/0000-0003-1620-6380
Symbiosis Institute of Digital and Telecom Management, India

C. G. Balaji
https://orcid.org/0000-0001-8870-1634
Symbiosis Institute of Digital and Telecom Management, India

ABSTRACT

The Internet of Medical Things (IoMT) is a subset of IoT, where the use of connected devices in medicine and healthcare has rampantly expanded in the past few years. IoMT comprises of smart medical instruments, wearable technologies, and sensors that synchronize, capture, and evaluate health information. Good authentication method is essential to provide a secure communication in the IoMT environment. Security threats such as unlawful access, invasion of the data, and jeopardiza-

DOI: 10.4018/979-8-3373-4332-7.ch005

tion of patients' safety are derivable if the authentication is weak or powerless. Thus, this chapter seeks to provide a qualitative analysis of the different type of authentications that can be used in IoMT. The goal is to introduce best practices in identifying proper and reliable authentication solutions that are fundamental in making IoMT safe and reliable. Therefore, with regard to the utilization of IoMT, it is apparent that awareness of the benefits and drawbacks of various mechanism enables appropriate decisions to be made for the protection of patient data and the successful operation of IoMT systems.

INTRODUCTION

As a branch of the Internet of Things, or IoT, the Internet of Medical Things or IoMT is a specialized category that has been rapidly evolving in recent years (Huang et al., 2023). While IoT covers a vast spectrum of interconnected devices, from mundane household appliances like smart refrigerators to complex urban traffic management systems, IoMT zeroes in on the intricate web of healthcare-related technologies. This niche focus encompasses a triad of critical elements: connected medical devices and appliances, wearable health tech, and overarching healthcare systems. It's an ambitious endeavor aimed at the collection, exchange, and analysis of medical data on an unprecedented scale. Imagine a scenario where a patient does a smartwatch, not just as a timepiece, but as a silent sentinel, vigilantly tracking their heart rate, blood pressure, and daily activities. This wealth of biometric data doesn't simply reside in the device; instead, it seamlessly cascades into an anonymous cloud application, where clinicians can scrutinize real-time trend data, potentially uncovering hidden patterns or alarming anomalies that might otherwise go unnoticed.

The IoMT environment, which has its branches spread out across the entire spectrum of the healthcare sector, brings a possibility of a new era of medical practice and personal health management. It is not just about connecting; it is about establishing interactions where patients, devices, and even healthcare systems must mutually interact. In this elaborate tango of datasets and diagnostics, IoMT simply watches and contributes. Consider the potential of IoMT in chronic disease management: a diabetic patient's monitor may not only notify them of the abnormal glucose level but also relay the same to the medical practitioners for immediate modification of the patient's regimen. Such an instantaneous specificity as the care, resulting from the data analysis, was once only seen in the sci-fi movies, but it is quickly becoming the new norm.

However, the proliferation of IoMT does bring its many problems and the issues of ethics into the conversation. While people rush towards a world where personal health information is shared down the Facebooks and Twitters of the world, the issues of personal privacy, protection of personal health information, and ownership of such data turn to be highly sensitive. While IoMT is comprised of a vast network of interconnected devices, this strength is also the network's weakness, as a cyberattack on this system could be apocalyptic. Moreover, the sheer volume of data generated by IoMT devices presents both an opportunity and a conundrum: how then filter this tidal wave of content? this is the key question whenever one is trying to move through the information age without being overwhelmed? What is important to note, however, is that we are at the edge of this revolution in the provisioning of healthcare services, through IoMT, and it is vital that as we chart these waters, we do not lose our patients' trust and choice-making capabilities in the process.

The IoMT Ecosystem

The components of the IoMT architecture in accordance with (Dhanvijay & Patil, 2019) are shown in Figure 1, and consist of the following layers. The IoMT **Devices Layer** includes wearable sensors as well as Implantable devices and Remote monitoring equipment which forms the central data providers in the appointed ecosystem. The **Edge Layer** includes local gateways and fog nodes that engage in data collection and data processing from IoMT devices as well as forwarding the information to the cloud layer. The **Network Layer** depicts the communication media through which data travels in the layers comprised of 5G, Wi-Fi, and Bluetooth among others. The **Cloud Layer** contains the storage area where raw data are stored, analysis engines, and AI/ ML algorithms that facilitate the processing of data. Thus, the tier of **Application Layer** provides patient applications, health care provider interfaces, and

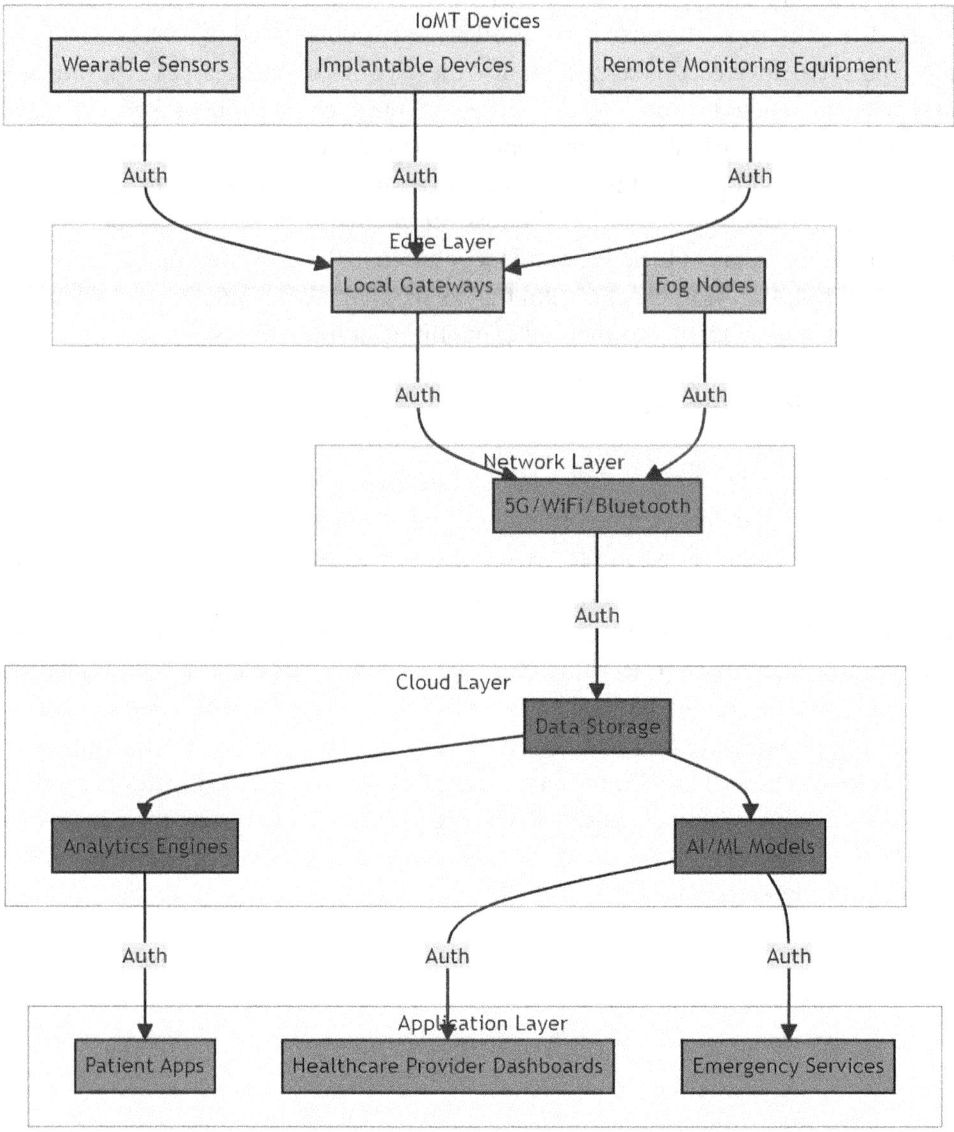

The arrows labeled with "Auth" indicate points where authentication is critical to ensure secure communication between different components of the ecosystem. This figure helps visualize the complex nature of the IoMT ecosystem and emphasizes the importance of implementing robust authentication mechanisms at various levels to maintain the security and integrity of sensitive medical data.

Distinct Characteristics of IoMT Devices

The category of IoMT subsequently differs from a broader IoT devices type because it targets healthcare sector only. Below are the most important features that can help in outlining the main distinguishing factors of IoMT. In contrast to IoT devices that are used daily for everyday conveniences, IoMT devices stay very close to the purpose of providing accurate measurement data (Razdan & Sharma, 2021). Indeed, inaccuracies of this data may be the difference between life and death in some medical decisions. For instance, pacemaker needs high precision, which measures the patient's heartbeat and adjusts the rhythm in the body tapping into the vast importance of precision in IoMT. IoMT devices are involved in managing a variety of patient information inclusive of the basic physiological data to the patient's treatment plans. Data security at this level becomes very important to ensure that this data is not accessed by unauthorized persons or exposed to breaches. This can be observed using encryption, secure datalink layers, tokens or biometrics, which are the pillars of any secure IoMT environment. While general IoT devices are used in many areas of consumer's lives (Lee, 2021), IoMT are limited to the application in healthcare. This includes everything in between wearable Electrocardiography (ECG) monitors that record the heart's activity, to smart insulin pumps that deliver the right amounts of medicine. This means that IoMT applications can range from patient monitoring and remote diagnosing to even delivering drugs into people's veins, giving a glimpse into how IoMT can revolutionize healthcare systems (Tariq, 2024).

IoMT devices need to conform to certain healthcare standards like Health Insurance Portability and Accountability Act (HIPAA) (Act, 1996) in the case of IoMT devices in the United States. There is another set of regulations for patient personal information and data security. For example, this is enforced by HIPAA where there are required protocols on how the patient health information (PHI) is to be handled, which IoMT devices have to meet for the patients' faith and data safety. This is one of the most critical barriers to the sustainable enhancement of the distributed ecosystem of IoMT devices and other IT systems in the healthcare facilities. Thus, an absence of unique communication platforms can cause interfacing problems, which results in the formation of barriers to data exchange in the context of the envisaged healthcare environment (Tariq, 2025).

Understanding the Differences Between IoMT and IoT

The availability of several IoT devices increases in different sectors due to comfort, efficiency, and pattern of control in different areas. However, there is the broader definition of the Internet of Things which finds IoMT nestled within, but with a special emphasis on healthcare. Here's what sets IoMT apart from IoT (Roberts,

2023), which is also given in Table 1 for quick reference. The ordinary IoT gadgets are versatile to various usages, IoMT is specialized for the medical industry. From the case of smartwatches that track heart activity to automated insulin delivery devices that provide accurate dosage of medication, IoMT solutions are all aimed and enhancing patients' well-being.

Table 1. Difference between IoT and IoMT

Aspect	IoT	IoMT
Focus	General applications across industries	Healthcare-specific applications
Purpose	Convenience, automation, and efficiency	Improved patient care and health outcomes
Design Priorities	User-friendliness, convenience	Accuracy, reliability, and security
Data Collection	Diverse data sources (e.g., sensors)	Medical data from devices and wearables
Challenges	Scalability, security, privacy	Regulatory compliance, interoperability

The user friendliness is highly emphasized in general in IoT devices which ensure that their devices are easy to use by consumers. On the other hand, IoMT focuses on the quality of the data, their reliability, and durable protection. The decision-making process in the medical facility relies mostly on the data captured by the IoMT devices, therefore, any small variation is consequential. Moreover, patients' privacy is at risk due to the nature of the collected data, so it is crucial to have robust protection and security within the IoMT environment. Standard IoT devices in our homes and surroundings collect different data from the sensors that are installed. On the other hand, most IoMT devices mainly capture health information right from the patient's own body. This information can be collected by regular health trackers or specialized health monitoring equipment, such as glucose meters, and integrates into a single health overview of a patient.

The major difficulties that concern general IoT are still relevant: scalability, security and privacy, but IoMT is confronted with some more issues. It is very important that regulatory compliance is achieved to protect the patient's information. Also, there is a need for interconnectivity of IoMT devices with different architectures incorporated in the targeted healthcare system and the existing IT systems for effective data sharing. These findings help to explain that IoMT is a set of very specific devices and the extent of their potential for revolutionizing healthcare. In this way, as IoMT is more developed, approaches to address the challenges that arise with its implementation will be critical to get the best out of it for the benefits of mankind.

Advantages of IoMT

1. **Enhanced Healthcare**: IoMT enables real-time monitoring, early diagnosis, and personalized treatment, leading to better patient outcomes.
2. **Remote Patient Monitoring**: Patients can be monitored remotely, reducing hospital visits and improving accessibility.
3. **Efficiency**: IoMT streamlines healthcare processes, minimizing paperwork and administrative tasks.
4. **Predictive Analytics**: Data-driven insights help predict disease trends and optimize resource allocation.

Disadvantages of IoMT

1. **Security Risks**: Protecting sensitive medical data is crucial; vulnerabilities can lead to privacy breaches.
2. **Interoperability Challenges**: Integrating diverse medical devices and systems remains complex.
3. **Regulatory Compliance**: Compliance with healthcare regulations (e.g., HIPAA) is essential.
4. **Cost and Implementation**: Deploying IoMT solutions can be expensive and require infrastructure upgrades.

The IoMT focuses exclusively on healthcare-related elements. Unlike the broader IoT, which encompasses everything from connected refrigerators to city-wide traffic systems, IoMT narrows its scope to precision medicine, patient care, and health monitoring. IoMT devices prioritize accuracy, reliability, and security over user-friendliness, ensuring that medical data remains confidential and trustworthy.

Based on the current trends in the IoMT solutions, medical devices, wearables and health care systems collaborate to revolutionize patient centric care. Wearable devices imply preventive health examination and monitoring such as fitness tracker, and glucose monitor. As enabled through IoMT, telemedicine is a form of consultation that does not have geographical restrictions and helps to increase the access to services. Finally, predictive analytics employ data from the previous years to predict the occurrence of diseases and the vulnerable communities, and to determine best use of resources. The IoMT bridges the gap between healthcare and technology, promising a future where well-being is not just a goal but a reality. Its impact on patient care, efficiency, and accessibility continues to evolve, shaping the healthcare landscape for the better. IoMT has become the backbone of the change in the healthcare system across each broad category and made patient care possible from distant care and improved chronic disease care to making practical data management a part of healthcare delivery (Tariq, 2024).

Remote Patient Monitoring (RPM)

Remote Patient Monitoring (RPM) is defined as the process of capturing health information from patients outside the conventional health care structures (Zohaib et al., 2024). They use IoMT devices in keeping track of the patient's physical signs/symptoms as well as their compliance to medication regimes. The features of RPM allow for monitoring the parameters such as blood glucose levels, blood pressure, and heart rate at any given point. This is because initial tracked values help in early detection of any deviation from the baseline and patients can remain at home while still being supervised hence help in avoiding hospital remission and general health care costs.

Chronic Disease Management

Chronic illness is one of the main focuses of health care and IoMT technology with its use of wearable devices (Dadkhah et al., 2021). Some examples of how mHealth is being implemented is through smart glucose monitors for managing diabetes, cardiac monitors to send constant data of heart conditions, and smart inhalers to effectively control asthma. All these wearables give patient specific care, engage the patient in his/her treatment, and avoid any worsening of the condition.

Data-Driven Healthcare

IoMT produces huge volumes of patients' data (Patil et al., 2023) such as wearable sensors data, electronic health records (EHRs) and genomic data. Data is analyzed by various sophisticated tools to derive information from it. The analysis of previous data enables general tendencies in diseases and risky population groups to be forecasted and resources be more efficiently used. IoMT also supports pervasive patient health monitoring (PPHM), drug discovery, and public health design (Tariq, 2025).

Challenges and Considerations

Security and privacy of patients' information is critical. There are still issues with integrating the developed solutions with existing health care systems and strict adherence to health compliance regulations such as HIPAA. The use of IoMT must be done ethically to ensure that there is adequate collection of data while at the same time respecting the rights of the patients about their consent and privacy.

As we will expand the usage of technologies further, only flagrant cooperation of healthcare providers, scientists, and IT specialists will define significant changes in the treatment outcomes. The IoMT or interconnected network of medical devices and applications creates the prospects of radical improvements to patient care and treatment systems. However, this convergence also creates new security threats that require some form of addressing in advance. One issue is that the IoMT devices are sometimes beset by numerous susceptibilities.

Security and privacy are core aspirations in IoMT structures (Thavavel et al., 2021). Although, integrated systems facilitate continuum of data across various systems thus improving healthcare provision, the issue of patient's privacy is of major concern. The verification mechanism ensures that only authorized people, or rather authorized personnel, can access the patient data, while encryption forms part of the mechanism of protecting such information.

Device integrity is also just as important to safeguarding (Koutras et al., 2020). All the connected IoMT devices must identify before they can request data, and before they can send data. This discourages the use of unauthorized devices to access the available networks in the organization. Secure booting procedures also authenticate the firmware of the device during booting while offering identities to devices assist in tracking their activities and responsibility.

Nevertheless, IoMT security has its issues. The ecosystem is in an elaborate one that consists of several devices, wearables, and gateways, thereby complicating the level of coordination across it. When multi-factor authentication (MFA) has been adopted, that means, a person's password (what you know), smart card (what you have) and fingerprint (who you are) all argue in favor of security. Role based access control involves the creation of roles defining the users' access rights and the objects those users must interact with. The role-based access control (RBAC) feature avails roles and other forms of grant in accordance with the data that the users are authorized to access. It means that there is always someone who pays attention to deviations and unauthorized intrusions.

Qualitative analysis (Lester et al., 2020) is the assessment of quality, characteristic, and attribute of a system or process with the use of qualitative data and not numerical data. Consequently, qualitative analysis has made a massive contribution in understanding the advantages and disadvantages of the different types of authentication protocols in IoMT security.

QUALITATIVE ANALYSIS IN IOMT SECURITY

Unlike quantitative approaches, qualitative analysis entails a process other than measurement and the use of numbers (Malamas et al., 2021). On the other hand, it is directed more towards the descriptive features like behaviour, use, and environment. Here's how it applies to evaluating authentication mechanisms for IoMT. Here is how it applies to evaluating authentication mechanisms for IoMT:

1. **Protocol Design and Implementation:** Qualitative analysis mainly focuses on the overall design principles and details of the implementation of the authentication protocols. Some of the attributes include simplicity relating to how easy it is to use as well as the clarity of the interface depicting good security practices.
2. **Usability and User Experience:** A good mechanism of authentication should be friendly to the users/ clients. The qualitative analysis focuses on factors such as simplicity of the implementation process, as well as interaction with a given technology by the users, including the healthcare practitioners and the patients.
3. **Threat Models and Scenarios:** Threat models unique to IoMTs focus of the study among the researchers. Quantitative assessment focuses on the notion of ensuring that the developed protocol also provides good solutions to most threats (for example, man-in-the-middle attacks, replay attacks) and in formulating good responses to other newly emerging threats.

Evaluating Authentication Mechanisms

This section deals with how qualitative analysis helps evaluate different authentication mechanisms: The taxonomy and literature review are centered on the categorization of the existing authentication schemes concerning their characteristics, the region of deployment – cloud, fog, and edge and the cryptographic techniques used, as stated by (Alsaeed & Nadeem, 2022).

Even though both quantitative and qualitative analysis can be employed to point out security flaws and vulnerability, several issues concerning the authentication protocols can be identified using qualitative analysis. For instance, there are schemes that have vulnerabilities such as, inability to withstand node compromise or replay attacks (Rahman & Jahankhani, 2021). They also describe problems associated with key management, session creation, and secure transmission. The Table 2 is a snapshot depicting different authentication considerations with regards to the IoMT and the strengths and weaknesses associated with them.

Table 2. IoMT authentication mechanisms

Mechanism	Strengths	Weaknesses
Password-based	Simple to implement	Vulnerable to brute force attack
	Widely understood by users	Often weak due to user habits
	Easily changeable	Susceptible to phishing
Certificate-based	Strong security guarantees	Complex setup
	Resilient to replay attacks	High computational overhead
Biometric-based	User-friendly	Privacy concerns (biometric data)
	Difficult to compromise	Limited scalability
Token-based (OTP)	Simplicity and low overhead	Vulnerable to device loss or theft
	Suitable for resource-constrained devices	Single point of failure (the token)

When examining usability, security, and real-life outcomes in a broader sense, researchers can make more relevant choices when considering or implementing authentication and authorization methods. Because IoMT is a continuously developing field, the quantifies are still causing improvements and guarantee the enhanced security for the connected medical instruments. The authentication mechanism comparison is given in in the following Table 3 which is self-explanatory.

Table 3. Authentication mechanisms comparison chart for IoMT

Criteria	Password-based	Biometric-based	Token-based	Certificate-based
Security Strength	••	••••	•••	•••••
Ease of Use	•••	•••••	••••	••
Implementation Complexity	•	•••	••	••••
Scalability	••	•••	••••	•••••
Resistance to Phishing	•	••••	•••	•••••
Suitability for IoMT Devices	••	•••	••••	•••••
Cost-effectiveness	•••••	••	••••	•••
User Privacy	••	•	•••	••••
Regulatory Compliance	••	•••	•••	•••••

Legend: • = Poor, •• = Fair, ••• = Good, •••• = Very Good, ••••• = Excellent

Qualitative Analysis of Authentication Mechanisms

The analysis of the type of authentication aims at assessing diverse approaches used to validate the user or the device, assessing their effectiveness and inefficiencies, and consideration of each strategy in a more complex setting notably in IoMT systems.

Security Attributes

When evaluating authentication mechanisms for IoMT communication, security attributes take center-stage (Svandova & Smutny, 2024). They describe the basic set of protection measures which guarantee the reliability, security and confidentiality of information shared between IoMT devices.

- **Confidentiality:** In the case of IoMT, confidentiality backs up a patient's information, in the sense that it prevents it from falling into the wrong hands. Suppose that information about the patient's heart rate, transmitted by a pacemaker, must be considered in this case? This means that there is a strong authentication to allow only the authorized health care personnel to access this data. If confidentiality is not maintained, then people might be able to listen in on this data and harm patient privacy or even discover some soft points and take advantage of it.
- **Integrity:** This attribute makes sure that information exchanged between point-of-origin and the destination IoMT device is not changed. As a subfield of medicine, data integrity is highly valued in IoMT since it enables correct diagnoses and treatments. For instance, a glucose monitoring system can transmit actual blood sugar level to a doctor at any given time. This is entailed in a strong authentication mechanism that provides integrity and means that the doctor receives correct information for important decisions such as when recommending a course of treatment.
- **Non-Repudiation:** This attribute creates responsibility since once the message is sent, the sender cannot come and deny having sent the message. The aspect of non-repudiation provides a key for clarity in the audit and the prevention of different scams in IoMT. Suppose an insulin pump gets a command to change the dosage. Non-repudiation means that only the entitled workers can launch such commands, which cannot be falsely attributed in a situation of mishaps or other undesirable circumstances.
- **Mutual Authentication:** This attribute does not just stop at authenticating the sender of the message. It also affirms that the device receiving the data is the rightful recipient of the data being received. When the communication is performed by devices that are constrained in terms of the available resources,

there is a need for M2M mutual authentication and avoid impersonation attacks. Let us assume that there is a pacemaker that has got firmware updates. The authentication ensures that the update is from the correct and trusted source, and that the pacemaker is the target, rather than allowing the malicious firmware that would corrupt the functionality of the device and perhaps put patient's lives at risk.

Another key attribute of IoMT security is that all the listed components are interdependent, and each contributes to the security of the communication. When choosing the strategies for authentication, relying on their capacities to protect the enumerated attributes, we can guarantee the secure data exchange in cases of IoMT that in its turn will save the patient's privacy and data, and by the same token, patients' health.

Balancing Security and Efficiency

Although stringent security properties are required for IoMT communication, attaining them must be done so, considering the resource-scarce medical devices. Whereas computing servers are powerful large appliances, most IoMT devices are portable with batteries that power them and are computationally and memory restricted. This delicateness of work-measure trade-off presents a major concern in the IoMT domain of authentication strategies.

Computational Overhead

To enable the authentication mechanisms, many algorithms demand functions such as encryption, decryption, and verifying digital signatures among others. Such operations may be costly in terms of time and computational resources in terms of CPU usage and amount of required memory. For resource-constrained IoMT devices, such as implantable sensors or wearable health monitors, the computational overhead (Osama & Nour, 2023) associated with complex authentication mechanisms can lead to:

- **Increased Power Consumption:** Crystal growth arising from enhanced cryptographic computation consumes more power hence affecting the device operational lifetime. The often-replacement or recharging cycle moreover affords not only additional expenses for batteries but also hampers the continuity of patients' care.
- **Performance Degradation:** If the exchange process that occurs during the authenticity check takes a lot of time and draws many resources of the device,

there will be problems with data transmission and response time. In critical situations of patient care, especially where data is vital and time sensitive, such delays could be disastrous e.g., continuous glucose monitoring.

Resource Efficiency

The introduction of authentication mechanisms for IoMT must be aimed at achieving resource savings. Here are some key considerations:

- **Lightweight Cryptographic Algorithms:** Since the authentication process requires less byte computations, using light-weight cryptographic algorithms with a lower number of computations will tend to take less power. Even though these algorithms might provide a little weaker security as compared to the heavier versions, the same might be justifiable as most devices are resource limited.
- **Pre-computation and Shared Secrets:** A few cryptographic parts can be pre-computed or employ pre-shared secrets to lower real-time computational load on the IoMT devices. This approach crowds some of the processing load to another server during enrolment of a device or management of keys.
- **Scalable and Flexible Mechanisms:** So, the authentication mechanisms should be elastic to allow scalability and fit to accommodate the technical differences in various IoMT devices in terms of their resource endowment. Ideally it should also be scalable in a way where it can be adjusted in strength of cryptography or the scheme of key management according to the limitations of the device in use.

Thinking through these issues, we can try to attain the idea of a presence of strong security, but, at the same time, the work is rationally organized concerning the consumption of resources. This shall help in guaranteeing that IoMT communication is secure while at the same time not stressing the functionality and battery life of important medical gadgets.

Accommodating Large-Scale IoMT Deployments

The growth in the usage of IoMT requires methods for authenticating that is flexible to support the large number of devices that could be connected to the IoT.

Challenges of Scalability

The fundamental problem that the key management and key verification-based solutions have while projecting for large scale IoMT environment is multi-folded. This kind of demand generally boosts with the addition of more gadgets; a situation that places enormous stress on central controllers and results in delays which compromise the real time characteristics of IoMT applications (Jolfaei et al., 2021). Third, a single server denies the distributed storage concept, which means that a server is a weak link in the security system; if the server is breached, all the data stored within a network may be at risk. Last but not the least, the administrative overhead in managing many devices and their credentials in centralized system grows immense.

Scalable Authentication Solutions

To address these challenges, several scalable authentication solutions have emerged:

A. **Decentralized Key Management (DKM):** Balancing the management tasks and deploying them across different devices or integrating the blockchain concept can reduce the load on a major server and enhance the organization's capacity (Ree et al., 2021). This decentralizes the load for scalable IoMT Security. In the period of large-scale IoMT integration, a single server that needs to store all the device credentials becomes a bottleneck and a serious problem for scalability and security. This is where DKM brings an attractive answer to all these issues. Just consider an instance of a traditional IoMT network asked to perform key management with the help of a central server. Every device that is present in the network starting from the pacemakers to the smart inhalers would need to authenticate itself and get the required cryptographic keys from the central authority. Moreover, the huge number of devices and their credentials are hard to manage which in turn consumes many resources. The DKM is the next logical step of dispersing the centralized control by distributing the key management operations between individual nodes throughout the network.

B. **Distributed Key Management Servers (DKMS):** In this approach, the main tasks of the management layer are divided between several servers located in different geographical locations. This way, each server is responsible for a certain number of devices which absolutely exclude overloading of a certain server and increases the scale of the system. Also, getting into one server would not lead to the penetration of the entire network as most people believe.

C. **Blockchain-based Key Management:** Application of blockchain technology, which is a distributed ledger system, is another efficient and most secure way of implementing DKM (Garg et al., 2020). Every device possesses its credentials that are safely stored on the blockchain and can be accessed only by the authorized subjects. Since blockchain is unalterable; these credentials cannot be manipulated by any individual including the attacker who may manage to compromise a node on the blockchain network.

Benefits for IoMT Deployments

Lightweight credentials in IoMT environments are helpful in the following ways according to (Norah et al., 2024). Better control and management of the network is realized because the credentialing is more manageable when they are small. Another advantage is the enhancement of the scalability in that credential with lightweight eliminates the storage space challenges which are realized when integrating other devices into the network.

Examples of Lightweight Credentials in IoMT

1. **Lightweight Elliptic Curve Certificates (LECC):** This standardized format provides smaller certificates compared to the normal X.509 certificates, hence making it good for use in the IoMT devices that are usually constrained.
2. **Identity-Based Encryption (IBE):** This approach may be applied in the IoMT settings where a master-slave situation occurs, for example, when a hospital server controls device identity. Devices can generate their certificates on- the flight thus avoiding the need for certificates to be pre-installed thus reducing the size of storage required.
3. **Hierarchical Authentication:** One of the possible solutions to address this problem is to introduce the hierarchical authentication system, in which devices submit their authentication requests to local gateways and, only then, connect to the server. This can become a problem in big-scale IoMT settings where tens of thousands of devices send data continuously, and centralized architectures could be easily a bottleneck

The Limitations of Centralized Authentication

Let's consider that there is a hospital network that needs to verify all the data received from numerous IoMT devices, from insulin pumps to smart mattresses. The latter, being a centralized approach, is associated with several difficulties. The scalability limit can be described by the cases when the central server receives a huge number of authentication requests and the real time communications limited. Furthermore, the central server has a high vulnerability because control is central; hence, the entire network could be at risk if the server is infiltrated, and patient's data become exposed from the millions of records that most hospitals collect. Another disadvantage concerns elevated latency, that stems from IoMT devices' data transferring to the central server for authentication, which may complicate real-time applications.

AUTHENTICATION MECHANISMS IN IoMT

Hierarchical Authentication

As a result of new technologies, hierarchical authentication organizes the IoMT networks in a hierarchical way and distributes the load of authentication along with offering multiple layers of security. IoMT devices communicate with nearby gateways, which should be deployed in areas of the network being considered. These gateways can be viewed as secure channels, as the devices provide authentication credentials to gain access and to exchange the information, thus lessening the burden on a chief server and boosting local security. This information can also be forwarded to the next higher level of analysis, that is the central server, after the data has been authenticated by the local gateway. The central server might also carry out other safety checks or control the access rights with the help of device permissions that will also enhance the network's security while maintaining data processing.

Benefits of Hierarchical Authentication for IoMT

Hierarchical authentication for IoMT (Zhang et al., 2023) has some advantages. Scalability is enhanced when the authentication is spread over the local gateways thus decreasing the dependence on the main server while accommodating the large number of devices though they take more time to respond. Better security is another benefit, since some layers are localized; even if the central server is breached, the attacker must also take control of a local gateway to eavesdrop on a device's connection. Additionally, reduced latency is a key benefit, as local data can be pre-authenticated, minimizing the need for each exchange to go through the main server. This reduction in latency enhances the response time of IoMT-based real-time applications.

Examples of Hierarchical Authentication in IoMT

1. **Smart Hospital (Woll & Tørresen, 2023)**: Suppose, there is a ward in a hospital with Local Gateway. All IoMT devices in the ward starting from the patient's monitor and the medication dispenser included must first validate with this local gateway before forwarding the data to the central hospital server. This illustration decreases the burden on the main server and protects the patient's record from potential threats.
2. **Remote Patient Monitoring** (Kegomoditswe, 2023): A patient would be capable of wearing a health monitoring device that transfers their physiological parameters to the nearest gateway, which could be in the patient's house. This gateway inputs, validates the device, pre-processes the information then passes a secure data connection to the doctor's server for analysis. The implementation of the hierarchical plan decreases latency and facilitates the secure transfer of information about the health state of the patient.

Thus, the hierarchical authentication can make the IoMT implementation maintain the scalability, security and real-time performance at a proper level. It helps in maintaining the confidentiality of information and at the same time, controlling large number of interconnected medical devices in large scale networks.

COMPARATIVE ANALYSIS OF AUTHENTICATION MECHANISMS

As the sphere of IoMT continues to expand, it becomes critical for healthcare organizations to focus on proper authentication to protect patients' information. However, resource scarcity of medical devices and the number of integrated devices in large-scale toppings present some issues. This section examines the comparison of various forms of authentication mechanisms in terms of attack on them and usability in IoMTs communication.

Security Attributes vs Attacks

Before diving into specific mechanisms, let's revisit the crucial security attributes and how they counter different attack types commonly encountered in IoMT, which is given in Table 4:

Table 4. Security attributes of different attack types commonly encountered in IoMT

Security Attribute	Description	Example Attack
Confidentiality	Ensures only authorized parties can access data	**Man-in-the-Middle Attack (MiM):** An attacker intercepts communication between two devices, potentially eavesdropping on sensitive patient data being transmitted from a pacemaker.
Integrity	Guarantees data remains unaltered during transmission	**Replay Attack:** An attacker captures and retransmits a legitimate authentication message to gain unauthorized access to an IoMT device, such as an insulin pump.
Non-Repudiation	Establishes accountability by ensuring the sender cannot deny sending a message later	**Spoofing Attack:** An attacker impersonates a legitimate device (e.g., a glucose sensor) to gain access to the network and potentially manipulate patient data.

Common Authentication Mechanisms

Let's explore some common authentication mechanisms and compare their effectiveness against the above discussed attacks and their suitability for IoMT deployments as given in Table 5, It is necessary to turn the lights on to the table and examine the content based on the type of mechanism and its consequences for IoMT security. Here's a breakdown of each approach:

Pre-Shared Key (PSK): It is for this reason that this mechanism is classified under what is known as Symmetric Key Cryptography. The device and the server that are involved in the said authentication process will contain the same secret key. When a device attempts a connection it first sends a message that is encrypted through the pre-shared key. The recipient is verified by comparing the received digital signature with a copy held in the server's database for the respective message once the message is decrypted using the same key. PSK is easy to adopt and incurs low computational complexity, thus it can be adopted in implementation for resource limited devices. If the shared key is lost or intercepted, then there is a possibility to have man-in-the-middle (MiM) attacks. Key management becomes difficult on a large scale as the number of devices in the network increases hence the need for good key management policies.

Table 5. Effectiveness of security mechanisms against the attacks and their suitability for IoMT

Mechanism	Description	Security Attributes	Resilience Against Attacks	Suitability for IoMT
Pre-Shared Key (PSK): Devices share a secret key beforehand for authentication.	Simple to implement, low computational overhead.	Confidentiality (if key remains secret).	Vulnerable to MiM attacks if the key is compromised.	Limited scalability, key management complexity grows with large deployments.
Public Key Infrastructure (PKI): Devices use digital certificates and public-key cryptography for authentication.	All security attributes.	Strong resilience against MiM and replay attacks.	Higher computational overhead compared to PSK. Requires a central authority for certificate management, less scalable for large deployments.	High Security Applications, Limited Deployment Size, Hybrid Approaches, Advancements in Lightweight PKI
Identity-Based Cryptography (IBC): Devices derive their credentials from a central authority on-demand.	Confidentiality, Integrity, Non-Repudiation.	Offers strong security attributes.	Requires a trusted central authority, vulnerable to a single point of failure.	Scalable for large deployments, reduces storage requirements on devices.
Lightweight Credentials: Utilize smaller key sizes and certificate formats.	Confidentiality (if key management is secure).	Suitable for resource-constrained devices.	May be more vulnerable to brute-force attacks compared to traditional PKI due to smaller key sizes.	Reduces storage requirements on devices, simplifies deployment in large networks.
Challenge-Response: Device responds to a random challenge from the server to prove its identity.	Confidentiality, Integrity.	Simple to implement, low computational overhead.	Vulnerable to replay attacks if the challenge-response exchange is not properly secured.	Suitable for resource-constrained devices but may not be strong enough for high-security IoMT applications.

Public Key Infrastructure (PKI): This one applies asymmetric key cryptography. There is a public and private key in every device where the public key is for encrypting the message and the private for decrypting it. Frequently, the devices employ the certificates obtained from the authentic central authority to authorize each other. A major advantage of PKI is that it has all three attributes of security i.e., confidentiality, integrity and non-repudiation - Proven to be very secure against both Man in the Middle and Replay attack. PKI introduces certificate authority, which becomes a significant point of attack for hackers. It is more elaborate than PSK because; therefore, it can be inconvenient for systems with numerous LPEs. This means that certificate management can be a problem in large environments as it negatively affects scalability.

Identity-Based Cryptography (IBC): This mechanism of a heavy central point is done without underlying trust but a different mechanism. Nevertheless, PFTs do not possess key pairs; instead, they obtain their credentials on the Internet in exchange for a quantity known as an identifier (ID), which is issued by the Certificate Authority (CA) where required. Identity-Based Cryptography (IBC) is not restricted by compatibility with a particular form factor and is highly scalable because certificates can be delivered without being embedded on a device in advance. It is space-saving because devices do not have to store key pairs and offers many of the same security features as PKI has to offer. Almost entirely, IBC uses many coordinators located throughout the network, which greatly contributes to additional risks given potential points of failure. By centralizing the authority, the threat means that an attack on the authorities puts the entire network in danger.

Lightweight Credentials: This approach solves the problem of PKI certificates, namely, the size of certificates and their requirements for resources. Elliptic Curve Cryptography is comparatively of smaller key size and have simple certificate structures when compared to the complex Hash-based Cryptography. The lightweight credentials are preferred and recommended for the low-power devices because they take less space. They cut down on the overall string length of the credentials used in large networks hence easy to deploy and manage. Security is somewhat inferior to standard PKI, and existing weaknesses include short keys compared to real-life PKI, which is cracked easily.

Challenge-Response: This easier process entails the server posing a random set of questions to that device. The device replies to the challenge with a message that has been encrypted using the shared secret key or its private key, based on the protocol in use. This response lets the server authenticate the device. Challenge-response has very low computational complexity and is easy to implement, thus suitable for the environment where resources are limited. Compared to other protection mechanisms it offers comparatively low level of security as well as can be vulnerable to replay attack in case the communication is not protected. It may not provide the utmost needed strength for the IoMT applications that needed strict identification.

It has been seen that the selection of an authentication mechanism has a direct influence on the level of protection, its expansiveness and its impact on the assets of an IoMT system. Thus, despite the advantages that PSK must provide, one must admit that the algorithm is not very robust. The case of PKI is that it offers a high level of security to organizations, but decisions involve a high level of management. Identity-Based Cryptography (IBC) and Lightweight Credentials (LWC) have potential for better scalability with little or no resources and Challenge-Response (CR) can be used as a very basic and low risk password method. The choice of the best mechanism requires a proper assessment of security requirements, on the other hand, the ability of the devices in the IoMT network, and the structure of the prevailing network.

The IoMT, involving connections of numerous connecting devices and applications in the medical field, offers great potential in improving patients' diagnoses and treatment. But it brings security issues which require prevention measures as the convergence advances. The first problem is that IoMT devices include elements that can be exploited by attackers. Such apparent weaknesses can be abused and lead to a patient's data confidentiality, data accuracy, and even patients' safety being at risk by hackers. If proper authentication is not developed then, unauthorized individuals can easily access the patient's details or harm the device operations. Security of data and information is highly sensitive in the field of IoMT. Although the flow of data among integrated systems in the delivery of healthcare is efficient, there are issues of privacy of the patients' data. The data authentication is also well developed, so that only authorized workers can access the patients records and the application of encrypted algorithms to protect data enables its safety.

Building trust in the condition of the device is also of great value. An IoMT device may involve in transmitting or receiving data should first go through the authentication process. High security ensures no other device in the network other than the approved ones gains access. Secure boot processes authenticate the firmware of the device when the booting of the device is being processed, while identity for devices is assigned to track their actions and reign in their activities. But there are challenges that are associated with security in IoMT. There are many devices, wearables, and gateways that make a co- ordination between such products a difficult task to do. Integration of MFA raises the security measures by offering the password factor, smart card factor, and the biometric factor all at once. Mandatory access control makes the users access only data relevant to their roles, and it allows it through role- based access control. Monitoring is constant: it identifies irregularities and violations as soon as they occur.

The framework of trust starts with authentication since it serves as the foundation of securing IoMT. Thus, with proper patient data protection and devices' reliability, it is possible to unleash the IoMT's full potential while minimizing hazards. The healthcare providers need to still be more cautious and keep monitoring and reading the practices of security to safeguard the patients and trust in this environment.

Key Findings from Qualitative Analysis

Based on the identified preliminarily insights of the study, the following critical findings of the present work in the context of IoMT authentication mechanisms could be summed up: Despite the presence of multiple types of IoMT authentication, which is certificate-based, biometric-based, and token-based, in general, multiple factors show the benefits and drawbacks of each, and, therefore, the choice should be made in consideration of certain parameters. Regarding trade-off, security is inversely proportional to its usability; for instance, certificate-based method is computationally intensive while the biometric method, though friendly to the user, is insecure. However, token-based approaches mainly have inherent security limitations although they are less complex compared to context- based methodologies. Especially in the case of resources applied in IoMT devices, e.g. wearables, it is essential to render the applied authentication as efficient as possible thereby considering aspects like battery lifetime, processing power and memory available. Also, there concerns touching privacy should be looked at when capturing biometric data including proper consent of the subjects and the storage of biometric templates.

Although this analysis offers useful information, the following sources of method bias should be noted: The work was confined to an area of authentication protocols only, thus further work in the direction should also be investigated other methods. Whereas many components of the analyzed IoMT environment were chosen based on static threat models, with the environments changing at a high pace, dynamic threat model, and adaptive authentication, remain an open issue. Moreover, the performative evaluation is generally weak in obtaining the users' perspective; conducting user studies patients, and manufacturers of the devices would contribute positively, as user preferences, acceptance and perceived security are important when the system is to be implemented in the real world. Compatibility issues have not been solved in terms of 'Interoperability' in the IoMT environment since different smart devices deployed by different manufacturers need to be interconnected and must work harmoniously. Otherwise, either standardization or cross-vendor solutions must be worked on to resolve this issue.

As IoMT continues to evolve, several trends and advancements shape the authentication landscape. These developments promise to enhance security, privacy, and user experience in the ever-evolving landscape of connected medical devices.

Zero-Trust Architectures

Zero-trust architectures (NIST, 2019) also disapprove the center-of-perimeter security model. Hence, it can be described that zero-trust principles deny inherent trust within a network, and any interaction is considered as suspicious. Here's how this concept applies to IoMT. In addition to extending the selective physical or logical access barriers, conventional perimeter security is insufficient as it is in IoMT-connected healthcare facilities, hospitals, clinics, and even patients' houses, a zero-trust security model is more suitable in this case. Namely, this approach claims that no device or a user should be considered secure by default because of their location or rank. In a Universal Trust and Access Control (UTAC) model, authentication is needed on every contact point: if a device wants to engage with a master server, or a clinician needs to get a patient's chart, or a wrist device is to connect with a hub.

Blockchain and Decentralized Identity

Cryptocurrencies are based on something known as blockchain (Mandrita, 2018), which is not solely for financial systems but has many other uses. In IoMT, blockchain-based solutions offer tamper-resistant identity management and data integrity. zero-trust involves dynamic access control, adjusting access rights based on the current context; for example, a nurse may have different access levels when using a workstation at the hospital versus a tablet during rounds. Immutable records, a key feature of blockchain, ensure that once data is recorded, it cannot be altered, which is crucial for maintaining accurate patient identities and treatment histories. Decentralized Identifiers (DIDs) facilitate self-sovereign identity, allowing patients to control their own identity information and share it selectively with healthcare providers. DIDs can also be linked to other data, such as proof of vaccination, without the need for a central authority. The application of blockchain in IoMT could lead to a patient-centric identity system, where each patient has a unique DID, enhancing the safety and efficiency of accessing health records across different providers and devices.

Continuous Authentication

Traditional authentication typically occurs at login, whereas continuous authentication monitors user behavior throughout a session. It utilizes behavioral biometrics, which consider patterns such as typing rhythm, mouse clicks, and steps as unique signatures, rather than relying solely on passwords or tokens. This approach reduces friction by enhancing security without disrupting the user experience, allowing users to remain logged in while the system continuously validates their identity. If the system detects deviations in behavior, such as erratic typing, it can prompt for reauthentication or trigger an alert, ensuring adaptive security.

These are the trends that will define the evolution of IoMT security in the future. To enhance and secure patient data, researchers, industries, and policymakers need to work together for the provision of strong and efficient, reliable and convenient authentication processes for the enhancement of the connected healthcare system.

CONCLUSION

Internet of Medical Things (IoMT) exists between the healthcare industry and the technology world while embracing patients' health status. Indeed, since there is a growing adoption of inter-operational connected healthcare devices, wearables, and telemedicine, the need for strong authentication systems cannot. However, it is crucial to uncover multiple layers of the authentication concept in IoMT and the way forward.

The patient data, that has been under Discussion in IoMT are patient records, temperature, treatments, past health history, and images. Proper privileges are accorded to the right people to guarantee that only valid people comprehend this information. Infringements of privacy can cause serious outcomes, meaning ruptured patient anonymity or even threats to individuals' lives. Thus, authentication becomes the first barrier of protection against the infamous threats of unauthorized access. IoMT devices such as smart insulin pumps, home telehealth devices; remote monitoring sensors are many form a network. The integrity of these devices is therefore relevant. In this method, unauthorized or compromised devices are prevented from joining the network. It provides security for the interconnected parts since there is confidence that they will not be tampered with by other parties.

In this research, it is evident that Health care professionals, patients, and the care givers have direct contact with IoMT systems daily. The methods of authentication must be secure but at the same time should be as convenient as possible. The organization is discouraged by cumbersome procedures required in authentication. On the other hand, smooth, comprehensible methods promote the implementation of the said measures and further, the improvement of the healthcare system. IoMT is used in environments such as hospitals, homes, ambulances, and even the wearables that the patients wear. There is a definite need for specialized authentication if the identified contexts are to be addressed. There are plenty such as protection strength and computational complexity, biometric and privacy, and scalability against simplicity. Currently many researchers encounter such issues. IoMT security can be described as dynamic in nature. Threats change times bring new threats which include new vulnerabilities, new avenues of attack, new aggressors. Innovation is constant, which means that Research & Development must go on without stoppage. Thus, the nature of the threats requires certain proactive idea and a search for new methods of the authentication. Zero-trust architectures have shifted the traditional concepts. They recommend that skepticism is always appropriate from anywhere and at any touch point whether it is online or offline. Blockchain for patient, apparent with tamper-resistant identity management and decentralized identifiers (DID), holds the security of patient identity. Continuous authentication best practice is a frequent approach that is found on using behavioral biometrics and does not interfere with constant usage.

The world of IoMT is constantly evolving and keeping it secure requires a team effort. Imagine researchers in universities brainstorming new ways to protect patient data, working together with companies developing the latest medical devices. Regulatory bodies would also be involved, making sure these innovations prioritize patient safety and privacy. It's like building a bridge – researchers design the blueprints, companies construct the sturdy pillars, and regulators ensure the bridge is safe to cross. Strong authentication isn't just a technical hurdle to jump over; it's the foundation of a secure IoMT ecosystem. It's about protecting lives, building trust with patients, and paving the way for a future where healthcare is seamlessly connected. We, as a community, need to stay alert, adapt to new challenges, and remain committed to securing IoMT for the well-being of everyone. After all, secure IoMT means healthier communities and a brighter future for healthcare.

REFERENCES

Act, A. (1996). Health Insurance Portability and Accountability Act of 1996, Pub. L. No. 104-191, 110 Stat. 1936. https://www.govinfo.gov/content/pkg/PLAW-104publ191/pdf/PLAW-104publ191.pdf

Ali, Z., Mahmood, S., Mansoor Ul Hassan, K., Daud, A., Alharbey, R., & Bukhari, A. (2024). A lightweight and secure authentication scheme for remote monitoring of patients in IoMT. *IEEE Access : Practical Innovations, Open Solutions*, *12*, 73004–73020. DOI: 10.1109/ACCESS.2024.3400400

Alsaeed, N., & Nadeem, F. (2022). Authentication in the Internet of Medical Things: Taxonomy, Review, and Open Issues. *Applied Sciences (Basel, Switzerland)*, *12*(15), 7487–7487. DOI: 10.3390/app12157487

Alsaeed, N., Nadeem, F., & Albalwy, F. (2024). A scalable and lightweight group authentication framework for Internet of Medical Things using integrated blockchain and fog computing, *Future Generation Computer Systems*, vol. 151, Pages 162–181, ISSN 0167-739X, DOI: DOI: 10.1016/j.future.2023.09.032

Banerjee, M., Lee, J., & Choo, K.-K. R. (2018). A blockchain future for internet of things security: a position paper, *Digital Communications and Networks*, vol. 4, no. 3, pp. 149–160, ISSN 2352-8648, DOI: DOI: 10.1016/j.dcan.2017.10.006

Boikanyo, K., Zungeru, A. M., Sigweni, B., Yahya, A., & Lebekwe, C. (2023). Remote patient monitoring systems: Applications, architecture, and challenges, *Scientific African*, vol. 20, pages. e01638, ISSN 2468–2276, DOI: DOI: 10.1016/j.sciaf.2023.e01638

Dadkhah, M., Mehraeen, M., Rahimnia, F., & Kimiafar, K. (2021). Use of Internet of Things for Chronic Disease Management: An Overview. *Journal of Medical Signals and Sensors*, *11*(2), 138–157. DOI: 10.4103/jmss.JMSS_13_20 PMID: 34268102

Dhanvijay, M. M., & Patil, S. C. (2019). Internet of Things: A survey of enabling technologies in healthcare and its applications. *Computer Networks*, *153*, 113–131. DOI: 10.1016/j.comnet.2019.03.006

Garg, N., Wazid, M., Das, A. K., Singh, D. P., Rodrigues, J. J. P. C., & Park, Y. (2020). BAKMP- IoMT: Design of Blockchain Enabled Authenticated Key Management Protocol for Internet of Medical Things Deployment. *IEEE Access : Practical Innovations, Open Solutions*, *8*, 95956–95977. DOI: 10.1109/ACCESS.2020.2995917

Huang, C., Wang, J., Wang, S., & Zhang, Y. (2023). Internet of medical things: A systematic review. *Neurocomputing*, *557*, 557. DOI: 10.1016/j.neucom.2023.126719

Jolfaei, A., Aghili, S., & Singelée, D. (2021). A Survey on Blockchain-based IoMT Systems: Towards Scalability. *IEEE Access*, PP, 1-1. DOI: 10.1109/ACCESS.2021.3117662

Koutras, D., Stergiopoulos, G., Dasaklis, T., Kotzanikolaou, P., Glynos, D., & Douligeris, C. (2020). Security in IoMT Communications: A Survey. *Sensors (Basel), 20*(17), 4828. Advance online publication. DOI: 10.3390/s20174828 PMID: 32859036

Lee, Y. (2021). Review of the Role of the Internet of Things (IoT) on the Consumer Market., 180–198. .DOI: 10.4018/978-1-7998-5326-8.ch009

Lester, J. N., Cho, Y., & Lochmiller, C. R. (2020). Learning to Do Qualitative Data Analysis: A Starting Point. *Human Resource Development Review*, *19*(1), 94–106. DOI: 10.1177/1534484320903890

Malamas, V., Chantzis, F., Dasaklis, T. K., Stergiopoulos, G., Kotzanikolaou, P., & Douligeris, C. (2021). Risk assessment methodologies for the internet of medical things: A survey and comparative appraisal. *IEEE Access: Practical Innovations, Open Solutions*, *9*, 40049–40075. DOI: 10.1109/ACCESS.2021.3064682

National Institute of Standards and Technology (NIST). (2019, December). Special Publication 800-207: Zero Trust Architecture (ZTA). National Institute of Standards and Technology. https://nvlpubs.nist.gov/nistpubs/SpecialPublications/NIST.SP.800-207.pdf

Khashan, O. A., & Khafajah, N. M. (2023). Efficient hybrid centralized and blockchain-based authentication architecture for heterogeneous IoT systems. *Journal of King Saud University. Computer and Information Sciences*, *35*(2), 726–739. DOI: 10.1016/j.jksuci.2023.01.011

Patil, S. D. D., Satpute, O. D., Mehta, P. S., & Mahajan, R. A. (2023). Enhancing Digital Healthcare with IoMT and Big Data: A Computational Intelligence Approach. In Chaki, N., Roy, D., N., Debnath, P., Saeed, and K., editors, *Proceedings of International Conference on Data Analytics and Insights, ICDAI 2023. ICDAI 2023*, volume 727. Springer. DOI: 10.1007/978-981-99-3878-0_18

Rahman, M., & Jahankhani, H. (2021). Security Vulnerabilities in Existing Security Mechanisms for IoMT and Potential Solutions for Mitigating Cyber-Attacks. In Jahankhani, H., Kendzierskyj, S., & Akhgar, B. (Eds.), *Information Security Technologies for Controlling Pandemics. Advanced Sciences and Technologies for Security Applications*. Springer., DOI: 10.1007/978-3-030-72120-6_12

Razdan, S., & Sharma, S. (2021). Internet of Medical Things (IoMT): Overview, Emerging Technologies, and Case Studies. *IETE Technical Review*, *39*(4), 775–788. DOI: 10.1080/02564602.2021.1927863

Ree, M., Vizár, D., Mantas, G., Bastos, J., Kassapoglou-Faist, C., & Rodriguez, J. (2021). A Key Management Framework to Secure IoMT-enabled Healthcare Systems. *2021 IEEE 26th International Workshop on Computer Aided Modeling and Design of Communication Links and Networks (CAMAD)*, 1-6. DOI: 10.1109/CAMAD52502.2021.9617796

Roberts, K. (2023). Internet of things vs. internet of medical things. Atlantic.Net. https://www.atlantic.net/life-sciences-pharma-biotech/internet-of-things-vs-internet-of-medical-things/

Svandova, K., & Smutny, Z. (2024). Internet of Medical Things Security Frameworks for Risk Assessment and Management: A Scoping Review. *Journal of Multidisciplinary Healthcare*, *17*, 2281–2301. DOI: 10.2147/JMDH.S459987 PMID: 38765613

Tariq, M. U. (2024). Social Innovations for Improving Healthcare. In Chandan, H. (Ed.), *Social Innovations in Education, Environment, and Healthcare* (pp. 302–317). IGI Global., DOI: 10.4018/979-8-3693-2569-8.ch015

Tariq, M. U. (2024). Leveraging AI for Entrepreneurial Innovation in Healthcare. In Özsungur, F. (Ed.), *Generating Entrepreneurial Ideas With AI* (pp. 192–216). IGI Global., DOI: 10.4018/979-8-3693-3498-0.ch009

Tariq, M. U. (2024). Application of blockchain and Internet of Things (IoT) in modern business. In Sinha, M., Bhandari, A., Priya, S., & Kabiraj, S. (Eds.), *Future of customer engagement through marketing intelligence* (pp. 66–94). IGI Global., DOI: 10.4018/979-8-3693-2367-0.ch004

Tariq, M. U. (2025). Gamification and Virtual Reality: Transforming Early Childhood Language Learning. In Manditereza, B. (Ed.), *Digital Pedagogy in Early Childhood Language Development* (pp. 25–52). IGI Global Scientific Publishing., DOI: 10.4018/979-8-3693-7437-5.ch002

Tariq, M. U. (2025). Leveraging Artificial Intelligence to Enhance Diversity and Drive Business Innovation. In Margaretha van der Poll, H., Andrew van der Poll, J., & Ngwakwe, C. (Eds.), *Diversity, AI, and Sustainability for Financial Growth* (pp. 79–96). IGI Global Scientific Publishing., DOI: 10.4018/979-8-3693-6011-8.ch004

Tariq, M. U. (2025). Quantum Algorithms and Predictive Analytics: Revolutionizing Consumer Insight. In Gajbhiye, P., Kannan, H., Rodriguez, R., & Rojas-Méndez, J. (Eds.), *The Quantum AI Era of Neuromarketing* (pp. 173–202). IGI Global Scientific Publishing., DOI: 10.4018/979-8-3693-7673-7.ch008

Vaiyapuri, T., Binbusayyis, A., & Varadarajan, V. (2021). Security, Privacy and Trust in IoMT Enabled Smart Healthcare System: A Systematic Review of Current and Future Trends [IJACSA]. *International Journal of Advanced Computer Science and Applications*, *12*(2). Advance online publication. DOI: 10.14569/IJACSA.2021.0120291

Woll, A., & Tørresen, J. (2023). What is a Smart Hospital? A Review of the Literature. In Duffy, V. G., Lehto, M., Yih, Y., & Proctor, R. W. (Eds.), *Human-Automation Interaction. Automation, Collaboration, & E-Services* (Vol. 10). Springer., DOI: 10.1007/978-3-031-10780-1_8

Zhang, H., He, S., Chen, J., Yang, K., Ran, P., & Chen, J. (2023). A Hierarchical Authentication System for Access Equipment in Internet of Things. *International Journal of Intelligent Systems*, *2023*(1), 1–11. DOI: 10.1155/2023/8899697

KEY TERMS AND DEFINITIONS

Certificate Authority (CA): An organization that gives out, administers and validates digital certificates that are used to confirm that a particular message has originated from a particular author in electronic communication.

Challenge-Response (CR): A server sends the client a challenge which the client must answer with a valid encrypted response, this makes it determined that the client is valid.

Elliptic Curve Cryptography (ECC): A kind of key exchange utilizing an elliptic curve defined over a finite field with relatively high security relative to other categories but with the keys that are considerably shorter.

Hash-based Keyed-Hashing (HL C): A technique of converting message digest in conjunction with a key utilizing hashing functions to maintain the originality of the data.

Identity-Based Cryptography (IBC): A cryptographic technique in which public keys are derived from an ID so that secure communication is achieved without the presence of PKI.

Lightweight Credentials (LWC): User authentication credentials that can be made relatively small, thus suitable for low power consuming devices with limited storages.

Man-in-the-Middle (MiM): An attack that is initiated with the purpose of intercepting the transmission of information between two parties and possibly even change the contents as well.

Predictive Analytics: This is the process of evaluating past events with the aim of predicting future trends in health, events like disease epidemics. It aids in the task of looking for populations that are vulnerable to such occurrences and the efficient utilization of resources in health care facilities to enhance decision making as well as preparedness.

Pre-Shared Key (PSK): A technique in which both the communicators agree and use the same secret key for encryption and decryption process.

Public Key Infrastructure (PKI): Software for the creation of programs that organize Digital Keys and Certificates for secure data transfer and verification with the use of Asymmetric Cryptography.

Remote Monitoring: This includes the monitoring of patients' vital signs, chronic diseases and the consumption of their prescriptions through IoMT devices on a consistent basis. Patients send data to their physicians, and they get timely intervention on the management of their health conditions.

Symmetric Key Cryptography (SKC): A technique that involves the utilization of the same key for the processing of data at the encryption and decryption phases, which emphasizes the aspect of security.

Telemedicine: A service-based solution that uses digital technology so that the patient can interact with practitioners and other healthcare providers face-to-face using video link or similar communication tools. It eliminates the geographical push and pulls and enhances the chances of receiving healthcare services amongst the people who frequently lack easy access to hospitals.

Wearable Devices: These are aware gadgets like the smart wristband for monitoring physical activity and smart glucose meter that a person uses to manage their health positively. They give instantaneous information on different health aspects, thus enabling users to manage their health.

Chapter 6
Comprehensive Modernity of Fish Smoking Techniques Using the IoT

Sabyasachi Pramanik
https://orcid.org/0000-0002-9431-8751
Haldia Institute of Technology, India

Soma Bag
Asadtala Nivedita Kanya Vidya Math, India

Atanu Roy
https://orcid.org/0000-0001-5269-9122
The Neotia University, India

Pranati Rakshit
https://orcid.org/0000-0001-5037-1326
JIS College of Engineering, India

ABSTRACT

The age-old method of smoking fish, which is popular in coastal areas, presents a number of difficulties since it requires a lot of labour, regular observation, and steady temperature maintenance. These problems often lead to inefficiency, uneven product quality, and even safety risks. Modern, dependable, and effective fish smoking techniques are desperately needed, since the smoked fish sector plays a vital role in supporting local economies in Indonesia's coastal regions. The goal of this project is to design an Internet of Things (IoT) integrated monitoring and control system for the smoked fish machine in order to overcome these problems. The goal

DOI: 10.4018/979-8-3373-4332-7.ch006

of this research was to create a machine monitoring and control system that would enable temperature monitoring, blower RPM management, and machine component on/off switching. The study's findings show that machine parts including a blower, servo motor, and light may be turned on by using IoT. Additionally, by connecting with a temperature sensor, IoT can track the machine's temperature in real time from a smartphone.

INTRODUCTION

Significant breakthroughs that have had a tremendous influence on human existence have resulted from technological developments (Cissé, M. et al. 2024). The development of digital technologies like the Internet of Things (IoT) is supported by the widespread ownership of smartphones because of the rapid pace of technological advancement. The "smart home" idea and Nest's release of a smartphone-controlled smart thermostat in 2010 helped to popularize the Internet of Things (IoT). Almost everyone has a smartphone these days, which makes the Internet of Things (IoT) idea possible by enabling things to connect to the internet and communicate on their own without the need for human interaction. These gadgets include automobiles, home appliances, industrial gear, and even ordinary items with sensors that can transmit and receive data. IoT's main purpose is to make life easier for people. IoT offers more efficiency and convenience by enabling the integration and remote control of several systems and devices. Through IoT-enabled mobile devices, users may remotely monitor and operate equipment including security apparatus, irrigation systems, and temperature regulators (Wang, J. et al. 2024). IoT reduces direct human participation and minimizes possible dangers by making it easier to use and operate equipment practically and by allowing remote operation and monitoring. One sector that might profit from IoT technology is the smoked fish business, which provides a living for many people in Indonesia's coastal regions. Now, stove-based traditional fish smoking techniques are used by most Indonesian smoked fish businesses. But when measured against more contemporary, regulated alternatives, these conventional techniques are often less efficient, effective, and sanitary. Conventional fish smoking takes longer to finish, needs constant monitoring, and often results in product degradation due to fluctuating temperatures. Modern technology has led to the development of more useful smoked fish machines. According to research by (Raoufi, P. et al. 2024), these devices may be connected to the Internet of Things to improve consumer ease. According to (Ahmid, M. et al. 2024), operators may remotely manage and monitor IoT-enabled smoked fish devices, enabling them to monitor the smoking process in real time. For Indonesia, a coastal country with an abundance of fisheries resources, the technical advancement of this sector is essential. It is anticipated that the adoption and integration of IoT into smoked fish machines would optimize operations, enabling customers to remotely monitor and manage the units using mobile devices in real time. Because users can carry out the required monitoring and control without having to physically be present at the site, IoT may save time and energy. Furthermore, it is possible to analyze the data produced by Internet of Things-enabled smoked fish devices to improve production efficiency and productivity.

The increasing use of information and communication technology in both business and daily life, together with current trends in information availability, makes it evident that civilization is on the cusp of a fourth industrial revolution. Today, concepts like "digital economy," "information society," and "internet of things" are recognized as the primary areas of state development that are essential to the competitiveness of the national economy. communication technology is the primary definition of the digital economy. Based on an experiment carried out at Murmansk State Technical University, the study given aims to create the idea of the Industrial Internet of Things (IIoT). Using a tiny drying device made specifically for drying and smoking fish and fish products, the idea of Things idea was put into practice. The suggested dewatering technique lowers the cost of electricity used to produce dried goods, prolongs the life of smoking and drying facilities at the expense of more effective coolant use, and enhances product quality by lessening the effects of relaxation-induced deformation in fish tissues.

The need for many essentials of life is increased by rapid economic expansion. It is inexpensive and simple to get, much like salmon, another popular animal protein source. On the other hand, fish soon deteriorate. Therefore, the public must be aware of fish preservation. Smoking is a common method of preserving fish. According to estimates from the Food and Agriculture Organization of the United Nations (FAO), smoking accounts for 30% of fish processing methods in tropical nations. Of course, this is a big number. As a result, the smoking technique for fish preservation must be perfected and continuously improved. Fumigation, also known as smoking, is the process by which volatile compounds from combustion enter fish, resulting in products with distinct flavors and scents and a long shelf life because antibacterial activity stops fish enzymatic activity, preventing fish from spoiling. The primary goal of smoking fish is to get the durability that the smoke produces. The second objective is to provide a unique scent. Smoking is a processing or preservation technique that uses a combination of drying treatment and heat generation to build compounds in the form of tar granules and steam. According to (Uhuo, E. et al. 2023), the smoked component adheres to the fish and dissolves in a layer of water on the fish's body, giving the final product a unique flavor and perfume as well as a golden or brownish hue. The most popular technique is using a chimney, which is seen to be unfriendly to the environment because of the heavy black smoke it produces. Most people still utilize time-consuming, traditional ways. Producing smoked fish with high competitiveness is very challenging due to poor production and smoking efficiency. This necessitates the development of contemporary fumigation technologies. Although there has been a lot of improvement in smoking machine types, the smoking cabinet design remains the same. Additionally, (Asiedu, B. et al. 2018) designed the fish smoking apparatus to enhance the manufacturing and hygienic conditions of smoked fish products. The outcome increases production capacity rather well, but it is inefficient in terms of smoking time and is still limited by the tool's size, requiring a lengthy smoking period. The amount of smoke, the fish's form, the smoking temperature, the humidity level in the smoking chamber, and the smoking device's design all have a significant impact on how successful the smoking process is. The author developed the SMOFIM (Smoke Fish Machine), a fish smoking innovation based on the issues. It will be a more efficient tool for smoking fish, more effective, eco-friendly, and built on cutting-edge Internet of Things (IoT) technology to raise living standards. The residents of Trisik Beach, Kulon Progo, to prevent the rotting of fish harvested by fisherman and to enable the use of smoked fish as a specialty dish in the Trisik coastal region. SMOFIM offers Android-based Internet of Things (IoT) automated control to make it easier to turn lights, electric motors, and blowers on and off. SMOFIM's energy source is a hybrid technology that uses solar photovoltaic, which is fuelled by sunshine, to run electric servo motors, blowers, and lighting. SMOFIM employs smoke filters to reduce CO2, CO, and HC pollution in addition to developing a fish smoker that is advanced and performs well. Filter paper made from leftover prawn shells is used as the filter. When chitosan membranes are mixed, it is known that tannins may precipitate automobile exhaust emissions. There are several organic substances in prawn waste, including protein (23-27%) and tannins. Shrimp waste contains chitosan, which is also utilized as a raw material to create membranes that are extensively employed for a variety of filtering or separation applications using natural chemical compounds from the burning of natural fuels. By burning, smoke

LITERATURE REVIEW

In the first research, (Phadke, M. et al. 2024) used DHT11 sensors to build a temperature monitoring system within an Internet of Things (IoT) framework. The system used a DHT11 sensor, an Ethernet shield, and an Arduino to provide data via a RESTful API to the AT&T M2X IoT platform. When significant variations in temperature and humidity were detected, the system could send alerts via email, SMS, and mobile push notifications thanks to the integration of the M2X platform's trigger capability with the IFTTT service. Significant differences in temperature and humidity were found across the sites, along with a strong negative association between the two variables. These findings underscore the need for better data centre layout design (Khanh, P. T. et al. 2024). In the second research, (Ray, A. M. et al. 2023) created an oven that used an Arduino Uno-based monitoring system to smoke fish using the cold smoking technique. In this work, a temperature monitoring system that could track the temperature within the smoke chamber of a smoked fish machine was introduced. An LCD that showed the current temperature conditions and a temperature sensor that could precisely measure the inside temperature were connected to an Arduino Uno microcontroller (Dhabliya, D. et al. 2024). Al Hudha et al.'s third project entailed creating a smoked fish machine with an Internet of Things-based control system (2019). The machine's IoT system was used to monitor temperature and turn on machine parts, which improved user access to the equipment's functionality. These parts may be used successfully and efficiently based on the conclusions from these three studies, which all examined the use of the Arduino Uno microcontroller, DHT11 temperature sensor, and Internet of Things in control and monitoring systems. As such, it is both possible and beneficial to include these technologies into a smoked fish machine. Murmansk Region is ranked third among the fish-processing regions by the Federal State Statistics Service, with Kamchatka Region coming in first thousand tones) and 699.1 thousand tones from Primorsky Krai. In the North-Western Federal District, where 1222.20 thousand tones of fish and fish products were produced in 2017 (figure 1), our area supplies over half of that output (42%). Figure 2 shows the dynamics of fish production in the Murmansk area and the Northwest Federal District. (Russian statistics compilation: Murmanskstat, 2017 45 p., Fisheries Industry of Murmansk area). But in recent years, the Murmansk region's fish and fish product production trends have slowed. In comparison to 2014, when production peaked, the amount of fish and fish products decreased from 563.8 thousand tones to 519.2 thousand tons in 2017. To put it another way, the Russian Federation's fisheries are a challenging industry to predict because of the wide range of factors that affect its growth, such as its production factor, material, and technological foundation.

This chapter mostly discusses the earlier research that has been conducted. As a result of humanity's constant attempts to make their job easier, automated smoker and grill control. The fabrication, modeling, and building of a grill and smoker are part of this project, along with material and pricing specifications. As easy to use as possible, the grill and smoker machine will provide improved heat radiation controllability, greater efficiency, lower cost, and better serviceability. The following standards apply to stainless steel 304: AMS 5513; ASTM A240; and ASTM A666. This provides specific details on the kind of stainless steel that was used for the design. The various families of FAs seem to be impacted by cooking. This is especially crucial in light of the suggested cooking techniques for the meat. "IoT-based home automation system using Node MCU ESP8266 module," by Suraj Tonage, Sandhya Yemul, Rajendra Jare, and Veena Patki, International Journal of Advanced Research and Development, volume 3, number 3, pages 332-333. This study describes a home management and monitoring system that uses an integrated micro-web server with IP connection to provide remote appliance access and control using an Android-based smartphone app. Temperature and current sensors have been included in the suggested control system to show the system's viability and efficiency.

The United States Environmental Protection Agency (US-EPA) and the European Union (EU) identified the potential for cardiovascular and cancerous disorders as health hazards linked to PAHs found in food items, such as smoked fish, in 2023 (Cisse, M. et al. 2023). Furthermore, PAHs were identified as known carcinogens by the International Agency for Research on Cancer (IARC) in 2012. Given that most Ghanaians eat smoked fish; this categorization might help explain the rising frequency of cancer diagnoses in the country (Ashisha, G. R. et al. 2023). Low market participation has resulted from the stringent monitoring and rejection of food goods, such as smoked salmon from poor nations, in international markets (Ngoc, T. T. H. et al. 2023). Because they utilize traditional ovens that depend on firewood, fishmongers often run the danger of suffering burns, respiratory problems, and headaches (Veeraiah, V. et al. 2023). Additionally, using firewood in these ovens may result in the loss of mangrove trees and other negative impacts on biodiversity (Roy, A. & Pramanik, S. 2023). Because of this, alternative ovens—especially ones with reduced PAH levels—have gained popularity. As a result, better fish smoking methods have been created, including the FAO-Thiaroye Technique, the Modern Landmark Smoker Oven, and the Ahotor Oven. Due to its fuel-efficient design and low PAH emissions, the Ahotor oven—which was first introduced by the West Africa Regional Fisheries Development Program (WARFP) in 2014—is an improved fish smoking technology (IFST). As a result, it offers fishmongers a safer and more cost-effective option than conventional ovens like Chorkor and Morrison, which are notorious for having high PAH levels. Ahotor ovens have been built in a few coastal villages in Ghana's Western and Central regions by the Cerath Development Organization (CDO) in partnership with the European Union-Ghana as part of the Power to the Fishers initiative (Cerath Development Organization, 2019). According to a recent research by (Ravishankar, C.N. et al. 2023), the use of better ovens for fish processing may enhance fishmongers' incomes by increasing market participation due to the reduced emission of PAHs. Fishmongers still often utilize traditional smoking ovens in Ghana, despite the CDO's significant efforts on the Ahotor campaign. Examining fishmongers' attitudes towards and preparedness to utilize the Ahotor ovens is essential to understanding their aims. Given that behavioral intention plays a major role in the diffusion of innovation, this could have a positive effect on that diffusion. A detailed analysis of their behavioral intentions about its use is required, given the limited use of the IFST in the fish smoking process, as shown by Ghanaian coastal fishmongers' use of Ahotor ovens. According to (Malode, K. et al. 2024), it is crucial to understand the connections between acceptability and user experience. In the context of the agricultural sector, including fisheries, the updated Unified Theory of Acceptance and Use of Technology (UTAUT-2) put forth by (Rawat, A. et al. 2024) provides a comprehensive framework for comprehending the elements that influence a person's intention to adopt any new technology or innovation (Veeraiah, V. et al. 2023). The seven dimensions of this model—performance expectancy, effort expectancy, social influence, facilitating conditions, habits, hedonic motivation, and price value—are essential for understanding how people make decisions about new innovations or technologies (Roy, A. et al. 2024). According to the sources mentioned by (Khanh, P. T. et al. 2024), the UTAUT2 model is well known for its ability to explain how new technology is adopted. The UTAUT-2 framework successfully clarifies the elements that influence people's behavioral intentions to accept cutting-edge technologies, according to a recent systematic review of theories on technology adoption (Pramanik, S. et al. 2020). According to this study, the UTAUT-2 model may provide policymakers and other stakeholders in the fisheries sector with useful insights into the variables that encourage and hinder fishmongers in Ghana's coastal villages from using Ahotor ovens. Nevertheless, prior research in Ghana has mostly addressed the Ahotor oven's scientific features without sufficiently considering the behavioral viewpoint. To our knowledge, the UTAUT-2 model has not been employed, despite the fact that there aren't many researches that concentrates on better technology adoption in the context of fisheries. This research provided a behavioral theoretical model based on the updated unified theory of acceptance and use of technology (UTAUT2) that was tailored to the Ahotor oven as IFST in Ghana to fill the current vacuum in the literature on fisheries.

CURRENT INFORMATION DEVELOPMENTS IN SOCIETY

The current state of civilization is clearly on the cusp of the fourth industrial revolution, as shown by the growth and pervasive usage of information and communication technologies in both business and daily life. These days, concepts like "information society," "internet of things," and "digital economy" are seen as the primary drivers of state growth that determine how competitive the national economy is. The World Bank defines the digital economy as a system of relationships centered on the use of digital information and communication technology. Without a question, the digital economy should have an impact on every aspect of government operations, but the manufacturing sector, which is the main driver of economic development, should be the first to undergo digital transformation (Tariq, 2025). Food safety and post-harvest loss are major issues in the worldwide food processing business, which includes Ghana's smoked fish sector. Previous studies conducted in Ghana have shown that traditional ovens used for fish smoking produce polycyclic aromatic hydrocarbons (PAHs), even though smoking fish improves quality and prolongs its shelf life, lowering the possibility of post-harvest loss. As a result, substantial concentrations of PAHs were found in Ghanaian smoked fish products. Therefore, it has been discovered that smoking fish in traditional ovens contains high quantities of PAHs, which are dangerous to human health. The United States Environmental Protection Agency (US-EPA) and the European Union (EU) disclosed in 2006 that PAHs found in food products, such as smoked fish, may contribute to the development of cardiovascular disorders and cancer. Furthermore, PAHs were identified as known carcinogens by the International Agency for Research on Cancer (IARC) in 2012. Given that most Ghanaians eat smoked fish; this categorization might help explain the rising frequency of cancer diagnoses in the country. This has led to stringent oversight and rejection of food items, such as smoked fish from developing nations, at global markets that involve evaluating complex theoretical frameworks and varying variables. However, SEM is seldom used in Ghanaian and other countries' fish processing technologies. This is a significant research void in the body of literature. This research sought to forecast the several parameters influencing Ghanaian fishmongers' usage of Ahotor ovens using SEM as an estimating approach. It is expected that the study's findings will help national and international organizations as well as policymakers develop programs and policies for the adoption of IFST, including the Ahotor oven, in the fish processing value chain to improve food safety conditions and lower post-harvest losses. A theoretical framework, the formulation of hypotheses, methodologies, findings, discussion, and conclusions are all included in the remaining portions of the article. Because they utilize traditional ovens that depend on firewood, fishmongers often run the danger of suffering burns, respiratory problems, and headaches. Additionally, using firewood in these ovens may result in the loss of mangrove trees and other negative impacts on biodiversity. Because of this, alternative ovens especially ones with reduced PAH levels have gained popularity.

As a result, better fish smoking methods have been created, including the FAO-Thiaroye Technique, the Modern Landmark Smoker Oven, and the Ahotor Oven. Due to its fuel-efficient design and low PAH emissions, the Ahotor oven—which was first introduced by the West Africa Regional Fisheries Development Program (WARFP) in 2014—is an improved fish smoking technology (IFST). As a result, it offers fishmongers a safer and more cost-effective option than conventional ovens like Chorkor and Morrison, which are notorious for having high PAH levels. Ahotor ovens have been built in a few coastal villages in Ghana's Western and Central regions by the Cerath Development Organization (CDO) in partnership with the European Union-Ghana as part of the Power to the Fishers initiative. According to a recent research by (Pramanik, S. & Raja, S. S. 2020), the use of better ovens for fish processing may enhance fishmongers' incomes by increasing market participation due to the reduced emission of PAHs. Fishmongers still often utilize traditional smoking ovens in Ghana, despite the CDO's significant efforts on the Ahotor campaign. Examining fishmongers' attitudes towards and preparedness to utilize the Ahotor ovens is essential to understanding their aims. Given that behavioral intention plays a major role in the diffusion of innovation, this could have a positive effect on the spread of such innovation. A detailed analysis of their behavioral intentions about its use is required, given the limited use of the IFST in the fish smoking process, as shown by Ghanaian coastal fishmongers' use of Ahotor ovens. According to (Martin et al. 2015), it is crucial to understand the connections between acceptability and user experience. In the context of the agricultural sector, including fisheries, the updated Unified Theory of Acceptance and Use of Technology (UTAUT-2) put forth by researchers provides a comprehensive framework for comprehending the elements that influence a person's intention to adopt any new technology or innovation. The seven dimensions of this model—performance expectancy, effort expectancy, social influence, facilitating conditions, habits, hedonic motivation, and price value—are essential for understanding how people make decisions about new innovations or technologies. According to the sources mentioned by (Pramanik, S. 2023), the UTAUT2 model is well known for its ability to explain how new technology is adopted. The UTAUT-2 framework well explains the elements that influence people's behavioral intentions to accept cutting-edge technologies, according to a recent systematic evaluation of theories surrounding technology adoption. According to this study, the UTAUT-2 model may provide policymakers and other stakeholders in the fisheries sector with useful insights into the variables that encourage and hinder fishmongers in Ghana's coastal villages from using Ahotor ovens. However, prior research in Ghana has mostly addressed the scientific features of the Ahotor oven without sufficiently considering the behavioral viewpoint. To our knowledge, the UTAUT-2 model has not been employed, despite the fact that there aren't much researches that concentrates on better technology adoption in the context of fisheries. This research provided a behavioral theoretical model based on the updated unified theory of acceptance and use of technology (UTAUT2) that was tailored to the Ahotor oven as IFST in Ghana to fill the current vacuum in the literature on fisheries. Additionally, recent research in the fisheries sector has shown how useful structural equation modeling (SEM) is as an accurate statistical technique for assessing the relationship

APPLICATION OF THE INDUSTRIAL INTERNET OF THINGS IDEA IN A DRYING FACILITY

For this reason, the fishing industry's technical and innovative growth is crucial. Therefore, the goal of our study is to demonstrate how to build the Industrial Internet of Things (IIoT) concept using a tiny drying machine intended to perform the drying, air-drying, and smoking processes for fish and fish products. The equipment is used by Murmansk State Technical University's fish processing facility. At the same time, our idea for creating IIoT was entirely original and did not make use of pre-made solutions. A Web application-based human-machine interface is part of the developed information system. It incorporates the features of mobile control and remote access, enables remote control of the drying unit's operation, and checks the data and condition of every sensor in the system. The online interface enables the drying unit operator to further extend the parameters of their workspace configuration and boost workshop productivity. It is important to note that the operator may visually see the technical operation of the little drying unit thanks to the online application receiving a video feed from the web camera. The MS SQL Server MSU Data database (Pramanik, S. 2024), which is intended to synchronize the work of all programs, is the central component of the information system while accessing a database in multi-user mode, fewer problems occur while connecting to a single resource (Figure 4). The following is the information system's fundamental idea: Information is gathered by the system core from human-machine interfaces, sensors, actuators, and controllers. Accurate and objective data on the technical process, its characteristics, and equipment operation time may be acquired thanks to the information gathered. By preventing unexpected downtime, equipment failures, unscheduled maintenance, and supply chain management interruptions, the knowledge gathered may help the business operate more effectively. The creation of our own software enables us to further modify the data exchange pool for integrating our workshop's technical equipment and other installations with the system. In this situation, it is very crucial that information be presented correctly and in a way that the user can comprehend. We are trying to update the core information system's data sets in the interchange formats of CSV (Pramanik, S. 2024), XML, and others in this respect. Production may be technologically and organizationally transformed thanks to the Industrial Internet of Things. It incorporates actual manufacturing, transportation, human, engineering, and other resources into almost infinitely scalable program-controlled virtual resource pools and the outcomes of their utilization at the management level.

Consequently, the Internet of Things was introduced on a wide scale based on a modest drying unit. Embedded sensors, software, and a data collecting and exchange pool were linked to computer networks and manufacturing facilities. Information equipment that guarantees remote controls of the technological process, its visualization, and, if required, remote interference with the system's operation was sent to the compact drying unit. The compact drying unit functions by taking into consideration the fundamental patterns of moisture mass transfer in capillary porous colloidal substances, both internally and externally. A multi-level control over the technical characteristics of the processed raw materials is implemented by the installation. The IIoT (Pramanik, S. 2024) idea is being developed in a constructive manner. The economic impact is already evident and enables us to prepare for upgrading our manufacturing line. From businesses that provide fish raw materials to businesses that are the ultimate consumers of fish products, we want to create a self-optimizing commodity turnover chain. As a result, a virtual business based on the production hall will be established and the transition to the lifetime information support principles (CALS-technology / Continuous Acquisition and lifetime Support) of fisheries goods will be implemented. Digital systems are linked to risk factors for the security of the data they handle since most of the information is sent over open communication channels. As a result, information security is given consideration in our project. The SSL (Secure Socket Layer) data transfer protocol is used to connect to the automation system server via user authentication to secure the large data (large DATA) sent inside the information system over the Internet.

RESEARCH METHODOLOGY

The phases of Internet of Things (IoT) system development as outlined by (Singh, D. et al. 2024) were used in this work, which followed an R&D approach (Figure 1).

Figure 1. IoT research and development stages

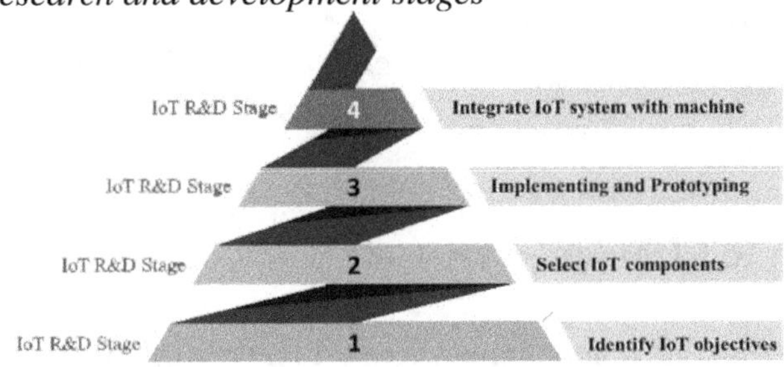

The four separate stages of IoT development that make up the research process are explained as follows:

1. Determine IoT Goals

It is critical to specify the main goals of IoT system development at this point. In general, the use of IoT in smoked fish machines is to provide an intuitive user interface that makes remote smoking room monitoring and control possible.

a) Controlling the blower's activation and deactivation, the machine's lights, and the servo motor that drives the machine's shaft are among the goals that need to be met.
b) A real-time temperature monitor that is located within the smoking tube in the fish smoking chamber.
c) Modulating the blower's Revolutions per Minute (RPM) to control the temperature within the combustion chamber.

These objectives show the visualized aims of the Internet of Things system that will be installed in the smoked fish machine for this project, as seen in the flowchart in Figure 2. Three goals are to monitor the blower's RPM, temperature, and machine components.

Figure 2. IoT objectives in this study

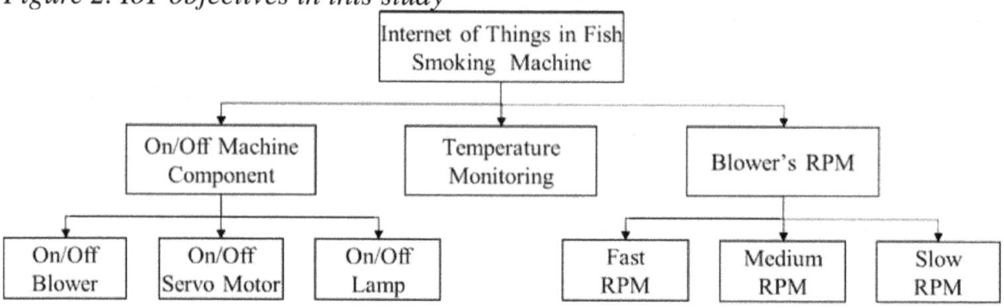

2. Choose an IoT Device

Selecting the required components was a critical next step after establishing the goals for the IoT system. Six crucial elements are identified by this study as being necessary for setting up the IoT system:

a) The microcontroller is the Internet of Things system's core processing unit, handling sensor data processing and providing control over other networked parts.

b) The temperature and humidity within the fish smoking chamber are continually and precisely measured in real time using the DHT11 temperature sensor.
c) For best performance, the IoT system is automatically connected with the blower's electrical system, which controls air circulation within the smoking chamber.
d) To link the Internet of Things system to the internet and allow remote data transmission and reception, an Ethernet (**Pramanik, S. 2024**) or WiFi module is needed.
e) Users may interact with and operate the Internet of Things system using the user interface on a mobile smartphone.

Following the identification of the main IoT system components in the fish smoking chamber, a thorough design was created that included both necessary and auxiliary parts. The electrical circuit design, as shown in Figure 3, shows the Internet of Things system's whole schematic and describes how different components—such as sensors, microcontrollers, communication modules, and output devices—are integrated.

Figure 3. IoT electrical system design plan

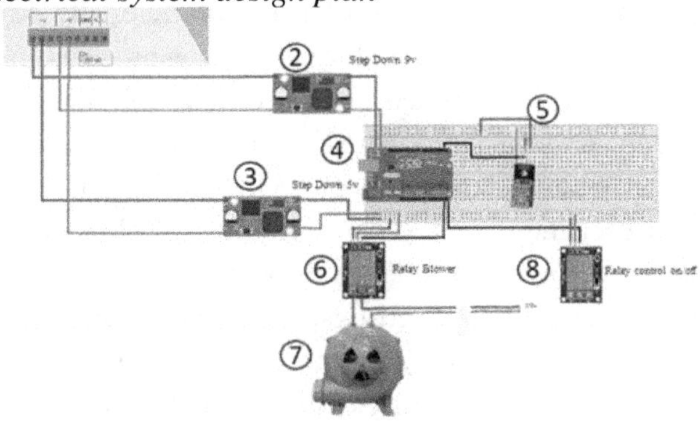

The elements below offer a thorough description of the characteristics of the IoT components shown in Figure 3, which includes a full schematic of those components.

Components of IoT Electrical Systems

1. 12 V microcontroller power supply
2. 9 V modulo step down
3. 9 V module step down
4. Arduino Uno microcontroller
5. DHT 11 Temperature Sensor

6. Blower relay with RPM setting
7. 2” 220 V blower
8. On/Off Relay Machine

3. Prototyping and Implementation

The process of implementing and prototyping comes next, after the choice of IoT components and system design completion. Implementing the hardware prototype and the software prototype are the two separate efforts that make up this step.

a) Implementation of Hardware Prototype

All IoT components are constructed and integrated in accordance with the finalized design requirements for the purpose of hardware prototype implementation. Connecting the Arduino Uno microcontroller to the DHT11 temperature sensor is the first step in the procedure. Then, software code is created to get temperature data in real time from the sensor and send it over the internet to the user interface. Furthermore, the microcontroller is integrated with the timer system and blower RPM. A matching computer code is written to automatically adjust the blower RPM depending on the desired temperature. The blower, servo motor, and other components may also be programmed to turn on and off using this technology. A WiFi (Manikandan, K. et al. 2024) or Ethernet module is inserted inside the microcontroller to provide internet connection for the Internet of Things system. The development of user interfaces, such as online or mobile apps, allows for the remote monitoring of temperature and humidity in addition to the management of fan speeds, timers, and temperature settings. After every component is integrated, the IoT system prototype is put through a thorough testing process. To verify that the device functions as planned and satisfies all functional requirements, this testing entails simulating a range of temperature settings and smoking situations.

Figure 4. Box panel and IoT component assembly

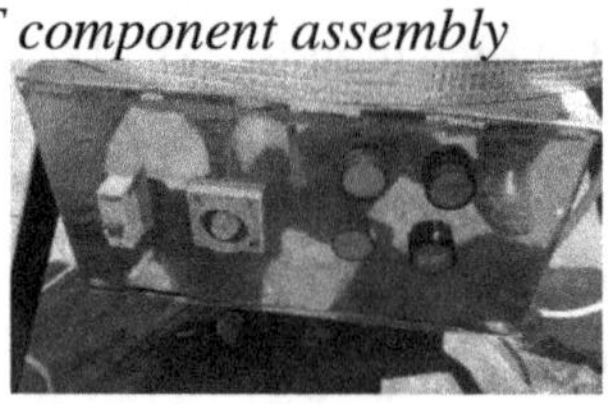

b) The control PCB circuit design, shown in Figure 4, is meant to be incorporated into the Internet of Things system to monitor and regulate the process of smoking fish. The machine's control panel has this electrical system installed, and both functionality and safety have been given serious consideration. A power supply, a step-down module, a DHT11 temperature sensor, a timer sensor, an Arduino Uno microcontroller, an IoT communication module, the machine's heating system, and the BlynkIoT platform are some of the essential parts of the design. The system's essential component, the DHT11 temperature sensor, measures the temperature in real time while smoking. It was chosen due to its precision in monitoring humidity and temperature as well as its ability to communicate digitally with the microcontroller. The Arduino Uno microcontroller enables smooth integration with the Internet of Things by facilitating wireless communication across a Wi-Fi network.
c) Implementing Software Prototypes

The oven temperature monitoring system, which uses the DHT11 sensor in combination with an Arduino Uno microcontroller and the Blynk platform, is coded as shown in Figure 5.

Figure 5. IoT coding on smoked fish machine

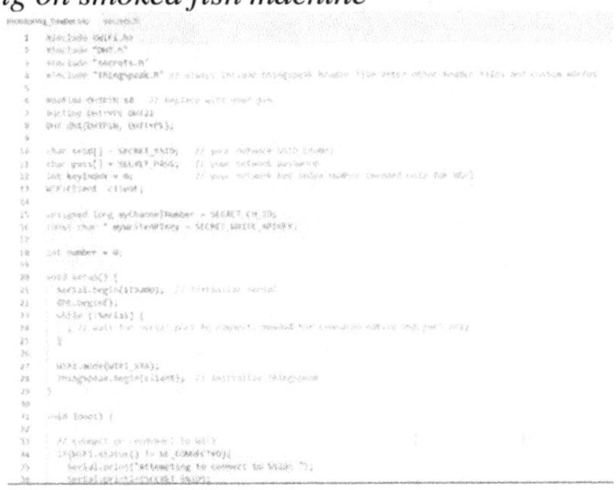

The program's functions include reading temperature data from the DHT11 sensor, sending it to the Blynk application for real-time monitoring, and displaying it on an LCD screen. Blynk makes it easier to see the current temperature and local time. It also can obtain local time using the Time Zone API.

Using a customized application, the user interface is intended to facilitate interaction with the system. Users may manage the blower's speed and RPM, as well as check the temperature throughout the smoking process, using this application. Users must log in using previously generated credentials to use the monitoring system. Figure 6 shows how the IoT system is integrated with the Blynk application and gives a visual depiction of the system's operation and status.

Figure 6. IoT user interface design display

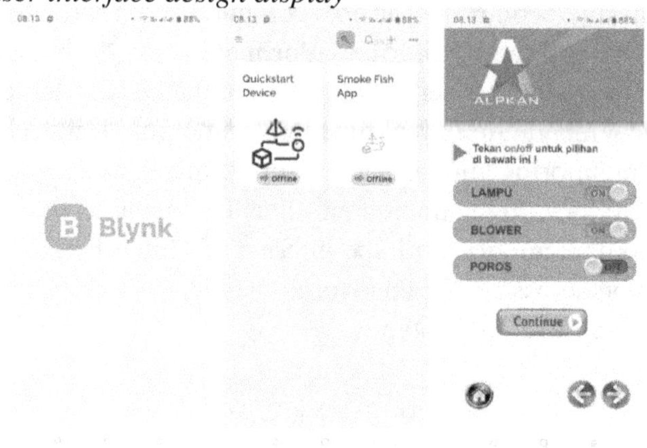

The design outcomes of the Blynk platform-based IoT monitoring and control system are shown in Figure 6. This phase is essential to the system's successful implementation. Using the Arduino Uno IoT connection module, the Blynk platform enables real-time data transfer from the temperature sensor to the Blynk cloud server. This data may then be seen and used using the Blynk online interface and mobile application. With features like historical temperature graphs and warnings for parameter deviations, Blynk's user interface provides an easy-to-use and intuitive display for real-time sensor monitoring. The Blynk interface also lets users establish goal temperatures and provides remote control of the blower's RPM via virtual buttons. Stable Wi-Fi connection is necessary for the IoT connectivity of the system. By keeping an eye on the consistency of data flow from the microcontroller to the Blynk platform, connectivity performance was evaluated. Designing the system with security in mind is essential, and the Blynk platform offers strong authentication and data encryption to secure connections between devices and cloud servers. Password security protects access to the Blynk user interface, guaranteeing that only authorized users may watch over and manage the system. The results of the testing verify that there is little to no major data loss during data transfer, which is efficient and stable.

4. Connect the Smoked Fish Machine to the Internet of Things

After the IoT system prototype for the smoking machine was successfully implemented, the next step was to integrate the system with the working machine. The IoT system had to connect with a smartphone and the smoked fish machine for this project. If the device is linked to a nearby Wi-Fi network and the user's smartphone is online, the Internet of Things system may be remotely monitored and controlled. When Wi-Fi isn't accessible, customers may use tethering (Mishra, S. et al. 2024), which allows a smartphone to connect to an IoT panel on a machine up to 30 meters away, as an alternate IoT connection option. The integration procedure between the components of the smoked fish machine and the IoT system is shown in Figure 7. This step is critical to guarantee the seamless operation of the smoking machine's IoT panel and the smartphone application for the Internet of Things. At this point, calibration tests were also carried out to confirm the reliability and correctness of the data readings.

Figure 7. Integrating IoT with smoking fish machine

RESULT AND DISCUSSION

Using the Internet of Things to Turn on and Off Machine Components

The operational sequence of the Internet of Things system applied to a smoked fish machine for component activation is shown in the flow diagram shown in Figure 8. The procedure begins with turning on the machine by hand. Power is then supplied using a supply unit that converts 220 V into 12 V. This is further controlled by a step-down converter, which lowers the voltage to 9 V and guarantees that the current stays within the Arduino Uno microcontroller's safe operating parameters. The smoked fish machines assigned IoT button must be pressed to enter the IoT mode once the device has been properly turned on. The user may operate the machine remotely after connecting it to their smartphone using the Internet of Things (IoT) protocol. Three components can be turned on and off more easily thanks to the Internet of Things system: the light, servo motor, and blower.

Figure 8. IoT system to activate components in smoked fish machine

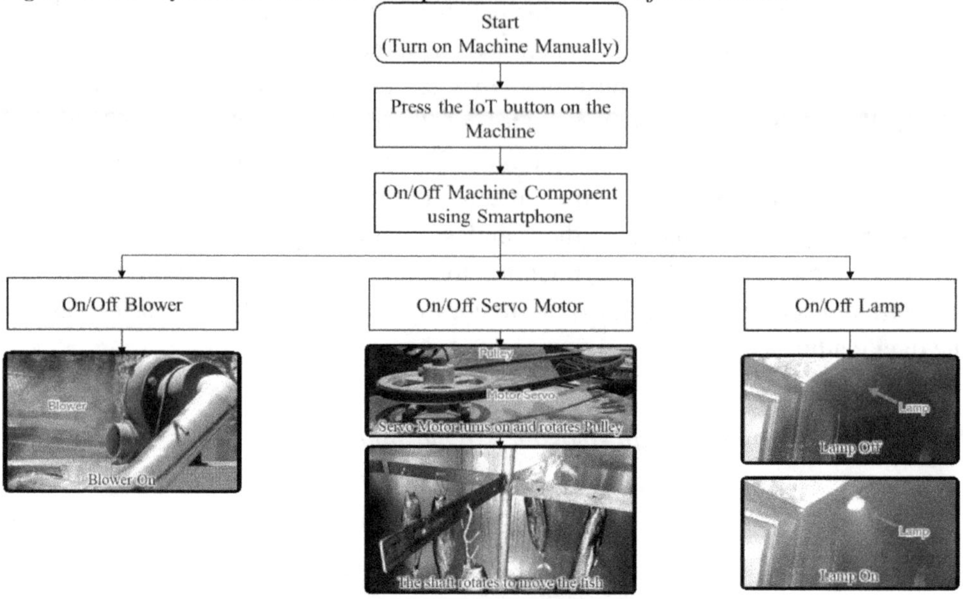

a) Temperature Monitoring IoT Performance Testing

By gathering data from the DHT11 temperature sensor, analyzing it, and sending it to the BlynkIoT platform via the connection module, the microcontroller plays a crucial part on the Internet of Things system. The temperature data that is gathered is monitored using the Blynk application. Additionally, the microprocessor controls the blower's RPM to keep it at the ideal setting for smoking. Furthermore, a timer mechanism is included to automatically stop smoking when the predetermined amount of time has passed. An extensive assessment was carried out to evaluate the effectiveness of the RPM blower monitoring and control system throughout the fish smoking process after the Internet of Things system was installed on the smoking machine. The primary goal of the first test was to assess how well the DHT11 temperature sensor tracked the parameters of the smoking process. The test results are shown in Table 1, which shows that the DHT11 temperature sensor performs well in terms of accuracy, with an average variance of 0.1-0.5°C when compared to the analogue thermometer that was used as a baseline. Ensuring that the temperature data collected properly matches the real circumstances inside the smoking machine is contingent upon this degree of precision.

Table 1. Temperature monitoring performance testing

bNo	Machine Temperature (°C)		Status Machine / Gas Solenoid Mode
	IoT Application	Analog Thermometer	
1	34.1	34.4	off
2	43.5	43.3	on / Fast RPM Blower
3	48.8	48.6	on / Fast RPM Blower
4	58.7	58.2	on / Fast RPM Blower
5	62.6	62.5	on / Fast RPM Blower
6	69.2	69	on / Fast RPM Blower
7	73.8	73.6	on / Fast RPM Blower
8	79.3	79	on / Medium RPM Blower
9	85.3	85.2	on / Medium RPM Blower
10	94.6	94.5	on / Slow RPM Blower
11	100.7	100.5	on / Slow RPM Blower

b) Using IoT to Regulate Blower's RPM

During the fish smoking process, a blower control system that is based on rotation per minute (RPM) is used to control the hot air circulation. To continually monitor the dryer's temperature, this solution combines a microcontroller and IoT technologies with a DHT11 temperature sensor. Testing was done to see how well this monitoring system maintained the ideal temperature during smoking. The temperature range that the blower system can withstand is 75–90°C. According to test findings, the temperature is successfully maintained by the RPM control system within the given range, with a maximum variance of 1°C. Furthermore, the system showed quick reaction to modifications in the smoking environment by quickly modifying the blower RPM in response to notable temperature variations in the smoking chamber. The smoking process is made more effective and efficient by this real-time parameter control.

Figure 9. Blower RPM control system installation

A tiny servo and a dimmer SCR are two of the fitted parts shown in Figure 9 that regulate the blower's RPM (Rotation per Minute). This device automatically adjusts power levels by integrating a servo motor and dimmer. By adjusting the dimmer lever, the servo motor improves blower efficiency by providing automatic RPM control without the need for human interaction. The RPM control system adjusts the smoking machine's air circulation according to the drying chamber's temperature. The RPM is lowered to 500 when the temperature rises over 100°C to protect the fish from overheating. The optimum temperature range for smoking is between 75 to 90°C, and the blower runs at a medium speed of 1,500 RPM to keep the temperature there. To efficiently boost the temperature in the smoking chamber when the outside temperature drops below 75°C, the system quickly replaces the cool air with hotter smoke by increasing the blower RPM to 2,500.

Figure 10. Working scheme for SCR dimmer and RPM blower control system design

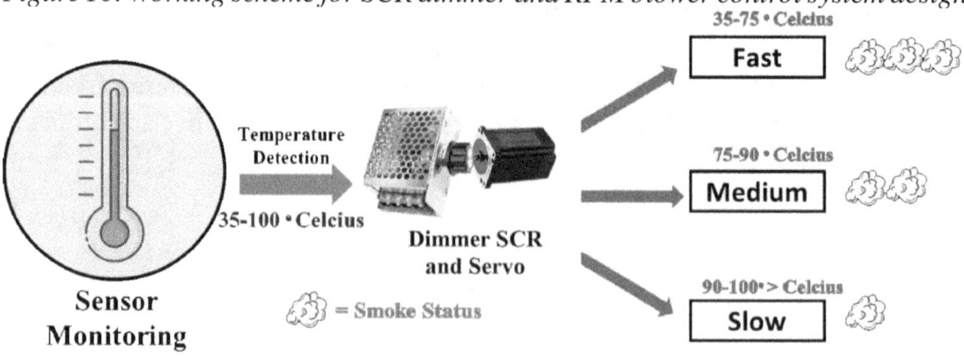

The RPM blower control system's operating scheme is shown in Figure 10. The purpose of this automated control system is to maximize and stabilize the heat produced by combustion smoke, guaranteeing that the temperature within the smoking processing tube stays continuously between 75 and 90 degrees Celsius. To control temperature variations throughout the smoking process, the Blower RPM Control System relates to the Internet of Things infrastructure. The temperature may be precisely adjusted thanks to its integration, rising or falling as needed. The system's architecture strives to effectively keep the smoking processing tube's temperature within the 75–90°C ideal range that has been shown to be beneficial in the literature (Mohammad, A. et al. 2024). The technology guarantees that the fish smoking process runs as effectively and efficiently as possible by automating the temperature regulating process.

CONCLUSION

An Internet of Things-integrated monitoring and control system for a smoked fish machine was successfully constructed and put into use by this research (Tariq, 2025; Tariq and Sergio, 2025). Users may remotely switch on or off the machine's components thanks to the IoT system's exact control over them. The device also makes it easier to monitor the temperature in the fish smoking chamber in real time. Additionally, it modifies the Revolutions per Minute (RPM) of the blower to control the temperature inside the combustion chamber. By adding an Internet of Things system to the machine, users may watch and manage the fish smoking process from a distance, doing away with the requirement for them to physically be present at the machine's location. Performance testing showed that when machine components such a blower, servo motor, and light are turned on via a smartphone interface, the Internet of Things system performs as intended with just a little delay of around one second. Additionally, testing showed that the DHT11 sensor, which is linked to the Internet of Things, has a very high degree of precision when detecting humidity and temperature, with an average difference of just 0.1 to 0.5°C when compared to an analogue thermometer. It was discovered that the blower RPM control system was also successful in keeping the temperature within the ideal range, with a maximum variation of 1°C, and that it was capable of effectively modifying hot air circulation according to the temperature.

Overall, the study's findings suggest that there is a lot of room for improvement in operational efficiency and control when IoT technology is used to smoked fish plants. To further optimize the system's efficiency, future development efforts might concentrate on incorporating other features like sophisticated data analysis and predictive maintenance systems.

REFERENCES

Ahmid, M., Kazar, O., & Barka, E. (2024). Internet of Things Overview: Architecture, Technologies, Application, and Challenges. In Boulila, W., Ahmad, J., Koubaa, A., Driss, M., & Farah, I. R. (Eds.), *Decision Making and Security Risk Management for IoT Environments. Advances in Information Security* (Vol. 106). Springer., DOI: 10.1007/978-3-031-47590-0_1

Ashisha, G. R., Anitha Mary, X., George, T., Martin Sagayam, K., & Unai Fernandez-Gamiz Hatıra Günerhan, M. N. (2023). Uddin and S. Pramanik, Analysis of Diabetes disease using Machine Learning Techniques: A Review. *Journal of Information Technology Management*.

Asiedu, B., Failler, P., & Beygens, Y. (2018). Ensuring food security: An analysis of the industrial smoking fishery sector of Ghana. *Agriculture & Food Security*, *7*(1), 38. DOI: 10.1186/s40066-018-0187-z

Cisse, M., Karamoko, G., Cisse, M., Chene, C., & Karoui, R. (2023). Determination of the quality of smoked sea herring (*Sardinella maderensis*) produced by three smoking processes practiced in Ivory Coast. *European Food Research and Technology*, *249*(11), 2989–3000. DOI: 10.1007/s00217-023-04343-5

Cissé, M., Karamoko, G., Cissé, M., Chèné, C., & Karoui, R. (2025). Quality assessment of Ivory Coast smoked herring (*Madeiran sardinella*) produced by three smoking processes by physico-chemical and mid infrared spectroscopy coupled with chemometric tools: A preliminary study. *European Food Research and Technology*. Advance online publication. DOI: 10.1007/s00217-025-04674-5

Dhabliya, D., Veeraiah, V., Dari, S. S., Raja Kumar, J. R., Dhabliya, R., Pramanik, S., & Gupta, A. (2024). *Virtual and Augmented Reality Applications for Healthcare and Goodness*. IGI Global., DOI: 10.4018/979-8-3693-1463-0.ch017

Khanh, P. T., Ngoc, T. T. H., & Pramanik, S. (2024). Evaluation of Machine Learning Models for Mapping Soil Salinity in Ben Tre Province, Vietnam. *Multimedia Tools and Applications*, *84*(1), 201–220. Advance online publication. DOI: 10.1007/s11042-024-18712-z

Khanh, P. T., Pramanik, S., & Ngoc, T. T. H. (2024). Soil Permeability of Sandy Loam and Clay Loam Soil in the Paddy Fields in An Giang Province in Vietnam. *Environmental Challenges*, *15*, 100907. Advance online publication. DOI: 10.1016/j.envc.2024.100907

Malode, K., Rajawat, B. S., Shankar, S. A., Kumar, R., Khubalkar, D., & Pramanik, S. (2024). *Mathematical Modeling of Carbon Sequestration in Agricultural Soils: Implications for Climate Change Mitigation Policies, in in Mathematical Modeling in Agriculture*. John Wiley & Sons., DOI: 10.1002/9781394233731.ch3

Manikandan, K., Nayagam, R. D., Namdev, A., Sudha, K., Patil, T., Gupta, A., & Pramanik, S. (2024). *Enhanced Deep Convolutional Neural Network for Breast Cancer Recurrence Prognosis*. IGI Global., DOI: 10.4018/979-8-3693-1463-0.ch007

Mishra, S., Namdev, K. P. A., Venkataramana, S., Pramanik, S., & Gupta, A. (2024). Analysis Model at the Sentence Level for Phishing Detection. In *Deep Learning, Reinforcement Learning, and the Rise of Intelligent Systems* (pp. 209–227). IGI Global., DOI: 10.4018/979-8-3693-1738-9.ch011

Mohammad, A., Asija, D., & Viral, R. (2024). Group Induction Motors' Speed Control via ACPWM-Based Single Speed Controller. In Kumar, R., Phanden, R. K., Tyagi, R. K., & Ramkumar, J. (Eds.), *Advances in Manufacturing and Materials. FLUTE 2023. Lecture Notes in Mechanical Engineering*. Springer., DOI: 10.1007/978-981-97-3173-2_30

Ngoc, T. T. H., Pramanik, S., & Khanh, P. T. (2023). The Relationship between Gender and Climate Change in Vietnam", The Seybold Report, DOI DOI: 10.17605/OSF.IO/KJBPT

Phadke, M., & Korde, M. (2024). IoT-Based Weather Monitoring System Using NodeMCU ESP8266. In Senjyu, T., So–, C., & Joshi, A. (Eds.), *Smart Trends in Computing and Communications. SmartCom 2024 2024. Lecture Notes in Networks and Systems* (Vol. 945). Springer., DOI: 10.1007/978-981-97-1320-2_18

Pramanik, S. (2023). An Adaptive Image Steganography Approach depending on Integer Wavelet Transform and Genetic Algorithm. *Multimedia Tools and Applications*, *82*(22), 34287–34319. Advance online publication. DOI: 10.1007/s11042-023-14505-y

Pramanik, S. (2024). Applications of 3D Printing in Contemporary Medicine, IGI Global, DOI: DOI: 10.4018/979-8-3693-1862-1.ch003

Pramanik, S. (2024). *Examining HRM Practices in the Contemporary Era With Regard to Technological Development and Innovation*. IGI Global., DOI: 10.4018/979-8-3693-3466-9.ch001

Pramanik, S. (2024). *Overcoming the Digital Frontier: Indonesia's NHS E-Health Plan and Medical Revolution From a Socio-Technical Perspective*. IGI Global., DOI: 10.4018/979-8-3693-1463-0.ch013

Pramanik, S. (2024). *Utilizing Waste Water for Sustainable Energy Generation via IoT-Integrated Technologies*. IGI Global., DOI: 10.4018/979-8-3693-1014-4.ch011

Pramanik, S., Singh, R. P., & Ghosh, R. (2020). Application of Bi-orthogonal Wavelet Transform and Genetic Algorithm in Image Steganography. *Multimedia Tools and Applications*, *79*(25-26), 17463–17482. Advance online publication. DOI: 10.1007/s11042-020-08676-1

Pramanik, S., & Suresh Raja, S. (2020). A Secured Image Steganography using Genetic Algorithm. *Advances in Mathematics: Scientific Journal*, *9*(7), 4533–4541. DOI: 10.37418/amsj.9.7.22

Raoufi, P., Hemmati, A., & Rahmani, A. M. (2024). Deep learning applications in the Internet of Things: A review, tools, and future directions. *Evolutionary Intelligence*, *17*(5-6), 3621–3654. DOI: 10.1007/s12065-024-00949-0

Ravishankar, C. N., & Elavarasan, K. (2023). Innovations in Fish Processing Technology. In Bansal, K. C., Lakra, W. S., & Pathak, H. (Eds.), *Transformation of Agri-Food Systems*. Springer., DOI: 10.1007/978-981-99-8014-7_16

Rawat, A., Ramachandran, T., Gupta, Y. C., Mishra, M. K., Michael, G., Pramanik, S. (2024). Simulating the Effects of Land Use Regulations on Agricultural Land Values, .DOI: 10.1002/9781394233731.ch14

Ray, M., Pramanik, S., Das, B., & Khanna, A. (2023). Hybrid Cryptography and Steganography Method to Provide Safe Data Transmission in IoT. *ICDAM*, *2023*, 513–524. Advance online publication. DOI: 10.1007/978-981-99-6550-2_39

Roy, A., Pramanik, S., Mitra, K. and Chakraborty, M. (2024). "Advanced hybrid neural network techniques for minimizing gas turbine emissions", *World Journal of Engineering*, Vol. ahead-of-print No. ahead-of-print. DOI: 10.1108/WJE-11-2023-0473

Roy and S. Pramanik. (2023). A Review of the Hydrogen Fuel Path to Emission Reduction in the Surface Transport Industry. *International Journal of Hydrogen Energy*. Advance online publication. DOI: 10.1016/j.ijhydene.2023.07.010

Singh, D., Jaison, F., Saxena, V., Sharma, S., Raj, S., Gupta, A., & Pramanik, S. (2024). *Cucumbers in PH Disease Monitoring using an IoT-based Mobile App*. CRC Press. DOI: 10.1201/9781003515883-9

Tariq, M. U. (2025). *AI-Driven Learning and Engagement in Higher Education*. IGI Global., DOI: 10.4018/979-8-3693-4074-5

Tariq, M. U. (2025). Innovative Business Models: Integrating Social, Economic, and Environmental Goals in Rural Manufacturing. In Prabhakar, A., Yu, P., & Erokhin, V. (Eds.), *Rural Social Entrepreneurship Development: Network-Based Manufacturing System Model* (pp. 147–174). IGI Global Scientific Publishing., DOI: 10.4018/979-8-3693-7515-0.ch005

Tariq, M. U. (2025). Revolutionizing Communication: EEG-Based Brain-Computer Interface for Speech and Mood Detection. In Prabhakar, A., Yu, P., & Erokhin, V. (Eds.), *Rural Social Entrepreneurship Development: Network-Based Manufacturing System Model* (pp. 237–264). IGI Global Scientific Publishing., DOI: 10.4018/979-8-3693-7515-0.ch008

Tariq, M. U., & Sergio, R. (Eds.). (2025). *Cases on AI-Driven Solutions to Environmental Challenges*. IGI Global, 2025. https://doi.org/DOI: 10.4018/979-8-3693-7483-2

Tariq, M. U., & Sergio, R. (Eds.). (2025). *Convergence of AI, Education, and Business for Sustainability*. IGI Global., DOI: 10.4018/979-8-3373-1917-9

Uhuo, E., Egba, S., Alaebo, P., Obike, C. A., & Ede, I. S. (2023). Evaluation of fertility hormones and biochemical stress initiated by toxicant in diet prepared with fish smoked with polyethylene (plastic) materials as a fuel source. *Comparative Clinical Pathology*, *32*(4), 527–537. DOI: 10.1007/s00580-023-03461-6

Veeraiah, V., Shiju, D. J., Ramesh, J. V. N., Ganesh, K. R., Pramanik, S., Pandey, D., & Gupta, A. (2023). *Healthcare Cloud Services in Image Processing, Thrust Technologies' Effect on Image Processing*. IGI Global.

Veeraiah, V., Shiju, D. J., Ramesh, J. V. N., Ganesh Kumar, R., Pramanik, S., Gupta, A., & Pandey, D. (2023). *Healthcare Cloud Services in Image Processing, Thrust Technologies' Effect on Image Processing*. IGI Global.

Wang, J., Feng, X., Yu, Y., Huang, H., Zheng, M., Xu, Y., Wu, J., Yang, Y., & Lu, J. (2024). Rapid temperature-responsive thermal regulator for safety management of battery modules. *Nature Energy*, *9*(8), 939–946. DOI: 10.1038/s41560-024-01535-5

KEY TERMS AND DEFINITION

IoT: The interconnected system of physical items, or "things," that are equipped with sensors, software, and other technologies to communicate and exchange data with other systems and devices over the internet is known as the Internet of Things (IoT).

Arduino: Artists, designers, amateurs, hackers, novices, and everyone else interested in making interactive things or settings were the target audience for the Arduino hardware and software. Buttons, LEDs, motors, speakers, GPS devices, cameras, the internet, and even your smartphone or TV may all be controlled with Arduino.

IIoT: The Industrial Internet of Things is referred to as IIoT. Data from industrial activities is gathered and analysed by this network of sensors, gadgets, and apps. IIoT is utilised to increase security and safety, save expenses, and increase efficiency.

Chapter 7
Leather Defect Classification in Footwear Manufacturing Industries

V. Mareeswari
https://orcid.org/0000-0002-2768-943X
Vellore Institute of Technology (VIT), Vellore, India

R. Vijayan
Vellore Institute of Technology (VIT), Vellore, India

Praveen Kumar S.
Vellore Institute of Technology (VIT), Vellore, India

Aravind P. Dhakshan
Vellore Institute of Technology (VIT), Vellore, India

ABSTRACT

Leather is a natural and durable material derived from tanning animal skins and hides. The value of leather is contingent upon its quality and surface condition, making it imperative to detect defects accurately. Traditionally, detecting leather defects has been labor-intensive and time-consuming, prone to human error and eye strain. This paper proposes a fully automatic defect detection system that employs the EfficientNetB0 neural network on the leather to provide defect-free leather. The system can classify leather patches into different classes, such as folding marks, grain off, growth marks, loose grain, non-defective, and pinhole. This proposed work suggests this defect identification software system can be employed with a few mechanical components such as fixing the camera in a conveyor system to identify the defects of moving leather pieces. This research contributes a practical and efficient

DOI: 10.4018/979-8-3373-4332-7.ch007

solution for enhancing quality control processes in footwear manufacturing industries, ultimately leading to streamlined production and improved product quality.

INTRODUCTION

Leather is a crucial raw material across various manufacturing industries, particularly in India, the world's second-largest exporter of leather garments and the fourth-largest exporter of leather goods globally. Leather has long been a vital material, valued for its durability and versatility in fashion, furniture, and automotive applications. Its production involves a meticulous process, starting with rawhide selection, cleaning, tanning, dyeing, and finishing. Defects can arise from animal living conditions (like scratches or insect bites) and processing stages (like cuts or blemishes from machinery). The global leather industry, particularly in India, faces challenges in maintaining quality due to increasing demand. Conventional defect detection methods, relying on visual inspection, are subjective and inefficient. Manual error checking remains an indispensable quality control step in the leather and textile industries.(Kumari et al., 2021). To address these issues, there's growing interest in developing deep learning techniques for automated defect detection in leather. Deep learning, a subset of artificial intelligence, is adept at recognizing patterns and can be trained to identify subtle defects in leather surfaces from annotated image datasets. This approach could enhance product quality and streamline production processes, but it requires overcoming challenges like dataset diversity and adapting to varying lighting and textures. Embracing these technologies is crucial for the leather industry to stay competitive and meet consumer expectations.

Due to the enduring popularity of leather products, the global leather industry plays a crucial role across multiple sectors, such as fashion, automotive, and furniture. However, maintaining consistent quality is vital for manufacturers to meet consumer expectations and protect brand reputation. Traditional methods of detecting defects rely on manual inspection, which is labor-intensive and subjective. As demand for leather increases globally, there is a growing need for more efficient and reliable defect detection techniques. Recent advances in deep learning, a branch of artificial intelligence, have opened new possibilities for automating defect detection processes. Deep learning algorithms (Mareeswari et al., 2022), especially convolutional neural networks (CNNs), have shown great promise in image recognition tasks. By analyzing annotated images, these models can accurately identify and categorize defects. (Vijayan, Mareeswari, & Jaswanth, 2024). Implementing deep learning in defect detection for the leather industry could streamline quality control processes and enhance overall product quality. However, this work proposes a fully automated defect detection system to address these challenges. By employing CNNs, the system analyzes finished leather to identify and flag any imperfections, ensuring the production of defect-free leather goods. In leather manufacturing, this automated approach promises to streamline processes, enhance efficiency, and mitigate the inherent challenges of manual defect detection. However, challenges like acquiring diverse datasets and optimizing algorithms must be overcome for successful implementation. By leveraging advanced technologies like CNNs, manufacturers can optimize quality control measures, ultimately bolstering product quality and competitiveness in the global market.

EfficientNet is a family of CNN architectures known for their outstanding performance and efficiency in image classification tasks. These models achieve remarkable accuracy when using fewer parameters than traditional architectures, making them suitable for deployment in resource-limited environments. Given the significance of defect detection in leather and the potential benefits of employing efficient deep-learning models, there is a need to explore the application of EfficientNet in this field. By utilizing EfficientNet for leather defect detection, we can enhance the accuracy and efficiency of defect identification while reducing computational resources and processing time. This sets the stage for proposing a system that utilizes EfficientNet architecture for leather defect detection, aiming to achieve precise and efficient detection of various defects in leather and improve quality control processes in the footwear manufacturing industry. Continuous monitoring by IoT devices and assessment of products by AI algorithms will maintain high standards. However, challenges remain, such as the need for extensive and diverse datasets to train robust models and optimize algorithms for detecting subtle defects under different lighting conditions and surface textures. Overcoming these challenges will require ongoing research and collaboration between academia and industry to develop more sophisticated and reliable defect detection systems for the leather industry. The review section comprehensively examines the existing research on developing deep learning-based defect detection systems for assessing leather quality. It discusses the advantages and limitations of deep learning approaches compared to traditional methods, highlighting recent advancements in image processing and computer vision techniques that have improved the effectiveness of these systems. The proposed method can generate a significant accuracy rate from 3600 leather images that belong to six different classes. The comparative analysis against baseline models underscores the superiority of the Efficient Net architecture in terms of accuracy and computational cost.

Additionally, the work explores potential challenges and future research directions in this field, aiming to provide insights and recommendations for advancing robust and scalable defect detection systems tailored to the needs of the leather industry. Deep learning algorithms detect leather defects in various ways, including enhanced model architectures, transfer learning from pre-trained models, advanced data augmentation techniques, multimodal approaches, real-time detection systems, edge computing, and IoT integration. These advancements could improve detection accuracy, efficiency, and scalability. Transfer learning from models trained on extensive datasets could enhance performance when training data for specific defect types is limited. Advanced data augmentation techniques could expand the training dataset and improve the model's ability to recognize various defects.

LITERATURE REVIEW

Image processing is a truly important field in computer vision. The type of images plays an important part in the case of identification, insulation, and decision-making processes in automated systems. With the ever-adding complex tasks given to these computer vision-predicated tools and the adding reliance on these systems, groups systems must be robust and have high accuracy and discovery rates to perform error-free and without manual intervention. Leather is a main material used in vesture and appurtenant industries for making fashion accessories, footwear, and coats, amongst a host of other products. It's generally a residual product made from the external covering of brutes analogous to skins and hides, generally attained as by-products of the meat industry. “Hide” is the external covering of a large beast, analogous to a buffalo or a cow, whereas “Skin” is that of an angel, goat, etc. Leather, technically defined, is a natural protein polymer that is treated with tanning agents to make it resistant to enzymatic attack and corruption, as well as to increase several physical features. Leather and leather products are increasingly gaining significance in the world. The major task of the leather processing industry is to convert the skin or hide into leather, from which quality products can be made. The leather, which is reused from raw to finished, is dried. The coming step is the visual examination and type. Blights may arise in the reused distance of leather due to various reasons, analogous as wear and tear and gash and incision by the machines, heat, etc. These scars may vary in size and shape depending on the type of defect, analogous as holes, scrapes, bubbles, torn pieces, etc. The type takes place according to the face defect area of the defect, defect area, position, and distance viscosity, amongst multitudinous other factors. The number of scars determines the price, with farther scars going at a lower price. Thus, type is a truly important process to be accepted to determine the profitable value of the product. The manual type process, which is generally used, is error-prone. Long working hours, motorist fatigue, and other reasons unhinge the effective type and quality assurance process. A simple type error causes a loss of trust amongst guests and greatly devalues the product, inferring significant losses for these leather-producing factories. The scars, scrapes, holes, bubbles, etc., thus need to be linked, located, and ultimately removed. Thus, defect discovery is also essential.

Quality control is critical, and defect detection can be approached in various ways by learning and the evolution of defect detection methods (Tariq, 2025). The types of learning used to detect the defects in leather are supervised learning, semi-supervised learning, and unsupervised learning. (Dinesh et al., 2018). The supervised learning methods involved training a model on a labeled dataset, where the input leather pieces are paired with the correct targeted results, either defect or no defect leather. This learning method requires a substantial amount of labeled data. The model learns to associate specific features in the images with the presence or absence of defects. Some example algorithms are Support Vector Machines (SVM) and CNNs. The semi-supervised learning method combines a small amount of labeled data with a larger amount of unlabeled data during training. The labeled data helps the model learn the initial features, and the unlabeled data helps in fine-tuning and improving the model's generalization. Some example algorithms are self-training and co-training. The unsupervised learning methods involved training a model on an unlabeled dataset. (Bergmann et al., 2021), where the model learns to identify patterns and anomalies without explicit instructions. The model clusters the data based on similarities and can identify outliers or anomalies that might indicate defects. Some example algorithms are K-means clustering, Principal Component Analysis (PCA), and Autoencoders. (Wei et al., 2021).

The defect detection methods have evolved from the pre-deep learning era to the deep learning era. Before the advent of deep learning, unsupervised algorithms were widely used for defect detection. These image-processing filtering methods relied on traditional image-processing techniques. The edge detection algorithms like Canny, Sobel, and Laplacian filters were used to detect edges and contours in leather images, helping to identify cracks or other linear defects. The simple thresholding techniques helped to segment leather images into defective and non-defective areas based on pixel intensity. (Gan et al., 2021). Morphological operations techniques like dilation, erosion, opening, and closing were used to enhance and detect specific shapes and structures in images (Tariq, 2025).

In the deep learning era, the approach to defect detection has significantly evolved. The convolutional neural networks have revolutionized image analysis by automatically learning relevant features from raw pixel data, making them highly effective for leather defect detection. In transfer learning, leveraging pre-trained models on large datasets and fine-tuning them for specific tasks has accelerated the development of robust defect detection systems. Deep learning models enable real-time defect detection on production lines, significantly improving efficiency and reducing manual inspection efforts. The proposed leather defect discovery system using advanced deep knowledge algorithms holds great potential to transform the leather sedulity. It can significantly improve effectiveness, reduce costs, and enhance product quality through automated quality control. This system can snappily and directly identify scars, streamlining product processes and saving time and resources. It also minimizes waste and rework, leading to cost reductions. Likewise, the harmonious adherence to quality morals and real-time defect discovery ensures advanced-quality leather products, fostering customer satisfaction and brand dedication. These benefits are set to drive the wide handover of advanced quality control technologies across the leather industry, enhancing the competitiveness of leather manufacturers. The type and defect discovery system that is posted should be suitable to directly classify the images and descry scars, if present in the images. With manual labor getting increasingly precious and error-prone, the use of independent type systems has been the most preferred, but least posted at present, due to low rigor and solidness. Thus, the system needs to be as independent and reliable as possible. Generally used ways for similar systems are image processing predicated ways. The introductory ways include ocean analysis, canny edge discovery, and thresholding. Still, these ways are inaccurate due to their impertinence in the identification of different scars than on the one it was developed and tuned for. But with the proliferation of deep learning and, most importantly, CNNs, image recognition tasks have been simplified. They also have excellent performance and high accuracy rates, which give good responsibility.

The proposed system (Bin Abdullah et al., 2024) Uses a double channel system, where one channel has an inception V3 convolutional neural network, which classifies the leather image as "good quality" or "defect quality", while the other channel has Mask R- CNN, which detects and locates the scars in the leather image and draws bounding boxes for candidate defect regions, when the affair of the first channel is "defect quality". Experimental results show that the system achieves a high performance with the type of CNN, achieving a 99% accuracy rate on the training dataset and on the evidence dataset. The defect discovery CNN achieves 99.6% defect discovery accuracy on the training dataset, 99% on the evidence dataset, and 99% accuracy on the test dataset. Thus, the system is proved to be suitable to directly classify leather images and identify and descry scars in images, where only those images which are classified as "defective", are transferred to the alternate channel defect discovery convolutional neural network for discovery of scars, thereby making the process important hastily.

A machine vision system, including CNN and ML classifiers, was used to localize and categorize leather defects by practicing Google Net, SqueezeNet, RestNet, and Alexnet. Alexnet outperforms shallow feed-forward neural networks, achieving 99.4% classification accuracy. (Moganam & Sathia Seelan, 2022). A comprehensive data set has a collection of 3600 leather images with various defects, such as folding marks, grain off, pinhole, growth markings, loose grain, and non-defective leather surfaces. LASCC (Leather and Stitching Color Classification) model (Pazzaglia et al., 2021) Producing 97.6% efficiency and ensuring high-quality leather products involves a crucial step as visually inspecting the leather to identify defects in Industry 4.0. This approach (Z. Chen et al., 2023) utilizes an ultra-high definition (UHD) leather imaging system and optimizes key network parameters, achieving a high classification accuracy of 94.6%. The sliding patch window technique is cropping the collected UHD leather images with a specific pixel size according to certain moving rules. The work demonstrates that the Residual Neural Network (ResNet) 50-2-based framework achieves effective, accurate, and efficient classification of leather surface defects, meeting the requirements of automatic production in the leather industry.

AI-based automated leather flaw identification technique uses pre-trained deep neural network models to classify potential fabric problems. The developed Deep Convolutional Neural Network (DCNN) classifier with VGG16 models, which employs transfer learning practicing the AlexNet pre-trained network, generated an extreme accuracy of 92.60% (Sandhya et al., 2021; Uddin et al., 2023). A custom VGG16 model (Uddin et al., 2023) Was employed for leather defect detection, demonstrating high accuracy with a 98% correct defect detection rate on the MVTEC anomaly detection dataset. They focused on MLD (Materials Learning from Data), CURE-TSR (Textures and Segmentation in Real-world images - Tanning Skins and Rough leathers for developing ensemble networks to identify industrial leather defects accurately. The goal of this work (Khanal et al., 2022a) Is to replace human leather inspection with fully automated fault detection based on cutting-edge machine vision techniques. The work uses Industry Technology 4.0, semantic segmentation, and deep learning techniques on the MVTEC leather dataset. The study reveals that Mobilenet outperforms Resnet50 in the intersection of union (IOU) by 95% in five classes' segmentation. This research (Gan et al., 2022) Introduces an efficient automated defect classification system that determines whether the sample patches contain defective segments. Classification accuracy obtained is 94% by using ANN classifiers. This experiment is restricted to the defect type of tick-bite defect comprised of 1600 pictures of defective leather and non-defective leather images. The research (S.-Y. Chen et al., 2021) Presents an automatic fault detection tool using deep learning techniques, including R-CNN, AlexNet, and U-Net, on a newly collected leather defect database of 522 reflective leather images. The strategy achieves a classification performance of 95% and segmentation tasks with an average IoU of 99.00% and pixel accuracy of 99.82%.

This work (S.-Y. Chen et al., 2021) Proposed the Hyperspectral Leather Defect Detection Algorithm (HLDDA), which incorporates deep learning and Hyperspectral Target Detection (HTD) approaches. This work analyzed five defects in wet-blue leather, including brand mask, rotten grain, rupture, insect bite, and scratches. They practiced 3d-UNet, 1d-CNN, R-CNN, and constrained energy minimization

(CEM) on the hyperspectral image (HSI). The contemporary vision transformer (ViT) architecture classifies leather surface defects through low-resolution image-based anomaly detection for defect localization, practiced in small-scale datasets and low-resolution leather surface photos. This model gives the ability to represent complex relationships and patterns in images precisely, which makes it a good option for identifying anomalies in the MVTec anomaly detection benchmarking dataset. (Smith et al., 2023).

These authors (Wang & Kyoung, 2023) Suggested boosting the leather flaw detection in small sample circumstances by a two-stage defect enhancement network. They practiced the Gabor filter, gray level co-occurrence matrix energy, and TDENet. The findings of the experiments demonstrate the capability of efficiently identifying various flaws in varying hues and textures in both the objective quantitative evaluation index and the visual detection effect. Using HSI, this study examined a variety of wet-blue leather flaws, such as brand masks, rotten grain, rupture, insect bites, and scratches.

This approach (Prunella et al., 2023) Ensures a thorough performance assessment with a variation of detail levels by an auto-encoder, automatic recognition, feature attention mechanism, CNN, and GAN. It highlights the need for big, high-quality labeled datasets and a variety of neural architectures to identify well-visible and subtle flaws. The employed dataset is newly collected and composed of 67 images with two different colors of leather and seven different colors of stitching. It also investigates the performance of edge detectors and threshold detectors while assessing the identification accuracy of the traditional machine learning method SVM in detecting leather surface defects. (Z. Chen et al., 2022). This work (Winiarti et al., 2018) proposes a leather classification method using a pre-trained deep CNN to extract rich features from tanning leather images. 1000 defective and non-defective leather images were used to train the model. It demonstrates good accuracy and precision in classifying tanning leather images into five categories using a pre-trained CNN for feature extraction and SVM for classification. A Learning Vector Quantization (LVQ) network compared the chrome-tanning and vegetable-tanning leather. The deep learning architectures YOLO (Pople & others, 2023) and CNN tailored (Aslam et al., 2019) the image classification, detection, and segmentation can be classified into natural, chemical, flaying, curing, and tanning defects. Many researchers present a detailed and methodical review of leather surface defects, their effects on leather quality grading, and automated visual inspection methods for leather defect inspection to improve productivity in industries.

Image capturing is one of the crucial phases of the entire inspection system in the field of quality inspection. Textile inspection systems are performing better because of the development and commercialization of advanced AI-based cameras. The two types of sensors used in machine vision are Complementary Metal

Oxide Semiconductor (CMOS) imagers and Charge-Coupled Devices (CCD). In addition to these, thermal and infrared cameras are employed for a variety of industrial machine vision applications. Color and grayscale cameras are also crucial considerations when choosing a camera, depending on the kinds of photographs being taken. Two- and three-dimensional cameras are two more picture kinds that are utilized for flaw identification. One popular visual AI defect detection method is computer vision-based leather fault detection. However, it is still difficult to discover a fully interoperable system in the state-of-the-art that combines computer vision technology with a hardware platform. (Khanal et al., 2022b). This is a component of a study that looks for various leather flaws, some of which are obvious when the leather isn't altered and others that become more noticeable when the material is altered. Certain artifacts are produced in the region where the flaws are located when this manipulation is coupled with a certain kind of light. A comprehensive, interoperable system for identifying different kinds of leather surface flaws was given in this article. The hardware for the interoperable system includes an image acquisition system, mechanical components for handling leather throughout the flaw detection process, and a software prototype that includes data visualization and image processing methods. The segmentation of the faults that are obvious without modification is the sole connection between the AI outcomes presented in this work. (Liong et al., 2020).

Defect detection is demonstrated at any stage of the leather product production process, including defects in natural leather. (Gan et al., 2023), defects in blue damp leather (Aslam et al., 2020) And defects in the completed products. Leather is significant in tenures of frugality and operation. Assuring the leather quality and effectively detecting its face blights are overcritical for the enterprises. In tanneries, face disfigurement discovery is traditionally carried out by mortal inspectors, which is labour-ferocious and frequently apt to mortal inaccuracy. For a high-value product like leather, reducing the periphery of inaccuracy is economically profitable. Still, mortal inspectors are frequently unreliable and inconsistent due to varying situations of experience. Thus, the want for spontaneous discovery of blights on the leather face is critical.

Still, an arising challenge within this sphere is multi-label disfigurement discovery, which involves relating multitudinous manners of blights in a single image or sample. Table 1 illustrates the manners and figures of blights detected in inquiries conducted on this motive. One of the primary terminations of this path is the class imbalance case, where the number of cases for each disfigurement type can be significantly nonidentical, leading to prejudiced models that favour the maturity class. Likewise, the swelled complication of multi-label models results in advanced computational charges, making them more delicate to train and optimize. The interpretability of these models can be limited, making it difficult to understand why a personal disfigurement was detected. In the other phase, the above inquiries were conducted in laboratory surroundings with restricted datasets, and delicacy classes can be anticipated to drop significantly in real working surroundings similar to tanneries. For illustration, the blights were detected in veritably fragile areas similar to 40 × 40 pixels, which influences the delicacy of the styles applied to voluminous measure blights. In extension, since the datasets exercised in the inquiries are the same for utmost inquiries, nonidentical effects may be attained in nonidentical leather slices. Our path focuses on developing an engine literacy- grounded disfigurement discovery system that can directly identify colourful blights in leather productions.

The main goal of this study (Ataç et al., 2024) is to exercise AI ways for the discovery of leather blights to shake mortal crimes. In the stage after the process of turning raw leathers into semi-finished leathers, it's levelled to descry blights on the leather by using AI-grounded image dissection ways on photos taken with a high-conclusion camera. These ways carry YOLOv8 and Mask R-CNN within the Detectron2 frame-RetinaNet and Faster R-CNN. A relative dissection of nonidentical DL styles was conducted to determine the most effective path for leather disfigurement discovery. The integration of the most prosperous DL system with the slicing-backed hyperactive-conclusion frame has been achieved, and with the modified frame, the discovery of fragile disfigurement manners has been bettered, thereby adding the delicacy classes.

In discrepancy to former inquiries that employed fragile photos(e.g., 16 × 16, 32 × 32) in ruled laboratory surroundings, this exploration assumed images collected in real-world tannery settings. Specially, the optimal SAHI frame size was determined to achieve the stylish delicacy classes, thereby icing the discovery of both fragile and voluminous blights. The jolt of image size on delicacy classes and processing time was completely delved and reported, offering precious perceptivity for unborn exploration in this sphere.

This study also explored the sequel of disfigurement type categorization on the interpretation of DL styles, fastening on six common or garden disfigurement manners, for which a dataset was created. Also, a secondary dataset was generated by combining all disfigurement manners into a single" disfigurement" order. The dataset exercised to train AI was collected from tanneries in the Gerede quarter of the Bolu fiefdom in Turkey. An aggregate of 125 photos of 9280 × 6944 pixels was exercised. Unless stated else, since DL algorithms in the styles exercised in the study condense the images during training, the pixel valuations of the images were downgraded 10 moments with a software developed for this study in Python programming language. The new image confines have been downgraded to 928 × 694 pixels. Still, during the operation, comparisons were made with high-conclusion images. The effects demonstrated that training DL models on collectively labelled disfigurement manners led to better interpretation than adding them into a single order, emphasizing the significance of refined disfigurement bracket for leather quality control.

Nonidentical DL algorithms can be applied to the dataset by utilizing Detectron2. (*GitHub - Facebookresearch/Detectron2: Detectron2 Is a Platform for Object Detection, Segmentation and Other Visual Recognition Tasks.*, n.d.) Frame. Detectron2 is a computer unreality archive developed by Facebook AI and employed in tasks similar to object discovery, segmentation, and pose estimation. Detectron2's Model Zoo is a depository of pre-trained models that druggies can work for colourful tasks. This depository contains multitudinous ready-made models trained on nonidentical datasets, similar to Faster R- CNN, RetinaNet, RPN & whirlwind R- CNN, and Mask R- CNN. By utilizing Detectron2's Model Zoo, druggies can work with important and accurate models without demanding to train them from scrape, which is especially salutary when training models on voluminous datasets that requires significant time and coffers. RetinaNet R101 3x and Faster R- CNN R101- FPN 3x were exercised. To apply the YOLOv8 (Vijayan et al., 2023) Algorithm to the datasets, the data were first uploaded to the Google Drive account. This step is needed for the data to be exercised on Google Colab. There are 2 manners of data in the Drive account. In the first type, there's data labelled for 6 manners of blights, and in the alternate type, there's data in which all disfigurement manners are labelled only as "disfigurement". The YOLOv8 model was exercised because it gave a mAP 50-95 rate of 52.9 in 43.7 million parameters. Image addition styles were exercised in all routines as they can diversify the training data and support the model to generalize better. These styles are blur extension, median blur extension, grayscale transformation, and discrepancy-limited adaptive histogram equalization. Adam was exercised as the optimizer with a literacy rate of 0.001. Due to its structure, YOLOv8 has resized the images to 640 × 640 pixels, with hyperparameters set to optimize object discovery interpretation.

RESEARCH QUESTIONS

The research questions are enumerated here by acquiring knowledge of the existing systems.

1. How does the choice of pre-processing techniques, such as image normalization or augmentation, impact the performance of the EfficientNet model in detecting leather defects?

Preprocessing techniques like image normalization can help the EfficientNet model by ensuring consistent input data ranges, which can improve convergence during training. Augmentation can also be beneficial, as it increases the diversity of the training data, potentially leading to better generalization and performance.

2. How do various hyper-parameters, including learning rate, batch size, and optimizer selection, affect how well the EfficientNet model identifies flaws in leather?

The learning rate can significantly impact the training process; a high learning rate may cause the model to overshoot optimal solutions, while a low learning rate may result in slow convergence. Batch size affects the noise in the gradient estimation, with larger batches providing a more stable estimate but potentially leading to slower convergence. The choice of optimizer (e.g., Adam, SGD) can also impact convergence speed and final performance.

3. How do the training dataset size and diversity influence the performance of the EfficientNet model in detecting leather defects?

A larger and more diverse training dataset can lead to better performance, as the model has more examples to learn from and can generalize better to unseen data. However, too much diversity can also introduce noise, so a balance is needed.

4. Can transfer learning techniques, such as fine-tuning or feature extraction, improve the performance of the EfficientNet model in detecting leather defects?

Transfer learning techniques can be beneficial, especially when the pre-trained model (like EfficientNet) has been trained on a large, general dataset. Fine-tuning the model on a specific dataset can help it learn relevant features for leather defect detection faster and with less data.

5. How does the efficiency of the EfficientNet model compare to other deep learning models, such as VGG or ResNet, in terms of accuracy and computational resources for leather defect detection?

EfficientNet is known for its superior performance compared with other models like VGG. (Vijayan, Mareeswari, & Shobana, 2024) and ResNet, especially in accuracy versus computational cost. It typically achieves higher accuracy with fewer parameters, making it a more efficient choice for resource-constrained environments.

AI models are more accurate than human inspectors in identifying flaws, which lowers the error margin. The production process can be accelerated by automated systems' rapid processing of huge quantities of leather. Manufacturers will profit by cutting waste and enhancing quality control. Regardless of human variability or weariness, AI systems consistently and reliably discover defects. Hence, AI and IoT(Venkatachala Appa Swamy et al., 2023) Support more environmentally responsive production methods by maximizing resource use and cutting waste (Banduka et al., 2024).

MATERIALS AND METHODS

An exciting application that combines cutting-edge technology to enhance quality control in the leather sector is the identification of leather defects using AI and IoT. The high-resolution cameras and sensors, among other IoT devices, take pictures and record information from leather surfaces. The cloud or edge computing systems receive the gathered data and process it. The proposed approach examines the photos to find flaws like bug bites, scrapes, holes, and stitch lines. Following the analysis, the system detects and categorizes flaws, giving operators immediate feedback or initiating automated procedures to eliminate faulty parts. Over time, the system's accuracy and efficiency are improved by ongoing learning and development based on fresh information and results.

EFFICIENTNET

The proposed approach practices the EfficientNet model to detect defects in leather. It is a series of CNN designs developed by Google Research in 2019. It's aimed at achieving strong performance while minimizing computational demands. Unlike earlier models that adjusted just one aspect of network size, EfficientNet scales depth, width, and resolution simultaneously. This technique, known as compound scaling, optimizes the network's dimensions using fixed coefficients. Combining these adjustments, EfficientNet achieves better accuracy with fewer parameters and computational resources than other CNN architectures. Key innovations include the "MBConv" block and (NAS) methods to identify the best scaling factors.

Figure 1 depicts the EfficientNet architecture. The key features of EfficientNet are: 1. Compound Scaling: EfficientNet uses scaling coefficients (phi, ϕ) to adjust network depth uniformly, width, and resolution, achieving a balance between model complexity and performance. 2. Depth-wise Separable Convolution: This technique involves a depth-wise convolution followed by a point-wise convolution, reducing parameters and computational cost without compromising performance. 3. Inverted Residual Blocks: Inspired by MobileNetV2, EfficientNet uses these blocks with linear bottlenecks and shortcut connections, enhancing parameter efficiency and gradient flow. 4. Efficient Attention: Advanced variants like EfficientNetV2 incorporate self-attention mechanisms to enhance model performance. 5. Efficient Model Scaling: EfficientNet introduces a novel method of scaling neural networks by balancing depth, width, and resolution, achieving superior performance with fewer parameters. 6. Model Architecture: EfficientNet typically adopts a MobileNetV2-based backbone network with modifications for efficiency, followed by classification layers. 7. State-of-the-Art Performance: EfficientNet models have demonstrated top-tier performance in image classification tasks like ImageNet while utilizing fewer computational resources than other models. 8. Variants: The EfficientNet family includes variants (e.g., B0 to B7) varying in depth, width, and resolution, each optimized for specific trade-offs between accuracy and efficiency.

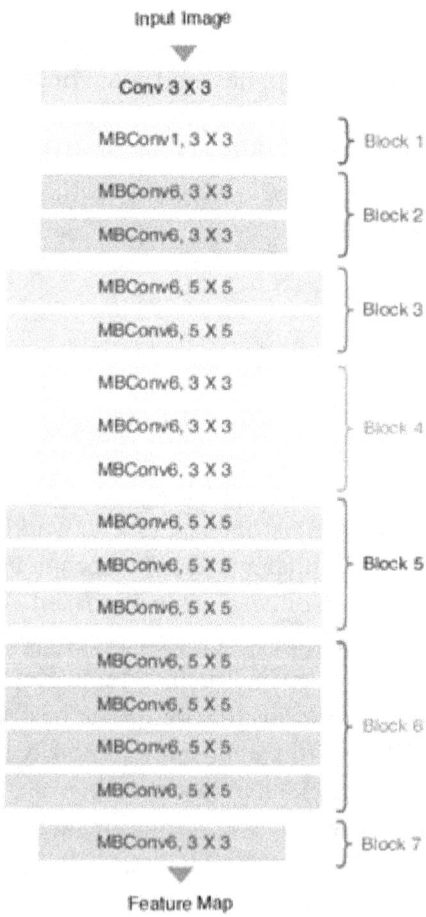

Experimental Settings

The layer description of the EfficientNet model is given in Table 1. The base model used is EfficientNetB0, a lightweight CNN recognized for its efficiency and effectiveness. It comes pre-trained on the ImageNet dataset, providing a foundation for learning intricate hierarchical features from images. Initializing the model with 'weights ='imagenet" enables it to benefit from the extensive knowledge acquired during training on ImageNet. Following the base model, the GlobalAveragePooling2D layer is applied to condense the spatial dimensions of the feature maps. This process calculates the average value for each feature map, resulting in a single value per channel. It helps in reducing model parameters and emphasizes essential features, enhancing the model's resilience to variations in input images.

Table 1. EfficientNet layer description

Layer No.	Layer Type	Output Size	Parameters
1	Input Layer	224x224x3	0
2	Conv2D	112x112x32	864
3	BatchNormalization	112x112x32	128
4	Swish Activation	112x112x32	0
5	MBConv1	112x112x32	320
N	Conv2D	7x7x1280	409600
N+1	BatchNormalization	7x7x1280	5120
N+2	Swish Activation	7x7x1280	0
N+3	GlobalAveragePooling2D	1280	0
N+4	Dense	1000	1281000
N+5	Softmax Activation	1000	0

The output from the GlobalAveragePooling2D layer is then fed into a dense layer comprising 1024 units with a Rectified Linear Unit (ReLU) activation function. This layer aids in learning intricate data patterns by applying a non-linear transformation to the input features. ReLU introduces non-linearity, empowering the model to capture more complex relationships within the data. Lastly, there's a dense layer with 6 units and a softmax activation function. This layer produces the final model output—a probability distribution across 6 classes of defects. Softmax ensures that output values range between 0 and 1, collectively summing up to 1, which is ideal for multi-class classification tasks.

The Rectified Linear Unit (ReLU) is a straightforward but powerful non-linear activation function extensively applied in neural networks. It calculates the output by taking the maximum between 0 and the input value. In formula terms, ReLU(x) = max(0, x). ReLU is favored due to its computational simplicity and ability to mitigate the vanishing gradient issue, enabling faster and more successful training of deep neural networks. It is useful in tasks like image processing and has become a common choice for activation functions in various neural network designs. The softmax activation function is generally used in the affair subcaste of a neural network for multi-class bracket tasks. It transforms the affair values into a probability distribution, where each value represents the probability of the input belonging to a particular class. The softmax function is defined as softmax(x_i) = exp(x_i)/ sum(exp(x_j)) for all j, where x_i is the input to the softmax function and the sum is over all classes. The affair of the softmax function is a vector where the values add up to 1, making it suitable for interpreting the laborers as chances. It is easier to interpret the model's prognostications, and cross-entropy loss facilitates training.

PROPOSED METHODOLOGY

The primary goal of this proposed system (see Figure 2) is to create a robust automated system using deep learning techniques to classify defects in leather products. Specifically, the work will investigate the effectiveness of the EfficientNet model, which is highly regarded for its performance in image classification tasks. This system aims to accurately identify and categorize common defects found in leather, such as scratches and cuts. The leather manufacturing process is intricate and can result in various defects that impact product quality and appearance. Manual defect inspection is time-consuming and prone to errors, as human inspectors may overlook subtle or difficult-to-detect defects. An automated defect classification system can significantly enhance efficiency and accuracy in defect detection within the leather industry, leading to improved product quality and reduced waste.

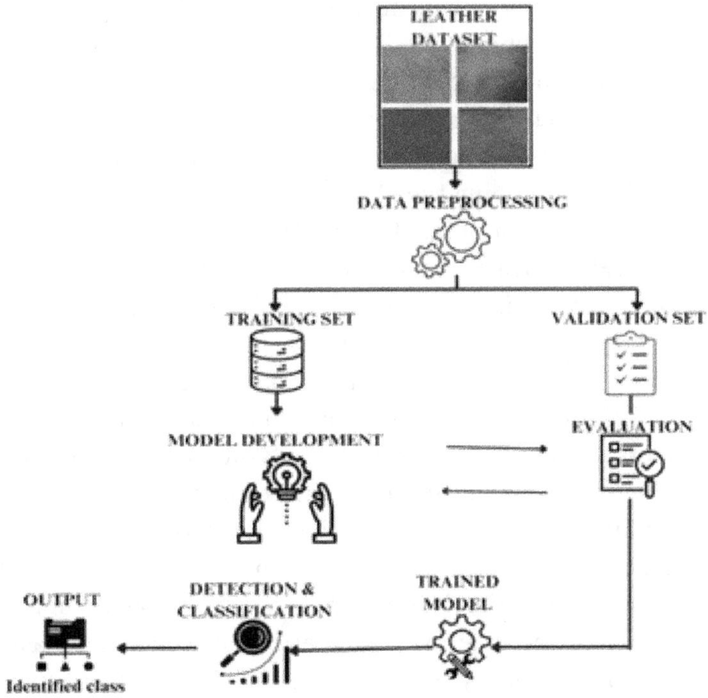

We collected the data from the Kaggle Low-resolution open-source Leather disfigurement Dataset repository. The dataset contains 3600 images with six classes, including folding marks, Grain out, Growth marks, loose imperfect, and pinholes. In a dataset, a training set is executed to make up a model, while a test (or evidence) set is to validate the model erected. Also, you have the complete training dataset. We can prize features and train to fit a model and so on. Once the model is attained, we can predict using the model attained on the training set. Some data may be used in a corroborative way, generally to corroborate that a given set of inputs to a given function produces some awaited result. Other data challenge the program's capability to respond to unusual, extreme, exceptional, or unexpected input.

Image preprocessing involves several essential steps to ensure the dataset is properly formatted and enhanced before training a model, ultimately improving its performance. The `ImageDataGenerator` class from Keras allows for real-time augmentation and normalization of the dataset. It contains the following operation: first, the `rescale` parameter within `ImageDataGenerator` is utilized to normalize pixel values of images, typically scaling them to a range between 0 and 1. This normalization is crucial for standardizing input data. The data augmentation step involves creating modified versions of existing images within the dataset, effectively expanding the dataset size. These modifications can include rotations, flips, shifts, and other transformations, enhancing the model's ability to generalize and perform well under varying conditions. The `target size` parameter specifies the dimensions to which all images are resized during loading. This uniform sizing ensures consistency across the dataset, which is necessary for effective model training. The `batch size` parameter defines the number of images in each training batch. Optimizing batch size can impact training speed and memory usage. The `class mode` parameter determines the format of label arrays returned by the generator. For instance, using `'categorical'` indicates that the generator will return one-hot encoded labels, suitable for models requiring categorical classification. Python is widely utilized and highly favored by data scientists worldwide, primarily due to its extensive libraries that facilitate specific data preprocessing tasks. Here are the key libraries used in our code: TensorFlow is an open-source machine learning framework developed by Google. It offers a comprehensive suite of tools, libraries, and community resources for constructing and deploying machine learning models. Keras is a high-level neural networks API that operates atop TensorFlow. It offers a user-friendly interface for building and training deep learning models, making it ideal for quick prototyping and experimentation. EfficientNet is a series of convolutional neural network models recognized for achieving state-of-the-art performance in image classification tasks. It stands out for its efficiency in terms of model size and computational demands. The ImageDataGenerator utility within Keras generates batches of augmented image data. It supports on-the-fly data augmentation during training and can enhance a model's ability to generalize to new data. Matplotlib is a Python plotting library employed in our code to visualize training and validation metrics (loss, accuracy, precision, and recall) across epochs. This visualization offers valuable insights into the model's performance.

The data is regularized for training and testing the model once drawn. We train the algorithm on the training data set when the data is also picked. The training procedure yields a training model based on algorithms, sense, and point values in the training set. The main goal of point birth is to align all values on a single scale. A training set develops a model's ideal parameters or the parameters it must learn from data. A machine literacy model's confirmation set is essential to its training process because it provides a means of estimating the model's performance on unobserved data and directs the training process toward improved conceptualization.

The model structure builds upon the EfficientNetB0 pre-trained model, known for its adeptness in extracting detailed features from images. It begins with this base model and incorporates a Global Average Pooling layer to condense spatial features. Next, a Dense layer with 1024 units and ReLU activation is added to facilitate learning intricate patterns. Finally, a dense layer with 6 units and soft-max activation is applied for the ultimate classification across six classes. The model is compiled using the Adam optimizer with a learning rate of 0.001 and categorical cross-entropy loss. It tracks accuracy, F1 score, precision, recall, and AUC at training. This architecture is encapsulated within a Model class, configured with the base model's input and the final classification layer output, making it ready for training on a specific dataset.

During the training phase, the code employs an ImageDataGenerator to generate augmented image batches from the training dataset. These batches are then used to train the model using the fit method while monitoring the progress with callbacks such as Model Checkpoint to save the best model weights based on validation accuracy and early stopping to stop training if validation accuracy plateaus. The model undergoes training for a set number of epochs, during which it processes the entire training dataset each time. Various metrics, including loss, accuracy, F1 score, precision, recall, and AUC, are calculated and recorded to assess the model's performance. Ultimately, this training phase aims to optimize the model's parameters to classify the input images accurately.

The code loads an image of a leather sample, then resizes it and adjusts its pixel values to prepare it for analysis. After that, the image is fed into the trained EfficientNetB0 model, which produces a probability distribution across different classes of leather defects. The predicted defect class is chosen as the one with the highest probability, and this is translated into a label that describes the type of defect, like folding marks or growth marks. Finally, the input image is displayed alongside the predicted defect label to visualize the classification outcome.

LAB ENVIRONMENT SETUP

We can build a lab with a few mechanical parts like a conveyor or moving camera system, a camera, a lighting system, a computer, and a display system.

A conveyor system is a quick and effective mechanical handling device employed in the suggested laboratory environment to move leather pieces while they are being inspected. The conveyor should be adjusted to travel smoothly and steadily, avoiding any jerks or disturbances that could interfere with taking pictures.

The camera system is installed above the conveyor's surface to capture the images of moving leather pieces in the conveyor. We should ensure the camera system is positioned to capture the entire width of the leather piece. We can adjust the focus and resolution to capture the detailed leather images. Hence, it identifies leather, captures its details, and transmits it to the computing devices.

The proposed system recommends a consistent and adequate lighting system to ensure the quality of pictures. It suggested the use of diffused lighting to avoid harsh shadows and reflections. Also, it should be properly placed to cover all the angles of the conveyor belt.

The computer system functionalities are to process the captured images, run the proposed deep-learning models, and display the results of the leather class. Subsequently, a sequence of image processing operations, including picture pre-processing, deep learning model training, and testing, will be performed to identify defects. Hence, we should ensure the computer has sufficient processing power, GPU capabilities to handle deep-learning tasks, and the required software installation and environment setup. For continuous model training and development, it is essential to effectively handle and store massive amounts of picture data. Finally, the display system is connected to the computer system to visualize the result of the leather defect process. It shows the real-time results of the moving leather pieces and feedback to guide further steps.

The proposed system process involves image pre-processing, which involves resizing, normalizing, and enhancing images for analysis. The deep-learning model is trained using labeled datasets of leather images, using techniques like CNN. The model is then tested on a separate set of images to evaluate its accuracy and performance. The trained model is then applied to real-time images from the conveyor system, analyzing the images and flagging any detected defects. The results are displayed on a screen, highlighting areas for further inspection.

To ensure the system's accuracy, the camera, illumination, and conveyor system must be regularly calibrated. Longevity and smooth functioning are guaranteed by routine maintenance of electronic systems and mechanical components. By combining these elements and procedures, the lab can quickly and accurately detect flaws in leather goods, guaranteeing premium goods and cutting down on human inspection work.

RESULTS AND DISCUSSION

The performance metrics are essential for assessing how well a model classifies images into their correct categories. Higher values in these metrics generally indicate better performance. Accuracy gives a broad view of overall correctness. The metrics like F1 score, precision, recall, and AUC provide more detailed insights, particularly in cases where the classes are imbalanced. Figures 3 – 6 visualize the proposed model's accuracy, loss, precision, and recall.

$$Precision = \frac{TP}{TP + FP}$$

A true positive (TP) denotes that the model correctly anticipated the number of sample positive classes as it is. A false negative (FN) represents that the model predicted the number of negative classes in the data and that the positive class is anticipated as a negative class. False positive (FP) represents that the model anticipated the number of positive class data and that the negative class is expected to be positive. A true negative (TN) denotes that the model anticipated the number of sample negative classes correctly and that the negative class is expected to be negative.

Recall is the rate of rightly prognosticated positive compliances to all compliances in the factual class.

$$\text{Recall} = \frac{TP}{TP + FN}$$

The recall metric is determined by adding up the true positives (defective samples correctly identified) and then dividing that sum by the total of true positives and false negatives (defective samples missed by the model). This calculation ensures that the recall value ranges from 0 to 1, with 1 indicating perfect recall (all defects are correctly identified) and 0 indicating no defects are detected. Including recall as an evaluation metric in the code helps assess the model's ability to spot defects, which is crucial for ensuring the quality and reliability of an automated leather defect detection system. High recall values suggest that the model effectively identifies most defects and supports maintaining high-quality standards in leather manufacturing processes.

Accuracy is the proportion of rightly classified images to the total number of images. While delicacy is a generally used metric, it can be deceiving when dealing with imbalanced datasets. For illustration, if one class dominates the dataset, a model that always predicts that class could still achieve high delicacy. Thus, it is important to consider other criteria, especially when dealing with imbalanced data.

$$Accuracy = \frac{TP + TN}{TP + FP + TN + FN}$$

The loss function is a critical element in machine learning model training, It evaluates how well the model performs by comparing its predictions against the actual labels in the training data. The code uses categorical cross-entropy loss, typical for multi-class classification tasks such as leather defect detection. During training, the loss is computed for each batch of data, and the average loss across all batches guides the updating of the model's weights. As training advances, the model strives to minimize this loss and improves performance in identifying and classifying leather defects.

The harmonious means of perfection and recall are furnishing a balance between the two criteria. The F1 score is a good metric to strike a balance between perfection and recall. It's the harmonious means of perfection and recall and provides a single value that summarizes the model's performance. A high F1 score indicates that the model has high perfection and high recall.

$$F1Score = 2* \frac{(Precision * Recall)}{(Precision + Recall)}$$

The rate of rightly prognosticated positive compliances to the total prognosticated positive compliances.

The proposed work involves training and testing different CNN models to select a model with higher accuracy and less loss percentage. The proposed analysis involves training many models and settings to add and reduce layers to improve accuracy. The result of the comparative analysis is given in Table 2.

Table 2. Comparative analysis of different models

Model	Accuracy
CNN model with 2 dropout layers	67%
CNN model with 6 layers and a dropout layer	69%
Sequential CNN with 8 layers	75%
Resnet Model	79%
EfficientNet-B0	90%

The EfficientNet-B0 model, with its 90% accuracy, stands out as a highly efficient and effective model for image classification tasks. Its strength is in its compound scaling technique, which uses a set of predefined scaling coefficients to scale all dimensions of depth, width, and resolution equally. This approach allows EfficientNet-B0 to achieve higher accuracy with fewer parameters than other models, making it computationally efficient. Additionally, EfficientNet-B0 has superior performance in various benchmarks, indicating its robustness and generalization capability across different datasets. Despite its high efficiency, EfficientNet-B0 may require more computational resources than simpler models. It might also be more challenging to interpret and analyze due to its complex architecture.

ResNet is known for its ability to train deep neural networks effectively by using residual blocks. It helps alleviate the vanishing gradient problem, enabling the training of deeper networks and producing 79% accuracy. ResNet models can be computationally expensive, especially as the depth of the network increases. They may also be prone to over-fitting if not regularized properly.

A sequential CNN with 8 layers can capture complex features in the data due to its deep architecture and produces 75% accuracy in results. It can learn hierarchical representations of the input images, which is beneficial for image classification tasks.

Deeper networks are more prone to over-fitting, especially if the dataset is not large enough. Training such models can also be computationally expensive.

Including a dropout layer in the CNN model with 6 layers helps reduce over-fitting by randomly dropping units during training. This regularization technique can improve the model's generalization performance with 69% accuracy. While dropouts can be effective, too many dropouts can lead to under-fitting. Finding the right dropout rate is crucial, and it may require experimentation.

The CNN model with two dropout layers and the model with multiple dropout layers enhance the model regularization, potentially reducing over-fitting and improving generalization. Similar to the previous model, finding the right dropout rates for each layer can be challenging. Using too many dropout layers or high dropout rates can hinder the model's learning ability.

The proposed model uses EfficientNet-B0, which stands out as the top performer in terms of accuracy. However, ResNet and other models also have strengths and weaknesses that should be considered based on the problem-specific context.

The proposed system suggests the proposed lab setup for the practical leather defect application. The lab setup build-up with the conveyor system, camera, and lighting would facilitate the acquisition of high-quality images of leather pieces. These images can be stored in the cloud or on the edge. Sequentially, the images will be processed using deep learning models to identify defects. Depending on the availability of labeled leather image data, we can choose between supervised, semi-supervised, or unsupervised learning methods for model training and deployment. The proposed system employs the AI algorithm as Efficient B0 and proves the accuracy. The proposed defect detection software can work together with the hardware components mentioned in Figure 7 and identify the defect of moving leather pieces over the conveyor system.

Researchers could make leather defect discovery systems using deep learning in the future. They could add advanced technologies like reinforcement learning (RL) and transfer learning (TL). RL helps the system make good opinions and adapt to new situations. TL uses knowledge from combined areas to enhance defect discovery accuracy. The systems could also be used in other industries besides leather manufacturing, like fabric or bus manufacturing. This would allow for automating quality control processes to boost effectiveness across different sectors. Exploring these avenues can enhance automated defect discovery systems' capabilities and versatility. This could lead to further industrialization, espousing the technology and driving invention.

CONCLUSION

The work aimed to create an automated system using the EfficientNetB0 deep learning model to detect defects in leather. The model was trained on a dataset containing images of various leather defects like folding marks, growth marks, pinholes, grain off, loose grains, and non-defective leather. The EfficientNetB0 model achieved an impressive 90% accuracy in identifying these defects, demonstrating its effectiveness in image classification tasks. Notably, the model showed efficient computational resources and training time, which are crucial for practical implementation. Its strong performance on new data suggests its suitability for industries where quality control is essential. This proposed system motivates us to practice this defect detection software with mechanical hardware components. To enhance the system, future studies could explore fine-tuning the model, using transfer learning, and refining data augmentation techniques to boost performance and reliability in real-world scenarios. Integrating this AI and IoT system into existing quality control processes could streamline leather manufacturing, minimize waste, and elevate overall product quality.

Future Work

Detecting leather defects using deep learning presents several exciting future pathways:

1. Enhanced Model Architectures: Continued research into specialized deep learning architectures optimized for leather defect image classification could significantly boost detection accuracy and efficiency.
2. Transfer Learning and Pretrained Models: Utilizing transfer learning from models pre-trained on extensive datasets could notably enhance performance, particularly when training data for specific defect types is limited.
3. Advanced Data Augmentation Techniques: Exploring sophisticated data augmentation methods could diversify and expand the training dataset, thereby improving the model's ability to recognize various defect types and variations.
4. Multimodal Approaches: Integrating other sensing technologies like spectroscopy or hyperspectral imaging alongside visual data could offer supplementary insights for more thorough defect detection.
5. Real-Time Detection Systems: Developing real-time defect detection systems that seamlessly integrate into manufacturing processes could provide immediate feedback for quality control purposes.

6. Edge Computing and IoT Integration: Designing lightweight models suitable for edge devices could enable distributed defect detection systems, enhancing scalability and operational efficiency.

REFERENCES

Aslam, M., Khan, T. M., Naqvi, S. S., Holmes, G., & Naffa, R. (2019). On the application of automated machine vision for leather defect inspection and grading: A survey. *IEEE Access : Practical Innovations, Open Solutions*, *7*, 176065–176086. DOI: 10.1109/ACCESS.2019.2957427

Aslam, M., Khan, T. M., Naqvi, S. S., Holmes, G., & Naffa, R. (2020). Ensemble convolutional neural networks with knowledge transfer for leather defect classification in industrial settings. *IEEE Access : Practical Innovations, Open Solutions*, *8*, 198600–198614. DOI: 10.1109/ACCESS.2020.3034731

Ataç, H. O., Kayabaşı, A., & Aslan, M. F. (2024). The study on multi-defect detection for leather using object detection techniques. *Collagen and Leather*, *6*(1), 1–12. DOI: 10.1186/s42825-024-00186-2

Banduka, N., Tomić, K., Živadinović, J., & Mladineo, M. (2024). Automated Dual-Side Leather Defect Detection and Classification Using YOLOv11: A Case Study in the Finished Leather Industry. *Processes (Basel, Switzerland)*, *12*(12), 2892. DOI: 10.3390/pr12122892

Bergmann, P., Batzner, K., Fauser, M., Sattlegger, D., & Steger, C. (2021). The MVTec anomaly detection dataset: A comprehensive real-world dataset for unsupervised anomaly detection. *International Journal of Computer Vision*, *129*(4), 1038–1059. DOI: 10.1007/s11263-020-01400-4

Bin Abdullah, A., Jawahar, M., Manogaran, N., Subbiah, G., Seeranagan, K., Balusamy, B., & Chengam Saravanan, A. (2024). Leather Image Quality Classification and Defect Detection System using Mask Region-based Convolution Neural Network Model. *IJACSA). International Journal of Advanced Computer Science and Applications*, *15*(4). www.ijacsa.thesai.org

Chen, S.-Y., Cheng, Y.-C., Yang, W.-L., & Wang, M.-Y. (2021). Surface defect detection of wet-blue leather using hyperspectral imaging. *IEEE Access : Practical Innovations, Open Solutions*, *9*, 127685–127702. DOI: 10.1109/ACCESS.2021.3112133

Chen, Z., Deng, J., Zhu, Q., Wang, H., & Chen, Y. (2022). A systematic review of machine-vision-based leather surface defect inspection. *Electronics (Basel)*, *11*(15), 2383. DOI: 10.3390/electronics11152383

Chen, Z., Xu, D., Deng, J., Chen, Y., & Li, C. (2023). Comparative study on deep-learning-based leather surface defect identification. *Measurement Science & Technology*, *35*(1), 15402. DOI: 10.1088/1361-6501/acfb9f

Dinesh, K. G., Arumugaraj, K., Santhosh, K. D., & Mareeswari, V. (2018). Prediction of cardiovascular disease using machine learning algorithms. *2018 International Conference on Current Trends towards Converging Technologies (ICCTCT)*, 1–7. DOI: 10.1109/ICCTCT.2018.8550857

Gan, Y. S., Chee, S.-S., Huang, Y.-C., Liong, S.-T., & Yau, W.-C. (2021). Automated leather defect inspection using statistical approach on image intensity. *Journal of Ambient Intelligence and Humanized Computing*, *12*(10), 9269–9285. DOI: 10.1007/s12652-020-02631-6

Gan, Y. S., Liong, S.-T., Zheng, D., Xia, Y., Wu, S., Lin, M., & Huang, Y.-C. (2023). Detection and localization of defects on natural leather surfaces. *Journal of Ambient Intelligence and Humanized Computing*, *14*(3), 1–15. DOI: 10.1007/s12652-021-03396-2

Gan, Y. S., Yau, W.-C., Liong, S.-T., & Chen, C.-C. (2022). Automated Classification System for Tick-Bite Defect on Leather. *Mathematical Problems in Engineering*, *2022*(1), 5549879. DOI: 10.1155/2022/5549879

Khanal, S. R., Silva, J., Magalhães, L., Soares, J., Gonzalez, D. G., Castilla, Y. C., & Ferreira, M. J. (2022a). Leather defect detection using semantic segmentation: A hardware platform and software prototype. *Procedia Computer Science*, *204*, 573–580. DOI: 10.1016/j.procs.2022.08.070

Khanal, S. R., Silva, J., Magalhães, L., Soares, J., Gonzalez, D. G., Castilla, Y. C., & Ferreira, M. J. (2022b). Leather defect detection using semantic segmentation: A hardware platform and software prototype. *Procedia Computer Science*, *204*, 573–580. DOI: 10.1016/j.procs.2022.08.070

Kumari, R., Bandara, G., & Dissanayake, M. B. (2021). Sylvester Matrix-Based Similarity Estimation Method for Automation of Defect Detection in Textile Fabrics. *Journal of Sensors*, *2021*(1), 6625421. DOI: 10.1155/2021/6625421

Liong, S.-T., Zheng, D., Huang, Y.-C., & Gan, Y. S. (2020). Leather defect classification and segmentation using deep learning architecture. *International Journal of Computer Integrated Manufacturing*, *33*(10–11), 1105–1117. DOI: 10.1080/0951192X.2020.1795928

Mareeswari, V., Vijayan, R., Sathiyamoorthy, E., & Ephzibah, E. P. (2022). A narrative review of medical image processing by deep learning models: Origin to COVID-19. *International Journal of Advanced Technology and Engineering Exploration*, *9*(90), 623.

Moganam, P. K., & Sathia Seelan, D. A. (2022). Deep learning and machine learning neural network approaches for multi-class leather texture defect classification and segmentation. *Journal of Leather Science and Engineering*, *4*(1), 7. DOI: 10.1186/s42825-022-00080-9

Pazzaglia, G., Martini, M., Rosati, R., Romeo, L., & Frontoni, E. (2021). A deep learning-based approach for automatic leather classification in industry 4.0. *Pattern Recognition. ICPR International Workshops and Challenges: Virtual Event,* January 10–15, 2021,*Proceedings, Part IV*, 662–674.

Pople, V. (2023). Public social distance monitoring system using object detection yolo deep learning algorithm. *SN Computer Science*, *4*(6), 1–21.

Prunella, M., Scardigno, R. M., Buongiorno, D., Brunetti, A., Longo, N., Carli, R., Dotoli, M., & Bevilacqua, V. (2023). Deep learning for automatic vision-based recognition of industrial surface defects: A survey. *IEEE Access : Practical Innovations, Open Solutions*, *11*, 43370–43423. DOI: 10.1109/ACCESS.2023.3271748

Sandhya, N., Sashikumar, N. M., Priyanka, M., Wenisch, S. M., & Kumarasamy, K. (2021). Automated fabric defect detection and classification: A deep learning approach. *Text Leath Rev*, *4*, 315–335. DOI: 10.31881/TLR.2021.24

Smith, A. D., Du, S., & Kurien, A. (2023). Vision transformers for anomaly detection and localisation in leather surface defect classification based on low-resolution images and a small dataset. *Applied Sciences (Basel, Switzerland)*, *13*(15), 8716. DOI: 10.3390/app13158716

Tariq, M. U. (2025). AI and the Ethical Crossroads: Navigating Privacy, Personhood, and Autonomy in the Digital Age. In Chakraborty, S. (Ed.), *Human Values, Ethics, and Dignity in the Age of Artificial Intelligence* (pp. 49–72). IGI Global Scientific Publishing., DOI: 10.4018/979-8-3693-9035-1.ch003

Tariq, M. U. (2025). Integrating ICT for Enhanced Organizational Resilience: Strategies and Case Studies. In Carvalho, L., Reis, L., Neves de Almeida Baptista Figueiredo, A., & Russo, N. (Eds.), *Leveraging Technology for Organizational Adaptability* (pp. 59–80). IGI Global Scientific Publishing., DOI: 10.4018/979-8-3693-6045-3.ch004

Tariq, M. U. (2025). Navigating Public Administration 5.0: Embracing AI for Smarter Governance. In Wongmahesak, K., Ahmad, J., & Mustanir, A. (Eds.), *Public Governance Practices in the Age of AI* (pp. 167–186). IGI Global Scientific Publishing., DOI: 10.4018/979-8-3693-9286-7.ch008

Tariq, M. U. (2025). Unlocking Competitive Edge: Harnessing the Power of Collaborative Innovation. In Martínez-Falcó, J., Sánchez-García, E., & Marco-Lajara, B. (Eds.), *Open Innovation Strategies for Effective Competitive Advantage* (pp. 315–344). IGI Global Scientific Publishing., DOI: 10.4018/979-8-3693-9266-9.ch011

Tariq, M. U. (2025). Bridging Generations: Digital Transformation and Its Multifaceted Impact Across Societal Sectors. In Anshari, M., Almunawar, M., & Ordóñez de Pablos, P. (Eds.), *Impacts of Digital Technologies Across Generations* (pp. 1–22). IGI Global Scientific Publishing., DOI: 10.4018/979-8-3693-6366-9.ch001

Uddin, S. M. I., Mozumder, M. A. I., & Kim, H.-C. (2023). *Automatic Leather Defect Detection Using Deep Learning.*

Venkatachala Appa Swamy, M., Periyasamy, J., Thangavel, M., Khan, S. B., Almusharraf, A., Santhanam, P., Ramaraj, V., & Elsisi, M. (2023). Design and development of IoT and deep ensemble learning based model for disease monitoring and prediction. *Diagnostics (Basel), 13*(11), 1942. DOI: 10.3390/diagnostics13111942 PMID: 37296794

Vijayan, R., Mareeswari, V., & Jaswanth, A. B. (2024). Triple Diseases Prediction System using Deep Learning Techniques. *2024 Second International Conference on Emerging Trends in Information Technology and Engineering (ICETITE)*, 1–7. DOI: 10.1109/ic-ETITE58242.2024.10493563

Vijayan, R., Mareeswari, V., & Pople, V. (2023). Public Social Distance Monitoring System Using Object Detection YOLO Deep Learning Algorithm. *SN Computer Science, 4*(6), 718. Advance online publication. DOI: 10.1007/s42979-023-02131-2

Vijayan, R., Mareeswari, V., & Shobana, V. (2024). Tomato Plant Leaf Disease Prediction and Suggestion Using Deep Learning. *International Conference on Power Engineering and Intelligent Systems (PEIS)*, 125–136. DOI: 10.1007/978-981-97-6714-4_10

Wang, F., & Kyoung, K. J. (2023). Leather defect detection method in clothing design based on TDENet. *IEEE Access : Practical Innovations, Open Solutions, 11*, 104890–104904. DOI: 10.1109/ACCESS.2023.3308493

Wei, W., Deng, D., Zeng, L., & Zhang, C. (2021). Real-time implementation of fabric defect detection based on variational automatic encoder with structure similarity. *Journal of Real-Time Image Processing, 18*(3), 807–823. DOI: 10.1007/s11554-020-01023-5

Winiarti, S., Prahara, A., Murinto, D. P. I., & Ismi, P. (2018). Pre-trained convolutional neural network for classification of tanning leather image. [CNN]. *Network (Bristol, England)*, *9*(1), 9.

KEY TERMS AND DEFINITIONS

Leather: A durable textile material that is more expensive than other textile materials because it is made from animal skin.

Leather Industries: The global manufacturing sector in importing and exporting leather materials and products. The leather products are used in footwear, clothing, furniture, and automotive manufacturing.

Leather Defect Detection: The Defect detection process identifies the defective leather in quality control.

Leather Classification: The Leather classification process categorizes the leather based on such features as quality, types of cut, made from types of animal, and leather finishing to make the different variety products.

Deep Learning: The subset of machine learning and uses artificial neural networks to learn from the large dataset by repeated training and testing to automate any complex systems.

Convolutional Neural Network (CNN): A type of deep learning algorithm that uses computer vision applications. It has multiple operational layers to analyze the features in image recognition, classification, and segmentation.

Efficient Net: The state-of-the-art model of the family of convolutional neural networks that maintain the computational efficiency and model complexity by scaling width, depth, and resolution in image processing tasks.

Chapter 8
Artificial Intelligence for Personalized Electric Vehicle Energy Management

Safad Ismail
https://orcid.org/0000-0001-5048-1920
Saintgits Engineering College, India

Harsha Vasudev
https://orcid.org/0000-0001-6491-458X
IIITDM Kurnool, India

ABSTRACT

The rapid advancement of Electric Vehicle (EV) technology presents opportunities and challenges in energy management. As EV adoption accelerates, optimized energy management strategies are needed for individual users. This chapter explores the potential of Artificial Intelligence (AI) in developing personalized energy management systems for EVs. Using Machine Learning (ML) techniques like Long Short-Term Memory (LSTM) algorithms, these systems dynamically adjust charging and discharging patterns based on real-time data, user preferences, and driving habits. Integrating these algorithms with sensors and connectivity enables analyzing data points such as route planning, traffic, weather, and grid demands. The proposed approach helps to extend battery life, reduce costs, and enhance sustainability. Case studies demonstrate AI's effectiveness in optimizing charging schedules, managing energy flow, and ensuring grid stability. The chapter also discusses AI adoption challenges, including data privacy, computational requirements, and cybersecurity.

DOI: 10.4018/979-8-3373-4332-7.ch008

INTRODUCTION

Electric vehicles (EV) provide a more environmentally friendly and sustainable option to conventional internal combustion engine vehicles for the automotive industry. Rechargeable batteries power the electric motors in EVs, which eliminate tailpipe emissions and lessen the reliance on fossil fuels. The components of Electric Vehicles are shown in Figure 1. For EVs to be widely adopted, their performance, cost-effectiveness, charging infrastructure, and energy efficiency are critical factors in addition to their safety, dependability, and economic viability. AI can significantly contribute to the improvement of these attributes by optimizing EV operations, enhancing user experience, and ensuring efficient energy use. Numerous studies have been conducted on the use of AI in EVs, and these applications can be broadly categorized into three areas, namely, range optimization; EV control system design and optimization; and EV battery design, fabrication, and management. In the context of smart energy management, AI can be used for energy Generation Forecasting Khan et al. 2023). The key challenges associated with the utilization of AI include ensuring data security Farayola et al., 2024), understanding the fundamentals of AI technology, maintaining robust cybersecurity, updating existing systems, and assessing the relationship between AI integration and financial performance. AI models can be applied to all forms of renewable energy, such as solar, wind, geothermal, hydro, ocean, bio, hydrogen, and hybrid energy (Sayed et al., 2023). Artificial Neural Networks (ANNs), Wavelet Neural Networks (WNNs), Decision Trees, Support Vector Machines (SVMs), Hybrid, and Ensemble are a few examples of AI model types. In general, hybrid ML models are quicker, more precise, and simpler to use. The need for a personalized energy management system is an approach where the energy is managed for achieving higher goals in an optimized manner. A complete understanding of the system's operation, the available resources, and the energy patterns is necessary for effective energy management. These requirements may alter overtime and in response to changing environmental conditions.

Figure 1. Components of electric vehicle

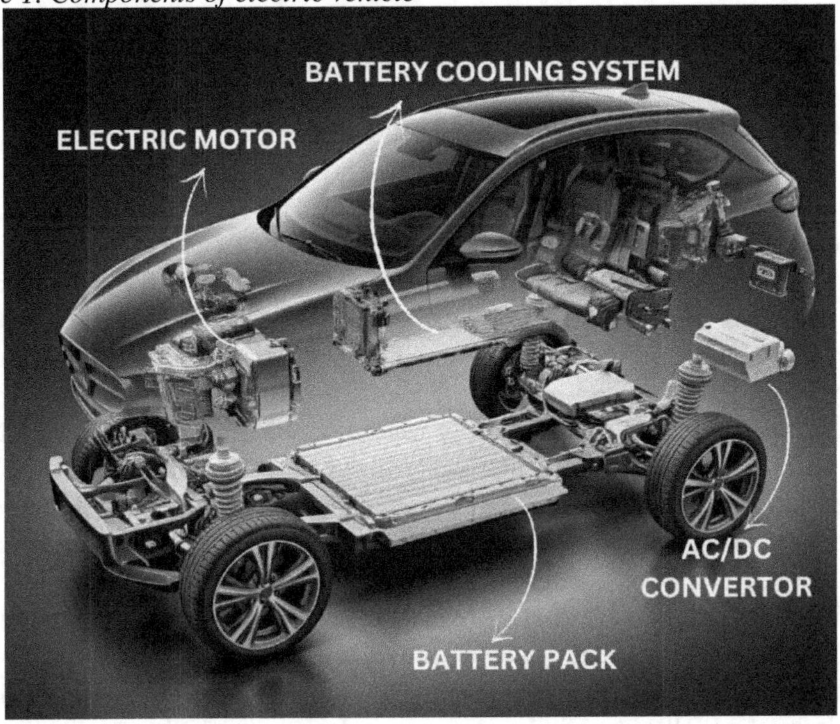

The main goal of this chapter is to examine how AI can revolutionize personalized energy management for EVs. To provide efficient energy solutions that are designed to meet the demands of each user, the chapter combines real-time data analytics with advanced AI approaches, including LSTM models. Through predictive modeling and dynamic adjustment techniques, the focus is to extend battery life, increase energy efficiency, and guarantee sustainable EV operations. This chapter addresses the issues and concerns in using AI for EV energy management. It also highlights the integration of AI with cutting-edge sensors, IoT technologies, and communication systems. The chapter is organized as follows. The related work section examines the role of ML, IoT integration, and energy management techniques for EVs. The Methodology section describes how LSTM models are used to forecast energy requirements and maximize battery utilization. With examples, the benefits of energy management discuss cost savings, battery life, and efficiency improvements. The predictive modelling and real-time data section focus on predicting energy requirements and making dynamic adjustments. The integration with sensors and IoT section highlights cutting-edge technologies for energy flow optimization. The challenges section covers cybersecurity, data privacy, and the computational needs for effective implementation. It also offers information about the real-world application of EV energy management systems. The chapter ends with a summary of the major developments and future possibilities in EV personalized energy management.

RELATED WORKS

To ensure an accurate and systematic assessment of the literature on the application of AI in personalized energy management for EVs from 2021 to 2024, the Preferred Reporting Items for Systematic Reviews and Meta Analyses (PRISMA) method was used. A systematic review following the PRISMA method selected 100 relevant studies from 450 initial records based on relevance, data availability, and research quality. Finally, 100 studies were incorporated into the study. These studies provide valuable insights into the application of ML, DL, and RL algorithms to optimize battery life, energy consumption, and personalized driving experiences like adaptive route planning and user-specific charging schedules. It was found that the use of AI in personalized energy management systems greatly improved energy efficiency by adapting to the unique needs and driving habits of EV users. The review showed the significant impact of AI in developing personalized energy management systems, showing how AI adjusts to the unique needs and driving patterns of each EV user, resulting in increased energy efficiency. The rigorous selection procedure made possible by the PRISMA technique improved the review findings' reliability and reproducibility and highlighted the vital role of AI in developing both general and personalized energy management systems for EVs. Several studies have examined the integration of AI techniques for optimizing energy management in EVs. Table 1 provides an overview of current studies describing the AI techniques used and important findings in relation to EV energy management. However, issues like data quality, computational complexity, and the scalability of AI systems in practical EV applications continue to be major obstacles to broader use. These obstacles must be addressed to fully realize the potential of AI in EV energy management.

Introduction to AI and Machine Learning

Artificial Intelligence (AI) and Machine Learning (ML) have emerged as key technologies, driving innovation across numerous industries in today's technology world. They are changing how we live and work, from enabling virtual assistants like Siri and Alexa to advancing financial forecasting and healthcare diagnostics Ismail et al., 2024). AI is the simulation of human intelligence in computers that are built to think and act like humans. AI systems are made to carry out tasks like speech recognition, visual perception, language translation, and decision-making that normally need human intelligence. AI is divided into Narrow AI, which focuses on specific tasks like facial recognition and chatbots, and General AI, which seeks to perform a wide range of human-like functions.

ML methods include supervised, Unsupervised, Reinforcement, Transfer, and Semi-Supervised Learning, each designed for different data processing approaches (see Fig 2).

- Supervised Learning: Labeled data is used by algorithms to learn and generate predictions.
- Unsupervised Learning: Algorithms identify patterns in data without labeled responses.
- Reinforcement Learning: Algorithms learn to make a series of decisions through interaction with the environment and feedback.
- Transfer Learning: The Application of information from one domain (source) to another that is similar but distinct (target).
- Self-Supervised Learning: The model generates its labels from the data, enabling it to predict certain data points based on other data points.
- Semi-Supervised Learning: Uses a small amount of labeled data and a large amount of unlabeled data.

Table 1. The role of AI in energy management for EVs (2021–2024)

Study	AI Technique	Application	Key Findings
Integrating Artificial Intelligence into Electric Vehicle Energy Management Hua et al. 2023)	Random Forest	Grid integration and optimization	AI facilitates effective EV integration into the grid by optimizing planning, forecasting, modeling, and enhancing energy system infrastructure.
Empirical Analysis of AI-based Energy Management in Electric Vehicles: A Case Study on Reinforcement Learning Hu et al.(2023)	Q-Learning	Energy management strategies in EVs	RL-based strategies outperform conventional methods in energy savings and real-time performance, particularly in hybrid and fuel-cell EVs.
Artificial Intelligence-Based Energy Management and Real-Time Optimization in Electric Vehicles, Pritima et al.(2022)	Gradient Boosting	Real-time energy optimization	AI enables the real-time optimization of energy consumption in EVs, leading to improved performance and efficiency.

continued on following page

Table 1. Continued

Study	AI Technique	Application	Key Findings
Optimizing Electric Vehicle Energy Consumption with Artificial Intelligence Kasaraneni et al.(2020)	SVM	Dynamic control and energy allocation	Advanced AI algorithms allow for dynamic control and allocation of energy resources within EV systems, enhancing performance.
A Systematic Review on the Integration of Artificial Intelligence in Energy Management Systems for Electric Vehicles Teimoori et al.(2022)	CNN	Energy efficiency and range improvement	Recent AI advancements significantly improve EV performance, energy efficiency, and range.
Towards Optimal Energy Management Strategy for Hybrid Electric Vehicle with Reinforcement Learning (Xu et al.(2020)	Deep Q-Networks (DQN)	Energy management in Hybrid EVs	RL-based strategies effectively learn optimal energy management, improving energy efficiency across various driving cycles.
A Review of AI Algorithms for Energy Management in E-Mobility Services Yan et al.(2023)	Genetic Algorithms	Addressing range anxiety and charge optimization	AI addresses challenges like range anxiety and charge optimization by enabling efficient energy management.
EnergAIze: Multi Agent Deep Deterministic Policy Gradient for Vehicle to Grid Energy Management Fonseca et al.(2024)	Multi-Agent Deep Deterministic Policy Gradient (MADDPG)	Vehicle-to-Grid (V2G) energy management	Multi-agent RL implements V2G technology, reducing peak loads and enhancing energy efficiency.
An Overview of Artificial Intelligence for Electric Vehicle Energy Management (Singh et al.(2023)	LSTM Networks	EV integration and energy management	AI offers solutions for challenges in EV energy management, providing optimization and efficiency improvements.
Managing Electric Vehicles in the Smart Grid Using Artificial Intelligence Ahmed et al.(2021)	Bayesian Networks	Smart Grid integration	AI optimizes energy storage and distribution in smart grids, reducing CO emissions, and supporting intelligent transportation systems.

Figure 2. Types of machine learning

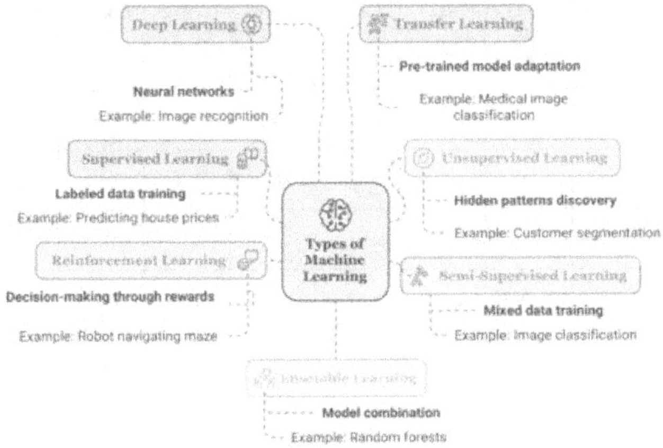

DATASET

The dataset used for the analysis of EV charging and discharging patterns Electric Vehicle Charging and Discharging Dataset 2024 comprises a set of technical attributes essential for modeling energy consumption dynamics. Each entry includes a Date and Time field, which provides granular time-series data for individual charging sessions or driving events, facilitating temporal analysis. Each EV user has an ID for personalized energy analysis. The battery level percentage metric denotes the state of charge before each session, which is critical for determining the necessity for charging or discharging actions. The dataset includes charging locations to assess usage patterns at different stations. Furthermore, the Charging Status attribute indicates whether the EV is in a charging or discharging state, thereby enhancing the clarity of operational modes during data collection. The Energy Consumed (kWh) and Energy Discharged (kWh) fields are used to record energy metrics measuring the energy flow throughout each session. Driving behavior is analyzed using distance and speed data. The Location (Lat, Long) coordinates give geo-location information, which enhances the contextual comprehension of charging patterns in various geographical areas. Finally, the Driving Behavior classification includes user driving styles like Normal, Aggressive, or Eco-Driving, which have a big impact on operational efficiency and patterns of energy use. A sample of the dataset is visualized in Table 2.

Table 2. Sample EV charging and discharging dataset

Date	Time	User ID	Battery Level (%)	Charging Station ID	Charging Status	Energy Consumed (kWh)	Energy Discharged (kWh)	Distance Driven (km)	Avg Speed (km/h)	Location (Lat, Long)	Driving Behavior
2024-01-01	08:00	U123	20	CS45	Charging	95.2	0	362.5	45	(40.7128, -74.0060)	Normal
2024-01-01	18:30	U123	80	CS33	Discharging	100.2	82.3	230.1	35	(40.7812, -73.9665)	Aggressive
2024-01-02	07:45	U124	25	CS45	Charging	116.5	83.6	390.0	50	(40.7060, -74.0086)	Normal
2024-01-02	12:00	U124	55	-	Discharging	110.3	101.1	270.5	40	(40.7506, -73.9934)	Eco-Driving
2024-01-03	19:00	U125	15	CS12	Charging	144.3	100.2	115.7	38	(40.7580, -73.9855)	Normal

METHODOLOGY

This chapter utilizes the LSTM model to analyze EV charging and discharging patterns, chosen for its effectiveness in processing sequential data and supported by prior research. To identify trends and produce precise forecasts, the LSTM model was trained using historical EV consumption data. This method guarantees a personalized energy management solution based on each user's particular charging habits.

Long Short-Term Memory- LSTM

We used the LSTM model to dynamically forecast EV charging and discharging trends. To fix missing values and standardize characteristics for model consistency, real-time data was first gathered and preprocessed. This includes battery levels, user preferences, driving behaviors, and environmental conditions. Factors such as trip frequency, duration, and peak hours were considered to improve model accuracy. The LSTM model architecture, as shown in Figure 3, was trained to detect temporal patterns and maximize charging and discharging predictions. The model updates predictions in real time, modifying charging and discharging based on user behavior, enhancing energy efficiency, and personalized EV management. The model is designed for real-time inference with minimal delay. It utilizes lightweight neural network architectures to reduce computational complexity, making it suitable for deployment in embedded EV systems. This approach ensures a balance between energy efficiency and predictive accuracy, optimizing performance for real-world applications. Mathematically, the LSTM gates are represented as follows:-

- Forget gate: $f_t = \sigma(W_f \cdot [h_{t-1}, x_t] + b_f)$ (1)

- Input gate: $i_t = \sigma(W_i \cdot [h_{t-1}, x_t] + b_i)$ (2)
- Cell candidate: $C\tilde{}_t = \tanh(W_C \cdot [h_{t-1}, x_t] + b_C)$ (3)
- Update cell state $Ct = ft * Ct-1 + it * C\tilde{}t$ (4)
- Output gate: $o_t = \sigma(W_o \cdot [h_{t-1}, x_t] + b_o)$ (5)
- Final hidden state: $h_t = o_t * \tanh(C_t)$ (6)

Figure 3. LSTM architecture

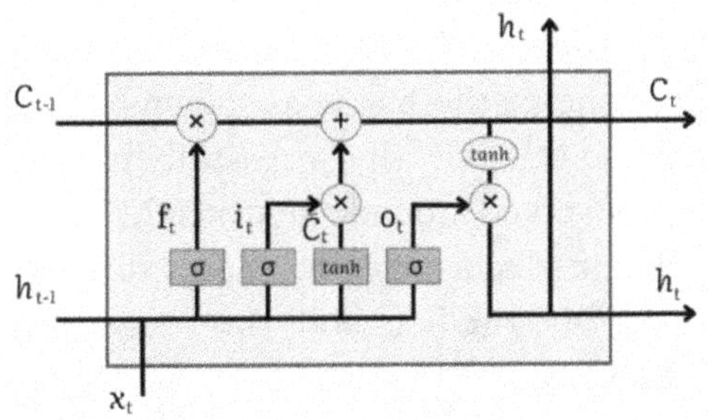

RESULTS

The prediction result of the LSTM Model for actual energy is shown in Figure 4. In the plot, the X-axis represents the Date ranging from '2020 to 2022' and the Y axes represent the consumed energy Load in kWh. The prediction graph shows the ability of the model to closely match actual energy used over time with predicted values. Energy consumption peaks, such as those seen in mid-2020 and early 2022, are well-represented, demonstrating the model's ability to predict high-demand periods. The model's accuracy in spotting patterns and trends is also demonstrated by the plot's small variances between expected and actual values throughout all phases. This result emphasizes the reliability of the LSTM model for energy management and demand prediction in EV systems.

Figure 4. LSTM prediction- energy consumed

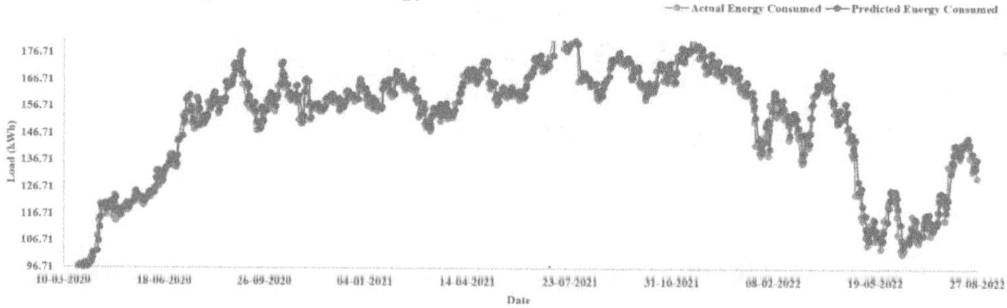

The predictions of the LSTM model accurately represent the peaks and trends in energy discharge across time, as shown in Figure 5. Notably, the model correctly predicts the times of high energy discharge, as shown by peaks in the plot, such as mid-2021 and late 2022. Additionally, it successfully tracks the downward trends, such as the sharp decline observed around August 2022 and the gradual recovery in subsequent months. In dynamic and changing energy discharge settings, this shows how well the model learns and predicts both short-term variations and long-term trends, guaranteeing reliability.

Figure 5. LSTM prediction- energy discharged

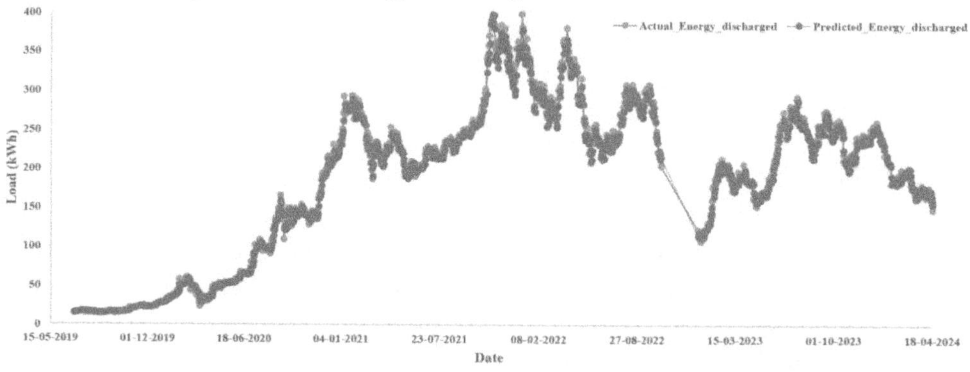

Table 3. Performance comparison of LSTM, ARIMA, and XGBoost for energy consumption prediction.

Model	RMSE ↓ (kWh)	MAE ↓ (kWh)	R^2 ↑
LSTM	2.110	1.568	0.92
ARIMA	4.256	3.325	0.85
XGBoost	2.045	1.770	0.88

Statistical evaluation shows that LSTM models outperform traditional methods in time-series forecasting by effectively capturing temporal relationships and complex data pattern

BENEFITS OF AI IN EV ENERGY MANAGEMENT

Electric Vehicles (EVs) represent a significant shift towards sustainable transportation, driven by advancements in battery technology, charging infrastructure, and regulatory support. However, the energy management demands and optimizing the performance of EVs present complex challenges that can be effectively addressed with AI technologies. The advantages of AI in improving EV energy management are shown in Figure 6.

Figure 6. Benefits of opting for AI in EV management

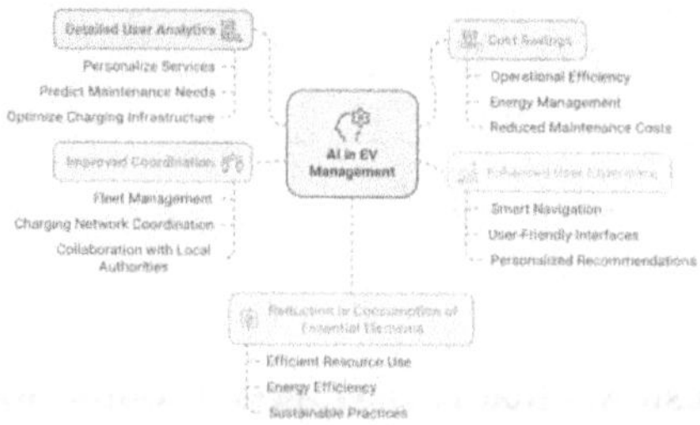

KEY AI TECHNIQUES: REINFORCEMENT LEARNING, NEURAL NETWORKS, PREDICTIVE ANALYTICS

AI-Driven Personalization in Energy Management

AI has become an innovator in the field of EV energy management by enabling customized energy optimization strategies that adjust to the preferences and behavior of each user. AI-driven personalization creates customized energy management solutions by using ML algorithms to evaluate data from several sources, such as driving habits, environmental factors, and the availability of charging infrastructure. The application of predictive analytics, which enables real-time energy consumption modifications depending on user behavior, is a crucial component of this strategy. AI systems, for example, can forecast a car's upcoming usage and adjust charging schedules, accordingly, reducing expenses and extending battery life Khan et al. 2021). Additionally, bidirectional energy flow is made possible by combining AI with vehicle-to-grid (V2G) technology, which allows EVs to release stored energy back to the grid during times of high demand, thus increasing energy efficiency Hanna et al. 2022). In addition to charging time optimization, customizing driving tactics is another aspect of personalization. AI can suggest driving practices that increase energy efficiency by evaluating past data, such as modifying acceleration patterns or making better use of regenerative braking. These personalized recommendations not only enhance the driving experience but also contribute to the overall sustainability of EV usage Liu et al. 2023).

Understanding User Preferences and Driving Habits

Understanding and adapting to user preferences and driving behaviors is essential for efficient energy management in EVs. By taking these things into account, energy consumption may be optimized, which enhances the vehicle's performance and user experience. It is possible to make more accurate range forecasts by identifying a person's driving patterns, including common routes, driving hours, speed changes, and acceleration-deceleration profiles Balsa et al. 2023).

Real-Time Data Analysis and Dynamic Adjustment

Maximizing energy consumption in EVs requires real-time data analysis and dynamic adjustments. This subsection presents case studies demonstrating their impact on vehicle performance and user experience.

- Real-time Data Analysis: The continuous tracking and analysis of data produced by the car and its driver is known as real-time data analysis. Numerous elements, including speed, acceleration, braking patterns, and even outside variables like traffic patterns and weather, can be included in this data. Using this data, EVs can make intelligent decisions that maximize energy efficiency. Wu et al.(2020) developed an eco-routing system that optimizes routes based on traffic, road gradient, and speed limits, reducing energy consumption by 51% compared to traditional navigation methods
- Dynamic Adjustment: This allows vehicles to modify energy-saving strategies, such as climate control, driving modes, and regenerative braking, based on real-time data. This idea is applied in real life in a study by Zhang et al.(2023). Based on the driver's braking intentions and the driving conditions at the time, the researchers created an intelligent regenerative braking technique that modifies the distribution of braking force Tang et al., 2023). The car dynamically controls regenerative braking to optimize energy recovery during deceleration events, like approaching a stoplight, by using a fuzzy logic control system. This braking method improves energy efficiency and provides a smoother driving experience.

- Personalized User Experience: Dynamic adjustment and real-time data analysis enhance energy efficiency while personalizing vehicle performance to individual driving habits.

The creation of P-ACC (Personalized-Adaptive Cruise Control) systems is a relevant example. A study by Zhao et al.(2023) addresses this concept. A P-ACC framework that learns and adjusts in real time to a driver's preferred following distance was presented by the researchers. The technology adapts the vehicle's acceleration and deceleration patterns to the driver's comfort level by examining the driver's actions during car-following scenarios. This customized modification optimizes speed profiles according to individual driving patterns, which not only improves driving comfort but also helps with energy efficiency (Tariq, 2025).

Predictive Modeling for Energy Needs and Battery Usage

For EVs, predictive modeling is essential to maximizing battery life and energy consumption. Through real-time data analysis and past driving patterns, these models can predict energy requirements, which improves vehicle performance and extends battery life. The creation of a probabilistic data-driven method for estimating Battery Electric Vehicle (BEV) trip-level energy consumption is a notable example. A model that integrates driver behavior characteristics, including average acceleration, average deceleration, and Relative Positive Acceleration (RPA) into NN with model uncertainty, was presented by researchers Maity et al.(2024). This approach produces more accurate energy forecasts by predicting a probability distribution for trip energy usage as opposed to a single-point estimate. The model outperformed other models in terms of accuracy, achieving a mean absolute percentage error of 9.3%.

The creation of a hybrid model that raises the prediction accuracy of energy consumption in BEVs is another significant advancement. This method uses several outside variables, like the weather, which significantly affect energy usage in addition to vehicle dynamics Huang et al. 2022). By considering these factors, the model can predict energy use more accurately, which helps EV drivers plan their routes and manage their energy more effectively. The significance of predictive modeling in controlling EV energy requirements and battery utilization is highlighted by these practical applications. These models help to create more effective and user-friendly electric transportation solutions by utilizing sophisticated data analysis techniques and considering a variety of influencing factors.

INTEGRATION OF AI WITH ADVANCED SENSORS AND CONNECTIVITY

The performance, safety, and efficiency of EVs are improved by combining AI with cutting-edge sensors and connections. This combination allows for real-time data processing and smarter decision-making.

The following subsection provides more details on these advancements.

Role of Sensors in Data Collection

In the realm of modern technology, sensors play a pivotal role in data acquisition, analysis, and utilization. With the advent of AI-based sensors and IoT-based sensors, the capabilities of data collection have been significantly enhanced. This section explores the functionalities and benefits of these advanced sensors, particularly in the context of EV energy management.

Figure 7. Advanced sensor integration in modern EVs

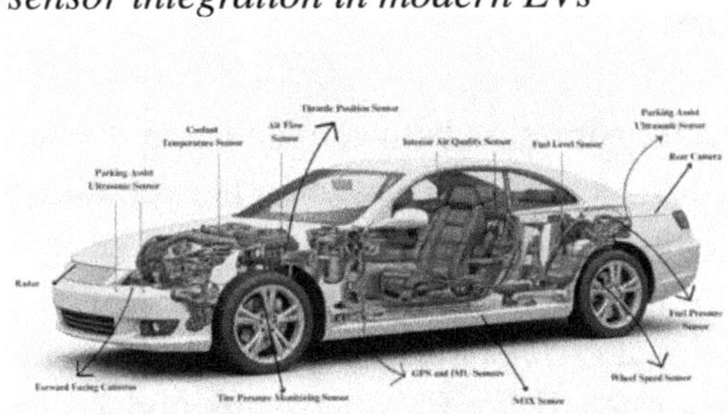

Data Acquisition and Analysis

AI-based sensors are instrumental in collecting vast amounts of data from various sources. By using ML algorithms, these sensors can analyze data in real time, provide critical insights into project progress, identify potential risks, and predict future trends. This real-time analysis enables proactive decision-making and enhances operational efficiency.

Predictive Maintenance

One of the significant advantages of AI sensors is their ability to predict equipment failures before they occur. By continuously monitoring the condition of components, AI sensors can identify signs of wear and tear, thus enabling timely maintenance. This predictive maintenance approach minimizes downtime, extends the lifespan of components, and reduces operational costs.

Safety Monitoring

AI sensors play a crucial role in ensuring safety by monitoring movements and site conditions. They can detect unsafe behaviors or hazardous conditions and trigger alerts to prevent accidents. As shown in Figure 7, the incorporation of these sensors into the EV infrastructure ensures effective monitoring of environmental conditions, vehicle performance, and user behavior. This real-time safety monitoring is essential in high-risk environments, contributing to a safer working environment (Tariq, 2025).

IoT-Based Sensors

With real-time data transmission between the car and external systems, such as charging stations, traffic control platforms, and other vehicles, the IoT is completely changing the role of EVs in a networked ecosystem. The foundation for more integrated transportation networks is laid by this connectivity, which improves user experience and operational efficiency. The ability of IoT to provide Over-the-Air (OTA) updates is a major benefit for EVs. Software updates can be remotely delivered, guaranteeing that EV systems always have the newest features, security patches, and improvements. This capacity is essential for autonomous driving technology, which needs frequent upgrades to improve safety, refine algorithms, and add new features. Predictive maintenance for electric vehicles is also made easier by IoT. IoT technologies can identify wear indicators or imminent problems by continuously monitoring different car systems and components, enabling proactive maintenance scheduling. This increases vehicle longevity and dependability while lowering the chance of unplanned failures. Additionally, by developing Electronic Manufacturing Services (EMS) for EVs, IoT is revolutionizing the manufacturing industry Roots EMS, 2023). IoT-enabled facilities in EMS manufacturing use predictive maintenance techniques, measure machine performance, and monitor production lines in real time. This high degree of automation lowers costs, increases productivity, and decreases manufacturing downtime, making EVs more accessible to consumers.

Figure 8. An overview of EV sensor placement

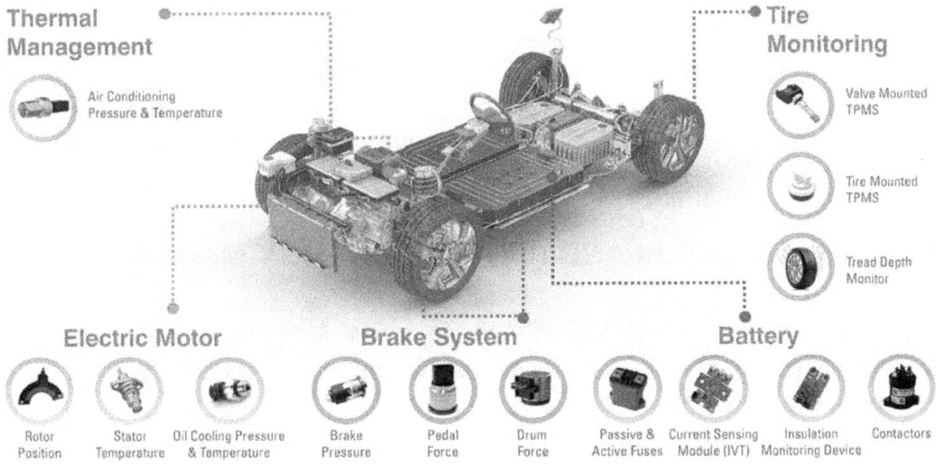

(Source: Sensata Technologies [EV Engineering, 2023)

Real-Time Monitoring

IoT sensors provide real-time monitoring of various parameters, such as EV energy usage. By optimizing performance and ensuring efficient energy management, these sensors contribute to the overall effectiveness of EV operations. The ability to monitor and adjust energy usage in real time is crucial for maximizing efficiency and sustainability. An outline of where EV sensors are located, highlighting electric motor control, tire and brake monitoring, and battery and thermal management systems, is shown in Fig 8.

Resource Management

IoT sensors are adept at tracking the usage of materials and energy. This capability is vital for reducing waste and enhancing sustainability. By providing detailed insights into resource consumption, IoT sensors help organizations implement more efficient and eco-friendly practices.

Automation

IoT devices enable the automation of numerous processes, such as the remote control of equipment and machinery. Automation not only improves efficiency but also enhances precision. The ability to control and adjust machinery remotely allows for more streamlined operations and reduces the need for manual intervention.

INTEGRATION OF AI AND IOT IN EV ENERGY MANAGEMENT

The integration of AI and IoT technologies offers numerous benefits in the realm of EV energy management. This section delves into the enhanced capabilities that arise from this integration.

Figure 9. Benefits of using IoT-based EMS

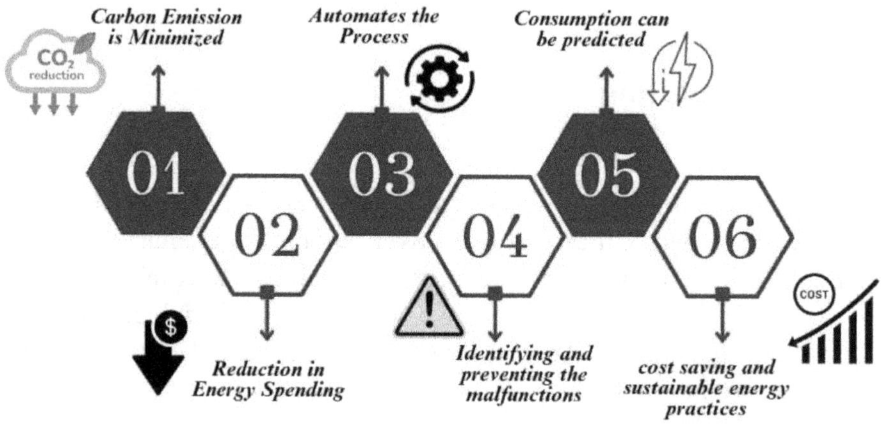

Enhanced Collaboration

By providing a unified platform for data sharing and analysis, the integration of AI and IoT facilitates better communication and collaboration among project stakeholders. This enhanced collaboration ensures that all parties have access to real-time data and insights, leading to more informed decision-making and improved project outcomes.

Improved Decision-Making

The combination of AI and IoT technologies offers data-driven insights that support informed decision-making. With access to accurate and timely data, stakeholders can make decisions that lead to improved project outcomes. The integration of these technologies ensures that decisions are based on comprehensive and up-to-date information (Tariq, 2025).

Sustainability

By optimizing resource use and reducing waste, AI and IoT technologies contribute to more sustainable manufacturing practices. The ability to monitor and adjust resource consumption in real time allows for more efficient operations, leading to a reduced environmental footprint and more sustainable practices (Tariq, 2025).

CONNECTIVITY TECHNOLOGIES AND DATA TRANSMISSION

Connectivity technologies and data transmission play a crucial role in the energy management of EVs. These technologies enable efficient communication between various components within the vehicle, between the vehicle and external infrastructure, and even between different vehicles.

In-Vehicle Communication Technologies

Different In-Vehicle Communication technologies include CAN, LIN, FlexRay, and Ethernet. A Controller Area Network (CAN)is a robust vehicle bus standard designed to allow microcontrollers and devices to communicate with each other without a host computer. Local Interconnect Network (LIN) is a low-cost, slower alternative to CAN used for simpler tasks where real-time performance is not critical. FlexRay provides high-speed and deterministic data transmission, used for Advanced Driver Assistance Systems (ADAS) and safety-critical applications. Ethernet can be used in modern vehicles for high-speed data transfer, particularly for infotainment systems and advanced diagnostics.

External Vehicle Communication Technologies

Cellular Networks (3G/4G/5G) are used for telematics, infotainment, Over-the-Air (OTA) updates, and vehicle-to-Everything (V2X) communication. 5G, in particular, offers low latency and high bandwidth, enabling real-time data transmission. Wi-Fi is used for connecting to home networks for OTA updates and data synchronization. They can be used in charging stations for communication between the vehicle and the charging infrastructure. Bluetooth facilitates short-range communication for in-car devices and smartphone connectivity. Dedicated Short-Range Communication (DSRC) is used for V2X communication, enabling vehicles to communicate with each other and with roadside infrastructure for safety and traffic management applications.

Vehicle-to-Everything (V2X) Communication

- Vehicle-to-vehicle (V2V): Allows vehicles to exchange information about their speed, position, and direction to improve safety and efficiency.
- Vehicle-to-Infrastructure (V2I): Enables communication between vehicles and road infrastructure, such as traffic lights and road signs, to optimize traffic flow and reduce congestion. • Vehicle-to-Grid (V2G): Allows EVs to interact with the power grid, enabling bidirectional energy flow where EVs can not only draw power from the grid but also supply energy back to it, supporting grid stability.

Figure 10. V2X communication technologies

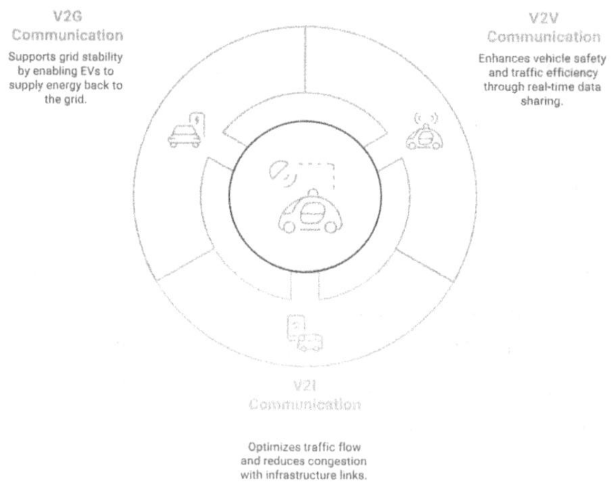

Data Transmission Protocols and Standards

Reliable and effective data transfer is crucial for smooth communication and improved system operation in the quickly changing world of EVs and linked infrastructure. In order to facilitate interoperability, security, and efficient data sharing inside EV networks, a number of protocols and standards—including MQTT, OPC UA, and ISO 15118—are essential.

- MQTT (Message Queuing Telemetry Transport) is a lightweight messaging protocol ideal for low-bandwidth, high-latency networks, commonly used in IoT applications Francis et al., 2024). • OPC UA (Open Platform Communications Unified Architecture): A machine-to-machine communication protocol for industrial automation, ensuring secure and reliable data exchange Tapia et al. 2023).
- ISO 15118: A standard for the communication interface between EVs and the charging infrastructure, supporting functions like plug-and-charge and smart charging Santos et al. 2024).

Figure 11. Data transmission protocols for EVs

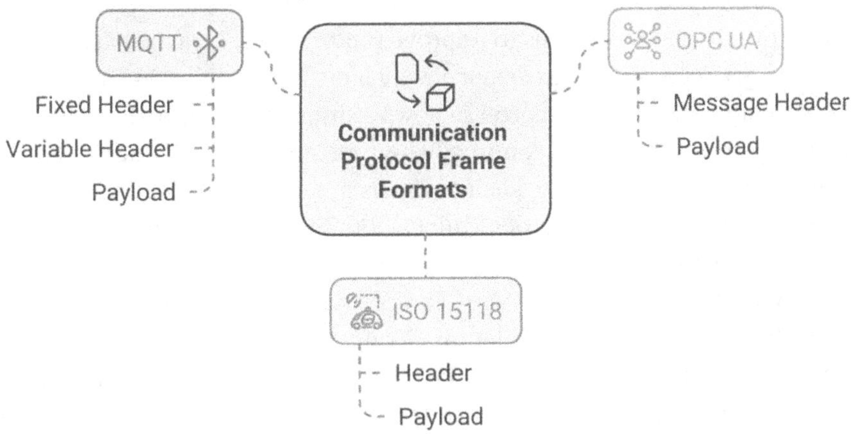

Energy Management Systems (EMS)

To maximize performance and extend battery life, efficient energy management is essential for EVs. A variety of techniques and technologies are utilized in EMS to track, optimize, and evaluate energy use. The Battery Management System (BMS), which guarantees the battery's safe and effective operation, energy management software that manages power distribution, and cloud-based solutions that use data analytics for improved efficiency and predictive maintenance are essential parts of EMS Gohil et al., 2022).

Integration and Utilization in Energy Management

EV efficiency and energy consumption optimization are greatly improved by advanced EMS. These systems use predictive modeling to forecast energy use and optimize charging schedules by analyzing both historical and current data. Energy conservation is ensured through adaptive control techniques, which enable dynamic adjustments to vehicle systems based on current conditions, such as traffic and weather. By using these data points, centralized management systems for commercial fleets effectively handle numerous vehicles. To optimize costs and minimize grid effect while adapting individual schedules and grid needs, smart charging solutions integrate home and public charging, significantly improving energy management Fowdur et al. 2024). By allowing EVs to return energy to the grid during periods of high demand, bidirectional charging, also known as Vehicle-to-Grid (V2G) technology, efficiently balances the load and offers financial incentives to car owners. Additionally, EVs' carbon impact is greatly decreased by renewable integration, which gives priority to charging from sustainable sources when possible. Real-time data is used by navigation systems to improve routes, forecast range based on driver behavior, and give eco-driving assistance, which provides advice and modifications to enhance energy efficiency. Figure 12 shows a simplified method for effective EV energy management and route planning. The first step involves finding important data points that affect energy use, such as environmental variables and vehicle performance indicators. The system determines both the elevation and route distance, as they have a great impact on battery usage. The charging stations are planned out along the route to avoid range anxiety. The predicted journey time in real time is determined by traffic and speed regulations. Traffic and road conditions are analyzed to reduce energy waste and prevent delays. In order to optimize planning, meteorological information such as wind and temperature is also included that affects battery efficiency. Finally, the integration with the power grid guarantees effective energy distribution and charging, avoiding overloads and lowering charging expenses. This approach provides an extensive framework for improving grid stability, driver convenience, and EV performance.

Figure 12. Data points for efficient route planning and energy management

CASE STUDY- SHENZHEN'S ELECTRIFICATION SUCCESS: AI-DRIVEN SMART CHARGING AND FLEET MANAGEMENT IN PUBLIC TRANSPORTATION

One of the EV capitals of the globe is Shenzhen, which lies in the southern Chinese province of Guangdong. The city is making quick advances toward EV adoption in private passenger cars and logistical trucks, and it has completely electrified its fleet of buses and taxis. By 2020, Shenzhen will have 63,000 private ride-hailing vehicles registered in the city, 24,000 taxis, and 19,000 electric buses. Approximately 80,000 of the 300,000 logistics vehicles in the fleet are electric. With about 4,300 of these vehicles on the road, Shenzhen is also in the early stages of electrifying its fleets of utility and sanitation vehicles. By the end of 2020, Shenzhen has more than 250,000 private passenger EVs, and by 2025, it expects to have 750,000 Hove A et al., 2024).

Figure 13. EV share by types in Shenzen

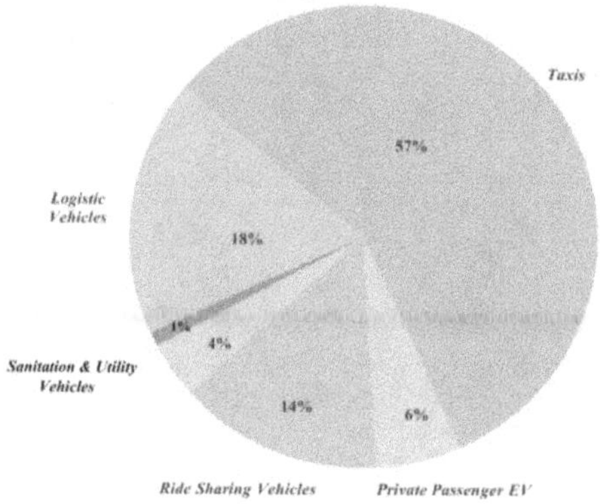

Regarding the infrastructure for charging, Shenzhen has developed an extensive EV charging network with over 5,000 stations and 83,000 charge points, supporting public and private EVs. The majority are bus fleet operators that build and run 120–150 kW chargers; occasionally, they share ownership with commercial charging businesses. To benefit from cheap time-of-use power pricing, most of the charging occurs at night, between 23:00 and 3:00 am. There is a little amount of midday bus pricing for longer routes. Although certain charge outlets are shared, Shenzhen's current bus charging infrastructure is adequate to meet the demand for bus charging on current routes and operations.

Shenzhen's Charging Strategy

Most public and commercial EVs complete charging within an hour, with taxis and ride-hailing services using HPC stations to reduce waiting time. This indicates that private vehicles have a greater degree of flexibility and don't require all-day operation. As a result, private automobiles can be charged at either central or dispersed locations. Commercial operators find centralized charging stations very appealing because they are often outfitted with 120–150 kW HPC equipment.

Table 4. Charging time by EV types in Shenzhen

EV Types	Charging Time	Charging within 60 min
Buses	23:00–3:00	45.90%
Taxis / Private ride-hailing vehicles	11:00–17:00 and 00:00–6:00	63.60%
Private passenger EVs	12:00–17:00 and 19:00–24:00	29.30%
Logistics vehicles	18:00–24:00	41%

Source: GIZ based on data from SAE 2022

The city's current fleet of vehicles can essentially be charged using Shenzhen's charging infrastructure, as shown in Table 4. However, charging station placement is based on grid capacity rather than user needs. This highlights the need for better coordination among stakeholders. The biggest disadvantage of Shenzhen's charging rollout thus far has been the separation of charging infrastructure planning by grid firms from the real needs of consumers, charging station owners, and operators. The building of charging infrastructure has been accelerated due to the emphasis on grid connection above other features, yet this has led to less-than-ideal locations when considering convenience or utilization.

INFRASTRUCTURE CHALLENGES IN AI ADOPTION FOR EV MAINTENANCE

The adoption of AI in the context of EVs and their charging infrastructure presents challenges such as data privacy, cyber security risks, and high implementation costs. Addressing ethical issues and guaranteeing compatibility across various platforms are also crucial factors. The following subsection lists several issues and difficulties.

Data Privacy and Security Concerns

Adopting AI technologies comes with several challenges and considerations, particularly regarding data privacy and security. Table 5 shows an in-depth look at these concerns and some strategies to address them.

Computational Requirements and Infrastructure

The computational requirements and infrastructure necessary for effective energy management in EVs are critical for optimizing performance, efficiency, and user experience. Here are the key aspects to consider:

1. Battery Management System (BMS) Sensors and Monitoring: EVs use sensors to track battery temperature, voltage, current, and state of charge (SoC). Continuous data processing helps to maintain safe operation and efficiency in charging and discharging. Safety measures prevent overheating, regulate cell balancing, and support battery performance. Safety Algorithms: Implementing algorithms that ensure safe operation of the battery, prevent overheating, and manage cell balancing.
2. Energy Management Software (EMS) Control Algorithms: Algorithms for optimizing energy consumption, including managing power distribution between the battery pack, electric motor, and auxiliary systems.
3. Predictive Modeling: Using historical and real-time data to predict energy demand and optimize energy usage during driving, charging, and vehicle-to-grid (V2G) operations. Integration with Vehicle Systems: Coordination with vehicle systems for real-time adjustments based on driving conditions, traffic, and user preferences.

Table 5. Challenges and considerations in adopting AI technologies

Category	Aspect	Description
Data Privacy Concerns Yanamala et al.(2023)	Collection and Storage of Personal Data	**Volume and Variety of Data**: AI systems often require vast amounts of sensitive data (e.g., location, health, financial data). **Data Minimization**: Limit data to only what's necessary for specific purposes. **Anonymization and Pseudonymization**: Techniques can protect identities but may sometimes be reversible.
	Consent and Transparency	**Informed Consent**: Users must be informed about data collection, usage, and sharing. **Transparent Practices**: Organizations should openly communicate data practices and AI algorithms.
Data Security Concerns Norris et al.(2023)	Cybersecurity Threats	**Data Breaches**: AI systems can be targets for attacks, risking exposure of sensitive data. **Encryption**: Strong encryption should protect data at rest and in transit. **Access Controls**: Authorization measures are essential to restrict access.
Regulatory and Compliance Challenges Padmanaban et al.(2024)	Global Regulations	**GDPR**: EU law on data privacy with strict compliance requirements. **CCPA**: California law offering data privacy rights. **Other Regional Laws**: Varying global regulations add complexity.

continued on following page

Table 5. Continued

Category	Aspect	Description
Ethical Considerations Smith et al.(2021)	Bias and Fairness	**Bias in Data**: AI may perpetuate biases in training data. **Fairness Audits**: These audits help to identify and mitigate model biases. **Diverse Training Data**: Using diverse datasets can help address biases.
Technical Challenges Brown et al.(2020)	Data Quality and Integrity	**Data Cleaning**: Ensuring high-quality data free from errors. **Data Integration**: Integrating multi-source data while maintaining consistency.
	Scalability and Performance	**Infrastructure Requirements**: Robust, scalable infrastructure is needed for AI systems. **Latency and Real-Time Processing**: Ensuring real-time, lag-free decision-making.

4. Connectivity and Communication Onboard Communication Systems: Integration with in-vehicle communication protocols like CAN, LIN, Ethernet, etc., for data exchange between components. External Communication: Connecting with external networks (e.g., cellular, Wi-Fi) for over-the-air (OTA) updates, remote diagnostics, and vehicle-to-infrastructure (V2I) communication.

Security Measures: Ensure secure communication to protect against cyber threats and unauthorized access to vehicle systems.

5. Computational Infrastructure Embedded Systems: EVs utilize embedded systems with microcontrollers and processors for real-time data processing and control. Edge Computing: Processing data locally in the vehicle to reduce latency and dependence on external networks, crucial for safety-critical applications. Cloud Integration: Leveraging cloud computing for advanced analytics, predictive maintenance, and fleet management, requiring robust connectivity and data security measures.
6. Charging Infrastructure Charging Controllers: Controllers manage the charging process, ensuring compatibility with different charging standards (e.g., AC charging, DC fast charging). Smart Charging Algorithms: Algorithms for optimizing charging schedules based on grid demand, energy tariffs, and user preferences. V2G Capabilities: Supporting bidirectional energy flow for V2G operations, where EVs can supply power back to the grid during peak demand periods.

7. User Interface and Experience Human-Machine Interface (HMI): Intuitive interfaces for users to monitor energy usage, manage charging schedules, and receive real-time feedback on driving efficiency. Integration with Infotainment: Incorporating energy-related information into infotainment systems for a seamless user experience.
8. Deployment Costs and Hardware Requirements: Scaling AI-driven IoT in EVs requires investments in sensors, edge computing, cloud infrastructure, and connectivity. Key hardware includes real-time sensors, AI-optimized chips (e.g., NVIDIA Jetson), and 5G/V2X modules for seamless data exchange. Costs also cover software development, cybersecurity, and ongoing maintenance. Optimizing AI for low-power embedded systems can help balance performance and cost-efficiency.

CHALLENGES AND CONSIDERATIONS

It is essential to supply enough memory and processing power for EVs to handle complicated algorithms and real-time data processing. Without sacrificing performance or safety, this capability must be matched with power consumption to optimize efficiency and range. As EV technology develops, especially with larger battery capacity and quicker charging times, scalable systems that enable the smooth incorporation of new features are crucial. To ensure a safe and dependable user experience, strong cybersecurity measures must also be put in place to safeguard private information and preserve the integrity of car systems against possible cyberattacks. Additionally, infrastructure challenges are crucial in AI-driven EV management. Reliable high-speed connectivity is needed for real-time energy tracking and optimization. Network disruptions can hinder predictive maintenance and resource management, making stable, low-latency communication essential for efficient EV operation.

Need for Robust Cyber Security Measures

Robust cybersecurity measures are crucial for EV energy management due to several key reasons.

1. Data Protection Personal and Vehicle Data: EVs collect location data, driving patterns, and user information, which must be protected from unauthorized access.
    - Homomorphic Encryption Enables computations on encrypted data without decryption, ensuring privacy.

- Data Anonymization: This removes personally identifiable information before analysis, safeguarding user data while allowing AI-driven insights.

2. Payment Information: Encryption and authentication safeguard financial data in EV charging transactions. Remote Access: Connectivity features in EVs allow for remote access and control over vehicle functions. Secure authentication and authorization mechanisms are vital to prevent unauthorized access by malicious actors.
3. Safety and Operational Integrity Vehicle Control Systems: In order to manage critical functions such as acceleration, braking, and steering, EVs rely on ECUs and communication networks. Protecting these systems from cyber threats is crucial to avoid safety risks.

Figure 14. EV charging cyber security

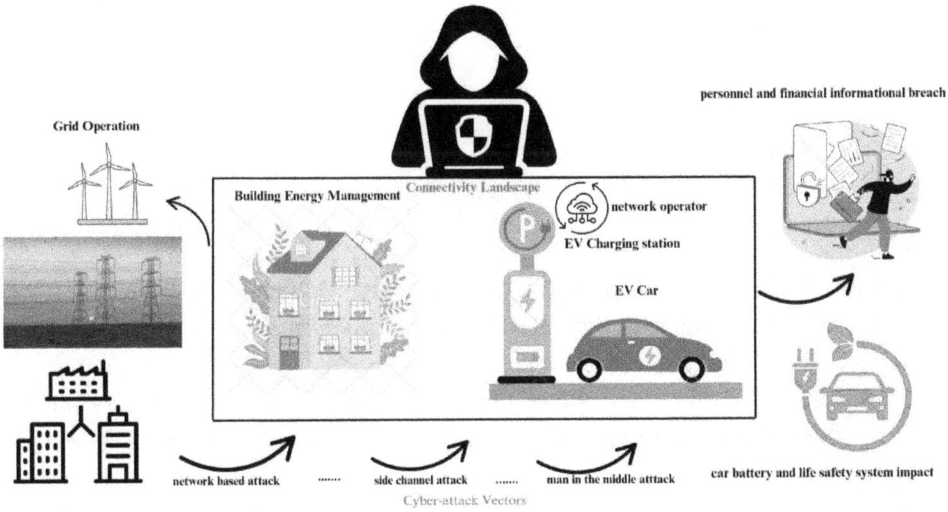

Figure 15. Malware transmission in the EV ecosystem

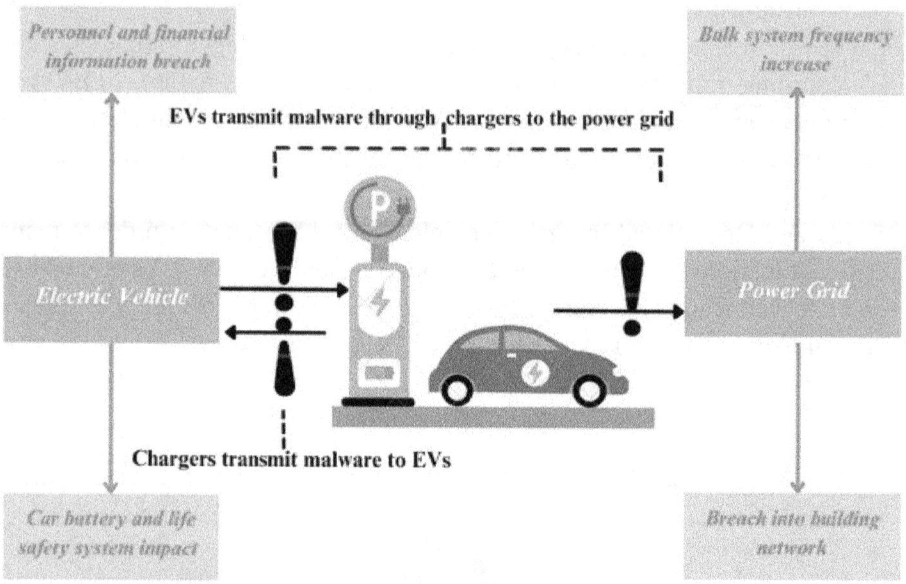

4. Battery Management: Cyberattacks on Battery Management Systems (BMS) can affect performance, safety, and stability. Protecting BMS from unauthorized access is essential for reliable energy management.
5. Firmware and Software Updates: To ensure that EVs receive timely patches for vulnerabilities and maintain the integrity of software systems, secure OTA updates are necessary.
6. Infrastructure Protection Charging Stations: EV charging infrastructure faces cyberattacks that can cause disruption or steal user data. To protect charging stations from malicious activities, secure communication protocols and robust authentication mechanisms are necessary. Grid Integration: Securing V2G technology prevents unauthorized access and ensures stable bidirectional energy flow.
7. Regulatory Compliance Data Privacy Regulations: Compliance with GDPR, CCPA, and ISO/SAE 21434 ensures data security in EVs. Industry Standards: Adhering to cybersecurity standards and best practices (e.g., ISO/SAE 21434 for automotive cybersecurity) helps ensure that EVs meet industry-recognized security requirements.

8. Threat Landscape Increased Connectivity: The growing number of connected EVs increases the attack surface for cyber threats. Malicious actors may exploit vulnerabilities in software, hardware, or communication networks to gain unauthorized access. Sophisticated Attacks: Cyber-attacks targeting automotive systems are becoming more sophisticated, requiring continuous monitoring, threat detection, and incident response capabilities to mitigate risks effectively. Mitigating Cybersecurity Risks
9. Encryption and Authentication: Strong encryption and authentication secure data transmission and access control.
10. Intrusion Detection Systems (IDS): IDS monitors network traffic for cyber threats.
11. Secure Software Development: Use secure coding practices and regularly test for vulnerabilities, like through penetration testing, to fix any weaknesses in the software.
12. User Awareness and Training: Educate users (drivers, fleet operators) about cybersecurity best practices and potential risks associated with EVs and charging infrastructure.
13. Collaboration and Information Sharing: Encourage collaboration among stakeholders (automakers, cybersecurity experts, and regulators) to share knowledge and enhance EV cybersecurity.

AI-DRIVEN SOLUTIONS FOR SECURE AND SUSTAINABLE EV CHARGING

AI-driven solutions improve sustainability, security, and efficiency through energy optimization and access control. To guarantee smooth and eco-friendly charging experiences, these technologies make use of real-time monitoring and predictive analytics. We will look over a few real-world case studies in the following subsection to discuss this.

Case Study: Tesla's Supercharger Network

The Tesla Supercharger Network consists of fast-charging stations strategically placed in cities and along important routes. These allow Tesla owners to recharge their cars in thirty minutes. The network has more than 20,000 charging stations Tesla, Inc., 2024). A breakdown of the Tesla Supercharger Network is shown in Figure 17. The adoption of the Supercharger ecosystem as the North American Charging Standard (NACS) is a significant move, as it permits access to non-Tesla EVs from major automakers such as Ford and GM. This strategic action increases

Tesla's market share and generates new, high-profit income sources. Because of its scalability and potential to change the EV charging landscape, industrial analysts have compared this evolution to the success of Amazon Web Services (AWS). The Tesla's Supercharger network employs AI to maximize charging efficiency, improve user experience, and maintain network security. To enable dynamic load balancing between charging stations, predictive analytics is used. This predicts energy consumption based on real-time traffic, vehicle usage, and weather data. ML models help to minimize grid overload and direct route decisions by modifying energy allocation. Tesla uses anomaly detection algorithms to detect and address cybersecurity risks and system issues. IoT device integration and V2G communication enhance resource management and real-time monitoring. These innovations make the Supercharger network more efficient, environmentally sustainable, and secure, setting a benchmark in EV charging technology.

Figure 16. Tesla supercharger network breakdown

ChargePoint's Technological Innovations in EV Charging

ChargePoint is a top supplier of EV charging solutions for fleet, business, and residential use, with more than 174,000 activated charging points worldwide. Its technology includes high-speed Level 3 DC chargers for quick public and fleet charging and Level 2 AC chargers for households and offices. ChargePoint's hardware and cloud-based software work together to provide seamless user experiences, dynamic energy optimization, and real-time charging station management. The 'Plug and Charge' function improves the customer experience by automating authentication and billing whenever a car connects to a charger. The company supports renewable energy sources like solar and wind by providing V2G technology. This allows automobiles to return energy to the grid during periods of high demand. Additionally, data analytics enable predictive maintenance, identifying charger issues early to improve reliability and reduce downtime. ChargePoint uses a subscription-based business model that combines its hardware with cloud services like energy management and station monitoring. This model supports expansion across businesses, government agencies, and individual users, establishing ChargePoint as a reliable and efficient EV charging provider.

Figure 17. ChargePoint's technological innovations

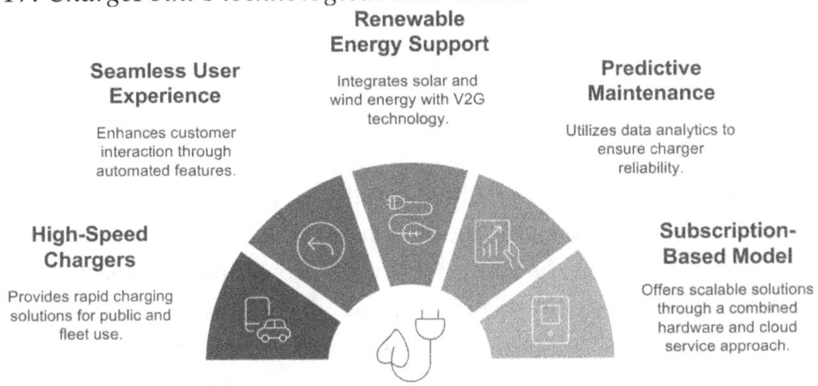

Enel X's AI-Driven Smart Charging Solutions in Italy

Enel X is a division of the Enel Group, founded in 2000. It is the world pioneer in sustainable energy solutions in Rome, Italy. The company specializes in energy efficiency and renewable energy technology with the goal of enabling everyone to have access to intelligent, clean energy. It uses AI technologies to meet Italy's renewable energy targets while ensuring safe, effective, and sustainable EV charging. One of their best-selling e-mobility products is the JuiceBox EV Charger'. JuiceBox offers several models to accommodate a range of charging requirements as shown in Table 6.

Table 6. Comparison of JuiceBox EV charging models

Model	Power Output	Ideal For	Price Range
JuiceBox 32	32-amp, 7.7 kW	Small EV batteries, daily commutes	\$500–\$600
JuiceBox 40	40-amp, 9.6 kW	Faster home charging, larger batteries	\$600–\$700
JuiceBox 48	48-amp, 11.5 kW	Rapid charging for large batteries, fleet use	\$700+

Figure 18. Enel X EV charging solutions

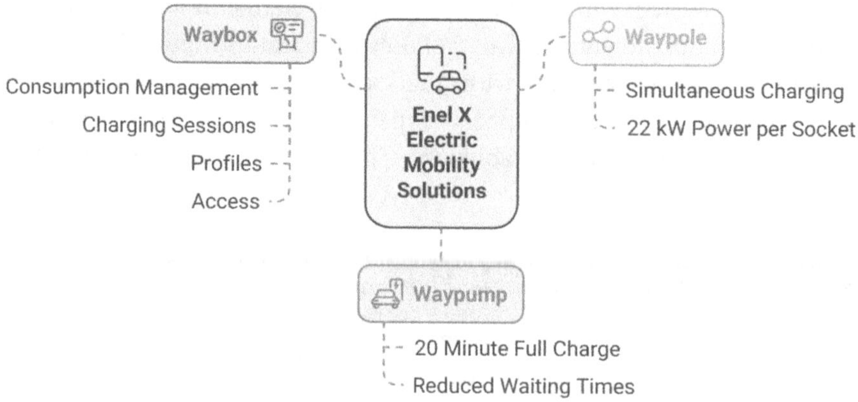

Enel X leverages advanced artificial intelligence within its JuiceNet platform to create an efficient and scalable electric vehicle (EV) charging ecosystem. The platform employs machine learning algorithms for real-time energy load management, optimizing charging schedules to align with off-peak grid demand and integrating renewable energy sources like solar and wind into the charging process. By dynamically balancing energy distribution across its network, Enel X mitigates grid overload risks, ensuring stability and reliability. The system's vehicle-to-grid (V2G) capabilities enable bidirectional energy flow, allowing EVs to act as decentralized storage units that can discharge energy back to the grid during peak demand periods, enhancing grid resilience. Additionally, predictive analytics monitor station performance and identify anomalies for proactive maintenance, minimizing downtime. With features like dynamic pricing optimization and user-specific charging recommendations delivered via a mobile interface, Enel X's AI-driven solutions represent a sophisticated response to the challenges of large-scale EV adoption and grid sustainability.

CONCLUSION

In this chapter, we explored how AI can transform personalized energy management for EVs. The combination of AI with advanced sensors and networks can analyze data like weather, grid needs, traffic, and route planning. This approach helps to create personalized energy plans that increase battery durability, lower energy expenses, and improve the sustainability of EV operations. AI can forecast user-specific driving behaviors, which ensures the EVs run as efficiently as possible in a variety of scenarios. Furthermore, through the integration of adaptive learning algorithms, these models can be improved over time, allowing energy management to be adjusted to the changing demands of both the environment and users. Recent developments in AI hardware, including neuromorphic computing, have the potential to speed up data processing and enhance energy-saving results. As a future research direction, transfer learning and federated learning can be explored to improve model adaptability, security, and personalization in AI-driven EV energy management.

The discussed case studies and real-world implementations highlight the effectiveness of AI in regulating energy flow, improving charging schedules, and maintaining grid stability. Looking ahead, additional improvements in AI models and computational frameworks will influence future developments in personalized EV energy management. The major areas of focus include enhancing algorithmic accuracy, integrating renewable energy sources into charging systems, and addressing important issues like data privacy, cyber security, and infrastructure scalability. These improvements will enable a more flexible and efficient EMCS, supporting the global transition to sustainable EV technologies. In summary, AI can completely transform EVs' personalized energy management. These technologies have the potential to be essential in the next generation of transportation systems, balancing user needs, energy efficiency, and environmental sustainability through the use of predictive modeling, real-time data analysis, and customized optimization methods.

REFERENCES

Electric Vehicle Charging and Discharging Dataset. (2024). *Kaggle Dataset.* Retrieved from https://www.kaggle.com/datasets/example-dataset-url

Engineering Online, E. V. (n.d.). *What role do advanced sensors play in EVs?* Retrieved from https://www.evengineeringonline.com/what-role-do-advanced-sensors-play-in-evs/

Farayola, O. A., Olorunfemi, O. L., & Shoetan, P. O.Oluwatoyin Ajoke FayayolaOluwabukunmi Latifat OlorunfemiPhilip Olaseni Shoetan. (2024). Data privacy and security in IT: A review of techniques and challenges. *Computer Science IT Research Journal, 5*(3), 606–615. DOI: 10.51594/csitrj.v5i3.909

Fonseca, T., Ferreira, L., Cabral, B., Severino, R., & Praca, I. (2024). *EnergAIze: Multi-agent deep deterministic policy gradient for vehicle-to-grid energy management.* arXiv preprint arXiv:2404.02361. DOI: 10.1109/SmartGridComm60555.2024.10738095

Fowdur, T. P., & Sanghan, A. (2024). *Achieving affordable and clean energy through AI and 5G powered Internet of Energy (IoE).* In *Artificial Intelligence, Engineering Systems and Sustainable Development* (pp. 97–108). Emerald Publishing Limited.

Francis, G. T., Souri, A., & Inanç, N. (2024). A hybrid intrusion detection approach based on message queuing telemetry transport (MQTT) protocol in industrial internet of things. *Transactions on Emerging Telecommunications Technologies, 35*(9), e5030. DOI: 10.1002/ett.5030

Gohil, H. H., Kachhara, V., & Kumar, R. (2022). A comprehensive review of battery management systems for electric vehicles: Current trends and future challenges. *Renewable & Sustainable Energy Reviews, 162*, 112419. DOI: 10.1016/j.rser.2022.112419

Hanna, J., Liu, Y., & Zhang, X. (2022). Integrating electric vehicles with smart grids: Opportunities and challenges. *Energy Reports, 8*, 123–135.

Hove, A. (n.d.). *Shenzhen charging roll-out has lessons for Berlin and other cities.* LinkedIn. Retrieved from https://www.linkedin.com/pulse/shenzen-charging-roll-out-has-lessons-berlin-othercities-anders-hove

Hu, J., Lin, Y., Li, J., Hou, Z., Zhao, D., Zhou, Q., . . . Zhang, Y. (2022). *Empirical analysis of AI-based energy management in electric vehicles: A case study on reinforcement learning.* arXiv preprint arXiv:2212.09154.

Hua, W., Mullen, D., Wahid, A., Sitabkhan, K., & Mason, K. (2023, September). *Integrating artificial intelligence into electric vehicle energy systems: A survey.* In *IFIP International Workshop on Artificial Intelligence for Knowledge Management* (pp. 92–115). Springer.

Huang, Y. W., Prehofer, C., Lindskog, W., Puts, R., Mosca, P., & Kauermann, G. (2022, December). *Predictive energy management for battery electric vehicles with hybrid models.* In *International Conference on Intelligent Transport Systems* (pp. 182–196). Springer.

Ismail, S., & Vasudev, H. (2024, September). *EIROB: An Emotionally Intelligent ROBot for EmpatheticSupport.* In *2024 IEEE International Conference on Signal Processing, Informatics, Communication and Energy Systems (SPICES)* (pp. 1–6). IEEE. DOI: 10.1109/SPICES62143.2024.10779758

Kasaraneni, R. K. (2020). AI-enhanced energy management systems for electric vehicles: Optimising battery performance and longevity. *Journal of Science and Technology*, *1*(1), 670–708.

Khan, A., Shah, S., & Majeed, A. (2021). Predictive analytics in electric vehicle charging management. *Journal of Cleaner Production*, *292*, 125–134.

Khan, S. U., Khan, N., Ullah, F. U. M., Kim, M. J., Lee, M. Y., & Baik, S. W. (2023). Towards intelligent building energy management: AI-based framework for power consumption and generation forecasting. *Energy and Building*, *279*, 112705. DOI: 10.1016/j.enbuild.2022.112705

Liu, R., Zhao, L., & Wang, T. (2023). Optimising energy consumption in electric vehicles through AI-driven driving recommendations. *IEEE Transactions on Intelligent Transportation Systems*, *24*(2), 456–469.

Maity, A., & Sarkar, S. (2024). Data-driven probabilistic energy consumption estimation for battery electric vehicles with model uncertainty. *International Journal of Green Energy*, *21*(9), 1986–2003. DOI: 10.1080/15435075.2023.2276174

Norris, D., & Mateczun, L. (2023). *Cybersecurity for local government: A primer.* UMBC Faculty Collection.

Padmanaban, H. (2024). Revolutionising regulatory reporting through AI/ML: Approaches for enhanced compliance and efficiency. [JAIGS]. *Journal of Artificial Intelligence General Science*, *2*(1), 71–90.

Pritima, D., Rani, S. S., Rajalakshmy, P., Kumar, K. V., & Krishnamoorthy, S. (2022). *Artificial intelligence-based energy management and real-time optimization in electric and hybrid electric vehicles*. In *E-Mobility: A New Era in Automotive Technology* (pp. 219–242). DOI: 10.1007/978-3-030-85424-9_12

Roots, E. M. S. (n.d.). *The role of AI and IoT in electric vehicles*. Retrieved from https://rootsems.com/the-role-of-ai-and-iot-in-electric-vehicles/

Santos, J. B., Francisco, A. M., Cabrita, C., Monteiro, J., Pacheco, A., & Cardoso, P. J. (2024). Development and implementation of a smart charging system for electric vehicles based on the ISO 15118 standard. *Energies*, *17*(12), 3045. DOI: 10.3390/en17123045

Sayed, E. T., Olabi, A. G., Elsaid, K., Al Radi, M., Semeraro, C., Doranehgard, M. H., Eltayeb, M. E., & Abdelkareem, M. A. (2023). Application of artificial intelligence techniques for modeling, optimizing, and controlling desalination systems powered by renewable energy resources. *Journal of Cleaner Production*, *413*, 137486. DOI: 10.1016/j.jclepro.2023.137486

Smith, J., & Johnson, M. (2021). Ethical implications of AI systems: Addressing bias and fairness. *Journal of Artificial Intelligence Ethics*, *15*(3), 233–245. DOI: 10.1234/jaie.2021.01345

Tang, M., & Zhang, X. (2023). Optimal regenerative braking control strategy for electric vehicles based on braking intention recognition and load estimation. *IEEE Transactions on Vehicular Technology*.

Tapia, E., Sastoque-Pinilla, L., Lopez-Novoa, U., Bediaga, I., & López de Lacalle, N. (2023). Assessing industrial communication protocols to bridge the gap between machine tools and software monitoring. *Sensors (Basel)*, *23*(12), 5694. DOI: 10.3390/s23125694 PMID: 37420858

Tariq, M. U. (2025). Revolutionizing Communication: EEG-Based Brain-Computer Interface for Speech and Mood Detection. In Prabhakar, A., Yu, P., & Erokhin, V. (Eds.), *Rural Social Entrepreneurship Development: Network-Based Manufacturing System Model* (pp. 237–264). IGI Global Scientific Publishing., DOI: 10.4018/979-8-3693-7515-0.ch008

Tariq, M. U. (2025). AI and the Ethical Crossroads: Navigating Privacy, Personhood, and Autonomy in the Digital Age. In Chakraborty, S. (Ed.), *Human Values, Ethics, and Dignity in the Age of Artificial Intelligence* (pp. 49–72). IGI Global Scientific Publishing., DOI: 10.4018/979-8-3693-9035-1.ch003

Tariq, M. U. (2025). Integrating ICT for Enhanced Organizational Resilience: Strategies and Case Studies. In Carvalho, L., Reis, L., Neves de Almeida Baptista Figueiredo, A., & Russo, N. (Eds.), *Leveraging Technology for Organizational Adaptability* (pp. 59–80). IGI Global Scientific Publishing., DOI: 10.4018/979-8-3693-6045-3.ch004

Tariq, M. U. (2025). Navigating Public Administration 5.0: Embracing AI for Smarter Governance. In Wongmahesak, K., Ahmad, J., & Mustanir, A. (Eds.), *Public Governance Practices in the Age of AI* (pp. 167–186). IGI Global Scientific Publishing., DOI: 10.4018/979-8-3693-9286-7.ch008

Tariq, M. U. (2025). AI-Enhanced Project-Based Learning: Revolutionizing Commerce Education With Generative AI. In ElSayary, A. (Ed.), *Prompt Engineering and Generative AI Applications for Teaching and Learning* (pp. 125–142). IGI Global Scientific Publishing., DOI: 10.4018/979-8-3693-7332-3.ch008

Teimoori, Z., & Yassine, A. (2022). A review on intelligent energy management systems for future electric vehicle transportation. *Sustainability (Basel)*, *14*(21), 14100. DOI: 10.3390/su142114100

Wu, G., Boriboonsomsin, K., & Barth, M. J. (2020). *Eco-routing navigation system for electric vehicles*. arXiv preprint arXiv:2008.09674.

Xu, B., Rathod, D., Zhang, D., Yebi, A., Zhang, X., Li, X., & Filipi, Z. (2020). Parametric study on reinforcement learning optimized energy management strategy for a hybrid electric vehicle. *Applied Energy*, *259*, 114200. DOI: 10.1016/j.apenergy.2019.114200

Yan, S., Shah, M. H., Li, J., O'Connor, N., & Liu, M. (2023, October). *A review on AI algorithms for energy management in E-mobility services*. In *2023 7th CAA International Conference on Vehicular Control and Intelligence (CVCI)* (pp. 1–8). IEEE. DOI: 10.1109/CVCI59596.2023.10397371

Yanamala, A. K. Y., & Suryadevara, S. (2023). Advances in data protection and artificial intelligence: Trends and challenges. *International Journal of Advanced Engineering Technologies and Innovations*, *1*(01), 294–319.

Zhao, Z., Liao, X., Abdelraouf, A., Han, K., Gupta, R., Barth, M. J., & Wu, G. (2023, October). *Real-time learning of driving gap preference for personalized adaptive cruise control*. In *2023 IEEE International Conference on Systems, Man, and Cybernetics (SMC)* (pp. 4675–4682). IEEE. DOI: 10.1109/SMC53992.2023.10394260

KEY TERMS AND DEFINITIONS

Range Optimisation: The use of AI algorithms to maximise the driving range of EVs by optimising energy consumption and driving efficiency.

EV Control System Design: AI-assisted development and tuning of control systems that govern EV functions such as motor control, braking, and regenerative systems.

Battery Management Systems (BMS): AI-powered systems that monitor, control, and optimise the performance, safety, and lifespan of EV batteries.

Energy Generation Forecasting: The application of AI models to predict energy generation patterns from renewable sources for smart charging and grid integration.

Hybrid AI Models: Combinations of multiple AI techniques (e.g., ANN + SVM) used for higher accuracy, efficiency, and adaptability in complex EV systems.

Chapter 9
Intelligent Mobility Assimilating IoT in Autonomous Vehicles:
Fostering Sustainable Cities and Communities

Bhupinder Singh
https://orcid.org/0009-0006-4779-2553
Sharda University, India

Kittisak Wongmahesak
https://orcid.org/0009-0000-2129-4691
North Bangkok University, Thailand

Saurabh Chandra
https://orcid.org/0000-0003-4172-9968
Bennett University, Greater Noida, India

ABSTRACT

Autonomous vehicle IoT integration enables intelligent mobility, which is a significant change in urban transportation and may sustain cities and communities. By utilizing IoT, autonomous cars can communicate with other vehicles, infrastructure and even pedestrians and optimize traffic flow while relieving the congestion on roads, which not only allows passengers to reach their destination faster but reduces environmental impact as well. Safety is increased by predictive capabilities built on advanced sensors, AI algorithms, and real-time data processing that helps prevent accidents. Reducing greenhouse gas emissions and air pollution by using renewable energy and electric propulsion systems, these vehicles provide cleaner air for urban

DOI: 10.4018/979-8-3373-4332-7.ch009

communities. IoT-powred shared mobility models decrease automobile ownership which further enhances effective resource allocation and equal access. Smart mobility solutions that solve problems like overpopulation, energy use, and climate change will bolster worldwide sustainability programs building more intelligent, resilient urban environments.

INTRODUCTION

Intelligent mobility is the advancement in transportation methods by combining advanced technologies, data processing, and efficient energy use to develop transport systems for efficiency, safety, and sustainability. With steeper urbanization, these vehicles represent a huge need for humans as environmental issues become more and more dire. Smart mobility responds to these challenges by developing integrated transport systems focusing on the user experience and reducing its ecological footprint. Intelligent mobility integrates all modes of transport, from conventional cars to newer approaches, including electric scooters and shared on-demand mobility. Such a comprehensive approach enables a flexible and responsive transport network catering to diverse user needs. Intelligent mobility uses digital technology and advanced transportation systems to improve the flow of people and freight in an urban setting. Smart mobility primarily depends on the Internet of Things (IoT), one of the most transformative elements in this transformation (Tan et al., 2018). This connectivity is crucial for creating smart transportation solutions that can adapt in real time to road conditions. IoT devices track traffic patterns, weather conditions, and vehicle performance to make real-time decisions based on facts. Providing this ability not only increases security but also maximizes the effectiveness of transport systems. Autonomous vehicles (AVs) represent one of their key components. Self-driving vehicles can travel without human intervention by using a combination of sensors, cameras, and advanced algorithms to guide their surroundings. Linking AVs with IoT looks feasible as both concepts collectively fill a unique technology niche. IoT is the backbone of AV technology, allowing vehicles and their environment to talk. This is realized with different kinds of V2X communication, which empowers the AVs to communicate with other vehicles (V2V), infrastructural elements (V2I), and pedestrians as well (like VANET) (Ji, 2021). IoT-enabled real-time traffic management, for instance, can drastically lessen congestion by optimizing the route according to live traffic conditions. They may then develop predictive maintenance systems to predict mechanical problems ahead of time and maintain the vehicle before it breaks down. Moreover, it can improve electric vehicles' efficiency by modifying driving patterns based on real-time data and energy optimization strategies. Intelligent mobility is also pertinent to sustainability beyond technological developments.

Using IoT-enabled AVs in urban transport systems can help reduce greenhouse gas emissions while increasing energy efficiency. Intelligent mobility can mitigate urban traffic congestion and pollution by promoting shared mobility solutions (e.g., ridesharing or car-sharing), thus reducing the need for private car ownership. Furthermore, intelligent mobility contributes to broader sustainability aims by promoting smart city projects that link transportation networks and other elements of urban infrastructure systems. By connecting all of this, it can work more efficiently to define the quality of life for residents. For example, intelligent traffic lights can modify their timings in response to real-time data received from linked cars to minimize waiting times and emissions at junctions. Smart mobility is an innovative way to solve modern transportation problems by integrating IoT and autonomous vehicles (Huang et al., 2021).

Intelligent mobility could also transform urban transport through integrated and efficient transport ecosystems with a heightened focus on safety and sustainable objectives. With cities growing and changing due to demographic switches and environmental challenges, moving towards intelligent mobility will be a vital next step for Smart, sustainable cities that serve all end users. How we leverage these innovations to transport ourselves is the key to our future, as sustainable development principles continue to serve society only when we make the effort. Defining Intelligent Mobility: Intelligent mobility is a range of strategies designed to maximize the efficiency of people and goods moving. It uses digital technology, sophisticated transport networks, and data analysis to build seamlessly integrated transport systems that maximize service efficiency and focus on users' needs. Intelligent mobility is all about improving user experience and minimizing environmental impact. One such paradigm for a seamless, interconnected driving experience is based on IoT, which would allow the data to be collected and processed in real time so that the vehicles, infrastructure, and users can interact with each other instantaneously. The idea is purposefully vague, encompassing everything from cars to new modes of transport, such as e-scooters and shared mobility. Its flexibility is required to solve the different urban mobility problems and have users with varying needs. With growing city congestion, intelligent mobility brings solutions focusing on user-friendliness, accessibility, and sustainability.

BACKGROUND

The concept of autonomous driving poised itself to the advanced stages of intelligent mobility, which today it would not have reached without the essential link partnering with IoT. AV technology relies on IoT as the driving force behind sensors, actuators, and communication smart device modules, which function independently while having interconnectivity. Autonomous vehicles (AVs) combined with the Internet of Things (IoT) technology in the transportation industry is the highway to a safer, more efficient, and sustainable mobility environment. AVs, enabled by recent advancements like artificial intelligence (AI), machine learning (ML), computer vision, and real-time data analytics, are revolutionizing urban mobility by minimizing human participation in driving. The integration of IoT-enabled sensors, GPS tracking, and cloud computing, along with the notion of vehicle-to-everything (V2X) communication, has bolstered communication in autonomous transportation, enabling vehicles, infrastructure, and pedestrians to communicate seamlessly. IoT in Autonomous Cars Collaboration via IoT Vehicle-to-Vehicle communication enables autonomous vehicles to get real-time updates about the traffic conditions and prevailing weather conditions and accident alerts too. Based on the data, the auto cars can prepare to take preventive action to avoid collision and maintain traffic flow. This interconnectivity also alleviates congestion, cuts down on journey time, and improves fuel efficiency, aligning with smart city programs and the international movement for sustainable urban transport.

IoT-powered autonomous vehicles improve road safety. IoT-powered autonomous vehicles improve road safety. AVs will save more lives than a human could by eliminating drunk driving, reckless behavior, distractions, and fatigue, with human error being responsible for at least 90% of road accidents worldwide. The IoT ecosystem operates to allow for real-time vehicle diagnostics, which ensures that AVs are running as needed, with mechanical issues, tire pressure changes, or braking system inefficiencies being identified before they can create malfunctions. Moreover, technologies like LiDAR (Light Detection and Ranging), radar, and AI-oriented perception systems allow AVs to detect pedestrians, cyclists, traffic lights, and street hazards, enriching decision-making processes and minimizing collision chances. IoT-connected self-driving cars can communicate with emergency services, sending accurate location information instantly in case of an accident, reducing response times and saving lives (Tang et al., 2021). Some of the key features that demonstrate this synergy are discussed below.

VEHICLE TO EVERYTHING (V2X) COMMUNICATION

This technology allows cars to speak with other vehicles and infrastructure elements (such as traffic lights) via short-range radio, improving traffic and safety. By exchanging real-time information about road and traffic conditions and possible dangers (V2X), a human driver or an automated system can improve its situational awareness. Apart from safety concerns, IoT-based autonomous vehicles are facilitating a paradigm shift in logistics and supply chain management. To optimize operations and cut costs, self-driving trucks and delivery drones are gaining traction with companies trying to accommodate a growing consumer appetite for faster, more reliable delivery. Telematics, fleet management software, and route optimization algorithms allow IoT-enabled AVs to find the most effective delivery paths, resulting in lower fuel consumption and carbon emissions. Efficient operations are enhanced by smart warehouses integrated with IoT sensors and AI-powered robotics technology, ensuring not only the smooth movement and storage of goods but also the seamless interaction between automated delivery vehicles and storage spaces. Within the ride-hailing and shared mobility industry, we see companies ranging from Waymo and Tesla to Uber and Cruise fielding autonomous fleets, meaning no human drivers, enhancing the ease of use for passengers while benefiting operational efficiency.

The integration of IoT with AVs also has excellent potential for smart city development. In order to ensure smoother traffic, governments and cities are focusing on intelligent transport systems (ITS), which integrate data from the Internet of Things (IoT)-based traffic monitoring, smart parking solutions, and adaptive signal control systems. For instance, IoT-enabled roadside infrastructure like- smart traffic lights, AI-based surveillance cameras, etc. communicate with AVs to control traffic density, ease congestion, prioritize emergency vehicles, and ensure pedestrian safety. AVs fitted with renewable energy sources, including solar panels and regenerative braking, reflect world efforts to minimize fossil fuel dependency and move towards zero-emission transportation.

But still, even having some advantages, there are lots of challenges to be solved to really enable autonomous vehicles and IoT. Data collectors will face significant cybersecurity threats, including hacking, data breaching, and unauthorized access to the vehicle systems, which will put the safety and privacy of both passengers and pedestrians at risk. IoT-linked AVs send and receive large amounts of data in real time, making them targets for cyberattacks that can impact the controls of the vehicle, navigation systems, or personal user information. Therefore, adopting strong cybersecurity designs, such as blockchain-driven information encoding, multi-factor validation, and intrusion prevention devices, is necessary to ensure the protection of AV communities. The regulatory environment and legal landscape around AV deployment are still convoluted; legislators must draft harmonized safety standards, ethical frameworks, and liability laws to address potential issues such as accidents, insurance claims, and culpability in the event of system malfunctions. (Tariq, 2024)

REAL-TIME TRAFFIC MANAGEMENT

Intelligent mobility systems can analyze data and optimize routes for the regional AV fleet. This feature saves travel time and could reduce congestion and emissions by guiding vehicles away from bottlenecks. The second major barrier is the cost of deployment of AV and IoT infrastructure. It's important to note that while major automobile companies and tech giants continue to pour billions of dollars into R&D for AVs, widespread adoption of fully autonomous transportation will only be realized after many upgrades and investments in infrastructure, 5G connectivity, and cloud computing capabilities. On the second hand, tar in rural locations with sparse access to the smart street networks may hit AV's gradual rollout, resulting in uneven transportation accessibility. The acceptance of and trust in the technology plays a vital role in this, as the majority of people are still reluctant to let an autonomous system take control of their vehicles due to unpredictability, reliability, and ethical issues that one must face while controlling an automobile (Tariq, 2025).

PREDICTIVE MAINTENANCE SYSTEM

IoT allows predictive analytics to predict when a vehicle needs to be maintained based on usage patterns and sensor data. By getting these proactive services done, owners can avoid breakdowns and increase the age of their vehicles. However, these challenges are not preventing IoT and autonomous vehicle technology from progressing. UDC, a chair of UCAV Technology Council, has mentioned the potential of having 5G and wide area mobile Edge Computing's potential to revolutionize the performance of autonomous vehicles, as well as quantum AI, and IoT can enhance a vehicle's agility, decision making, and intelligence in executing tasks, including complex ones. City grids revolutionize commuting and help create a smarter, “green” Earth.

ENERGY OPTIMIZATION

Speed, route, and many other factors can be tailored to optimize energy use. Again, an AV with IoT technology has the potential to adjust driving patterns according to real-time traffic conditions or energy availability. This is especially crucial for electric vehicles (EVs), as a vehicle's range and sustainability are directly tied to how efficiently it uses energy (Dai et al., 2018). However, with accelerating global investments in AV research, partnerships between automakers, technology companies, policymakers, and urban planners are key to building the regulatory framework that will facilitate the integration of AVs into the 21st-century transportation system. IoT-based Autonomous Mobility Future: A connected, intelligent, and self-sufficient system. Short of folly, if implemented, this expert technology revolution could overhaul.

SIGNIFICANCE OF SUSTAINABLE DEVELOPMENT CONCERNING IOT IN AUTONOMOUS VEHICLES

Autonomous Vehicles (AVs) have the potential to contribute significantly to sustainable development goals (SDGs) through the integration of the IoT, enabling smart and responsible mobility solutions. The urgency of sustainable transportation solutions arises from pressing challenges of climate change, urban congestion, and environmental degradation and plays an essential role in reducing carbon emissions, conserving energy, and optimizing mobility. The deployment of IoT-based AVs represents a major forward surplus for sustainable development (SD) by improving road safety, traffic management, fuel efficiency and smart urban planning, thereby fulfilling the SDG 11 (Sustainable Cities and Communities), SDG 13 (Climate Action), and SDG 9 (Industry, Innovation and Infrastructure). However, IoT-enabled AVs leverage real-time data analytics, AI-powered decision-making, and vehicle-infrastructure interconnectivity to combat environmental pollution, optimize road efficiency, and propel smart mobility, recasting the future of eco-friendly urban transportation (Tariq and Sergio, 2025).

The most important environmental benefit of IoT-powered AVs is their potential to provide a drastic reduction in GHG emissions. Conventional vehicles emit more than 20% of the global carbon, playing a major role in poor air quality and the acceleration of climate change. Electric autonomous vehicles (EAVs), which rely on the Internet of Things (IoT), provide an eco-friendly solution to traditional fossil-fuel-driven automobiles, decreasing reliance on non-renewable energy sources while encouraging the further development of technologies that utilize renewable energy sources, such as solar-powered charging stations and regenerative braking systems. Smart city infrastructures enable intelligent traffic management and connect routing systems that help AVs find the best routes for energy efficiency and traffic avoidance and reduce idle wait points through predictive routing using real-time data processing. AVs also communicate with smart traffic lights and dynamic road signs via vehicle-to-infrastructure (V2I) communication, facilitating smoother traffic flow by minimizing idling at intersections and avoiding over-consumption of fuel caused by stop-and-go traffic.

More than just reducing emissions, IoT-enabled AVs improve road safety and traffic efficiency, facilitating the transition to sustainable urban mobility. Road accidents resulting from human errors such as distracted driving, fatigue, and speeding lead to millions of deaths and injuries every year. IoT integrated with AI and machine learning constructs automated driving functions that minimize human-related risks and guarantee safer and predictable driving behavior. IoT-based AVs rely on LiDAR, radar, GPS, and AI-enabled perception systems to identify pedestrians, cyclists, and possible hazards and then swiftly change course to avoid accidents. In addition, vehicle-to-vehicle (V2V) communication enables AVs to share data about speed, lane changes, and emergency braking, which reduces multi-vehicle collisions and helps ensure that traffic moves in a coordinated manner. Aside from reducing car accidents, the connected network of intelligent AVs lowers the economic loss because of road accidents, medical spending, and damage caused to the infrastructure. IoT-enabled autonomous vehicles represent more than a technological innovation; they are essential to progress toward sustainable development goals. Inbound and Shutterstock AVs help sustainability in a few different ways:

Lowering Emissions: By enabling shared mobility solutions (e.g., ride-sharing or car-sharing services), AVs can significantly decrease road usage through privately owned vehicles. Fewer vehicles imply less greenhouse gas emissions, which enhances air quality.

Improving Energy Efficiency: By moving towards electric vehicles, intelligent mobility transforms emissions generated by combustion engines into something much lower. In addition, energy usage can be further reduced by leveraging IoT technologies to affect fleet driving patterns.

Optimizing Urban Mobility: Intelligent mobility systems envision creating smart cities by integrating transportation solutions. Integrating public transportation and private options like AVS can make it easier to use both while reducing individuals' dependence on cars. It can potentially allow cities to evolve into green spaces that are also pedestrian-oriented and optimized for public transport.

Smart mobility promotes shared mobility through enabling platforms for shared transport services, thus reducing personal vehicle ownership. This manner of shifting not only reduces the traffic in brownies but also helps promote solutions that are more sustainable and community-oriented transport.

Intelligent mobility is an encompassing vision for today's transportation that combines intelligent machines and sustainable approaches. The seamless integration of IoT with autonomous vehicles facilitates such implementations, paving the way for clean, safe, and efficient smart cities. Smart mobility then opens new sights to develop modern towns and connect with our growing populations in a smarter way that every citizen enjoys traveling within their city center while mitigating its negative impact on the urban environment (Luettin et al., 2022).

AUTONOMOUS VEHICLES AND COMPONENTS OF IOT

The key aspect of sustainable transportation is IoT in fleet management and logistics. Autonomous, IoT-based fleet management systems play a key role in reducing carbon emissions and resource waste in the logistics industry. Self-driving trucks and AI-enabled delivery drones, already enabled with real-time GPS tracking, predictive analytics, and automated route optimization, potentially increase supply chain efficiency, often by curtailing fuel use, shortening delivery times, and cutting down idle truck hours. IoT devices such as sensors can be used to monitor vehicle health and performance. Smart warehouses, when integrated with automated AV fleets, allow for better and efficient inventory management, load optimization, and last-mile delivery while reducing energy consumption and operational inefficiencies.

Autonomous ride-sharing and public transportation services based on IoT play a vital role in achieving sustainable urban mobility to solve traffic congestion, carbon footprint, and resource utilization. This results in higher vehicle density, underutilized parking spaces, and increased emissions. Though AV-based ride-sharing platforms, enabled by IoT-enabled mobility-as-a-service (MaaS) solutions that are based on vehicle pooling and shared mobility that are significantly reducing the number of cars on road. AI projects smart public transport systems, achieving passenger distribution and wait time minimization, along with being less reliant on fuel with AI-generated transit schedules and autonomous electric buses. Moreover, encouraging a green commuting culture through intelligent ticketing, real-time tracking & dynamic routing optimizations powered by IoT-based solutions will lead to higher public transport reliability and increased accessibility.

The other critical way IoT in AVs is impacting sustainable development is by changing the face of urban infrastructure. These intelligent transport system (ITS)-enabled smart cities feature smart traffic lights, instantaneous air quality monitoring, and AI congestion prediction models that develop a dynamic system that guarantees optimal use of roads with little environmental disturbance. AVs with Internet of Things (IoT) integration talk to smart parking systems to guide drivers to available parking spots, eliminating the fuel wasted idling and emissions from cars looking for parking. Furthermore, sensors are embedded in roads and bridges to monitor structural integrity, traffic density, and environmental factors, report IoT sensors, where data can then be used for sustainable urban planning.

However, the implementation of IoT-integrated autonomous vehicles has its challenges that need processing systematically before these technologies can attain development goals. Cybersecurity and data privacy There is a chief concern. IoT-enabled AVs constantly capture, process, and communicate massive amounts of data in real time, increasing their susceptibility to hacking, data breaches, and unauthorized access. Cyberattacks on AVs may result in vehicle malfunctions, disruptions to navigation, or even threats to safety. As such, IoT-powered AV networks must be protected with strong cybersecurity frameworks, blockchain-based data encryption, and AI- or machine learning-driven anomaly detection systems.

SENSORS AND ACTUATORS

At the core of all autonomous vehicles, sensors and actuators are crucial components that participate in perception and interaction with the surroundings. High-tech sensors, including cameras, LiDAR, radar, and ultrasonic instruments, amass teraflops of data on the environment around the vehicle. Likewise, Cameras use visual information to identify traffic Signs, Lane Markings, Pedestrians, and Vehicles. LiDAR systems use laser light to create 3D maps of the environment. Radar can detect objects far and near at varying speeds; Ultrasonic Sensors are primarily used for near-distance detection, such as when parking. Based on the information given by sensor inputs, actuators are deployed to bring different types of changes to the vehicle's running. They control steering, acceleration, and braking functions. With these sensors and actuators, AVs can perceive their environment accurately and respond appropriately when the conditions change (Halilaj et al., 2021; 2022).

Communication Networks must communicate continuously for autonomous vehicles to perform their tasks seamlessly. V2X or Vehicle-to-Everything communication refers to the interconnectedness between various devices and the vehicle, including wirelessly connected vehicles with other vehicles (Vehicle-to-Vehicle or V2V for short), infrastructure (Vehicle-to-Infrastructure or V2I), and a cellular network (Vehicle-to-Network or V2N). This means that AVs can communicate with other nearby vehicles, sharing details of their speed, direction, and intentions through V2V communication. The data exchange provides context and clarifies the driving environment, which helps predict or avoid dangerous situations. V2I communication is an interaction between AVs and infrastructure elements like traffic lights or road signs, as well as sensors installed inside materials on the roadway. This allows cars to receive information from the internet about current traffic, obstacles ahead, and future stop signs. For instance, an AV could change its speed based on how far a traffic light is from the vehicle or find another route if there are reports of an accident ahead. Communication of vehicles from cellular networks or cloud services for data processing and storage capabilities is called V2N communication. This connectivity is crucial for real-time traffic, weather, and other environmental variables influencing driving decisions.

Cloud Computing is essential for centralized data storage and processing of autonomous vehicles. A considerable amount of data is generated from sensors, which need to be analyzed well in time for decision-making. This data processing power is available to you via cloud platforms. The individual AVs can analyze historical data and real-time inputs using sophisticated algorithms, but they will benefit from some of the processing being performed in the cloud. This centralization enables advanced predictive analytics that improve navigational assistance. Using cloud-based systems can, for example, track and analyze traffic patterns over time to recommend routes that maximize speed while avoiding congested areas (Halilaj et al., 2023). Moreover, cloud computing allows us to learn continuously; as more data is delivered from other vehicles on the road, Algorithms can further be improved, passing time.

Artificial intelligence is a foundational technology that underlies many functions in autonomous vehicles. Conversely, AI uses algorithms to process data captured by sensors and complete movement-related tasks like navigation, object detection, and predictive modeling. Machine learning helps AVs learn over time based on past experiences and modify their behaviors. AVs utilize AI to find the fastest routes for navigation, considering the real-time traffic situation and any road closure or construction zones. Object detection algorithms enable vehicles to identify pedestrians, cyclists, and obstacles accurately, which is vital for split-second driving decisions that keep cars from colliding with others. AI predictive analytics take AV further, imagining possible hazards before they develop. An autonomous car, for instance, could learn from patterns in driving behavior or environmental conditions to anticipate when a driverless vehicle will have to decelerate or change lanes due to an approaching obstruction.

Also, integrating sensors and actuators, reliable communication networks, cloud computing capabilities, and advanced AI algorithms has allowed AVs to run safely in complicated arenas. The continued evolution of these technologies will provide an impetus for more intelligent transport systems that can improve mobility with lower congestion levels and environmental impacts. All these interacting components are crucial to the future of autonomous vehicles, working together and communicating across an IoT framework. This enhances the functionality of each vehicle and serves broader transportation objectives like enhanced safety on roads and emission reduction via driving pattern optimization. These elements will be vital to an expansion into self-driving cars that we sprint towards in the next decade, and getting to grips with them is necessary if we want to maximize their benefits in making our transport systems sustainable. Connected nature IoT is revolutionizing the autonomous vehicles (AVs) arena by onboarding multiple new functionalities that improve navigability, prediction, and energy management.

IOT-DRIVEN FUNCTIONALITIES IN AVS

High Cost Another challenge is related to the cost of AV infrastructure and IoT deployment. Autonomous driving and intelligent transport systems, based on IoT technology, will need a massive investment in 5G networks, AI cloud computing, and intelligent road infrastructure. Less developed and rural areas might be digitally disconnected with limited road networks where the transition and adoption of IoT-driven AV might become a long-term process, potentially creating inequities in sustainable mobility access. To tackle this issue, policymakers, technology firms, and governments need to work hand in hand to create affordable and scalable solutions that can provide equal access to smart transportation systems, irrespective of the geographical region. The widespread adoption of autonomous and IoT-powered transportation hinges on public acceptance and trust, which, yet again, are two of the major barriers. Many people worry about AV reliability, ethical dilemmas during emergencies, and the loss of driving jobs. This is where effective public awareness campaigns, transparent regulatory policies, and pilot programs that demonstrate the environmental and safety benefits of AVs will help bridge the gap in acceptance between technological advancements and society.

The future of sustainable development, however, will be closely linked with innovations available through IoT and autonomous vehicles. The integration of AI, IoT, 5G, and renewable energy technologies into AVs will only grow the momentum toward more sustainable, intelligent mobility systems as these technologies themselves develop. Collectively, governments, corporations, and research institutions must formulate ethical, inclusive, and environmentally sound guidelines by which the equitable distribution of IoT-driven AV solutions may progress. In conclusion, the success of sustainable transportation lies in achieving digital inclusion (entire society must be connected and contribute to making smarter decisions), having electric and green energy solutions as priority (entire world is looking for ways to stay green so this must be an initial agenda), coupling the use of IoT and AI in these solutions in an ethical way (everything IoT tends to be expensive so ethics comes in picture here as well, but people need to be convinced). IoT-enabled autonomous vehicles serve as essential facilitators toward sustainable development, resonating with worldwide initiatives to downsize pollution, streamline mobility, and build intelligent urban environments. With real-time data connectivity, intelligent decision-making, and eco-friendly innovations, AVs help create safer roads, more efficient transportation systems, and cleaner cities. Addressing technical, regulatory, and social challenges will be critical to ensuring IoT and AVs realize their promise as transformative forces for global sustainability. With the future of intelligent transportation on the horizon, the data-driven age is poised to splendor investment in AI, IoT, sustainable mobility, and will pave the way for a sustainable world of tomorrow will involve an environmental conservation, urban development, and booming smart cities in which, IoT-integrated AVs will act as an inseparable part.

AUTONOMOUS NAVIGATION

Navigation is among the most important applications of IoT in autonomous vehicles. Thanks to IoT technologies, route optimization, obstacle avoidance, and adaptive cruising have become straightforward for an AV to help it move in complicated environments more accurately. These vehicles consistently collect information about their environment through sensors, including, but not limited to, cameras, LiDAR (that uses lasers), and radar. The gathered data is then processed by sophisticated algorithms that help the vehicle decide how it should drive. For example, an AV approaching an intersection can access V2I (Vehicle-to-Infrastructure) communication to get real-time information about traffic signals and road conditions. This means it can speed up or even change lanes if it notices something in its way, which will be at least a red light. In addition, IoT allows vehicles to communicate with each other (V2V or Vehicle-to-Vehicle communication), exchanging essential information such as speed, direction, and intention. This connectivity not only increases situational awareness but also provides for vehicle-to-vehicle coordination, whereby multiple vehicles can plan paths to improve the safety of movement on roadways. Furthermore, with the help of IoT, adaptive cruising becomes possible by allowing an AV to adjust its speed according to traffic conditions. An example would be when a vehicle receives information that congestion is ahead through V2V or V2I communication and automatically begins to slow or reroute away from the potential delay. Being able to vary and respond dynamically improves the experience of traveling for passengers and the general efficiency of traffic flow (Elahi et al., 2020).

Predictive Analytics: IoT also fuels one more significant functionality in autonomous vehicles, i.e., predictive analytics. By analyzing large sets of data collected by onboard sensors and external databases, AVs can gain insight into such traffic patterns or the conditions for maintenance. This ability is critical to improving operational efficiency and extending vehicle life. Trend forecasting: IoT-powered systems can mine historical data in conjunction with real-time information to predict traffic congestion trends at certain times of the day or week. If they recognize these patterns, AV can pick the best and fastest routes to travel, which will take less time and save a lot of fuel. Further, predictive analytics will assist in identifying risks before they reach the escalated stage. For example, if the car detects (using V2V) that a neighboring vehicle is suddenly braking hard, it may suspect that a traffic jam is developing ahead and plot an alternate route. IoT also helps monitor the health of vehicle parts in real time, which is an essential aspect of vehicle maintenance. Sensors can be attached to things, from the engine temperature and fluid levels to tire pressure and battery status. By continually analyzing this data, predictive maintenance algorithms can identify when a component will fail or need service (Lecue, 2020). This approach prevents breakdowns and lowers maintenance costs, as work can be conducted before significant problems occur.

Smart Energy Management: Another domain of significant contribution includes efficient management of energy, which IoT supplements well in autonomous vehicles. As EVs are adopted, energy management has become a key enabler in optimizing performance and range. IoT can provide dynamic resource allocation to reduce energy consumption according to real-time conditions. For example, if you have IoT embedded into the AV, it can even check its energy consumption pattern and adjust its driving accordingly. For instance, if the car detects an upcoming steep hill or congested traffic conditions needing additional power consumption, it can adapt acceleration and speed to save energy. Also, via V2N (Vehicle-to-Network), an AV can determine whether a charging station near its path is open and how much the energy costs are. That ability includes the option to plan charging stops along longer trips cost-effectively. IoT enables optimal route planning depending upon energy efficiency concerning terrain and traffic conditions for an AV's best path (Kim et al., 2021).

For instance, if a route takes the vehicle through hilly terrain, energy can be saved by taking a flatter route with less resistance. Fine point: IoT provides innovative features for autonomous vehicles, changing the way they understand the road ahead, predict when maintenance is needed, and manage energy consumption. High-tech sensors enable autonomous driving through real-time data analysis while communicating with other cars and infrastructure components. Predictive analytics increase operational efficiency by predicting upcoming traffic patterns and determining maintenance needs before a minor issue becomes a more significant concern. Among these transitions, the revolution in the Internet of Things (IoT) enabled autonomous vehicles (AVs), that offer minute and impeccable features to eliminate the significant challenges in urban mobility. Decreased traffic congestion, public and shared transportation facilitation, and greater mobility for all population segments will lead to greener cities and communities using IoT-centric AVs.

IMPACT ON SUSTAINABLE CITIES AND COMMUNITIES

The most crucial impact is reducing traffic congestion. Static routes and schedules mean traditional trucks often fuel gridlock in urban environments. On the other hand, IoT-AVs can talk to each other and infrastructure systems, improving traffic flow. AVs can be informed in real time about the status of traffic, accidents, or road closures, for example, via V2X (Vehicle-to-Everything) communication. With this information, they can alter their real-time routes to avoid congested zones, effectively reducing delays. The resultant impact is a downtrend in the overall traffic volume, thus ensuring easy travel for all road users. Further, IoT-based intelligent traffic management systems also focus on prioritizing specific vehicle types during peak hours. For example, traffic signals can be configured to prioritize buses or ambulances, reaching their destinations on time and minimizing waiting times for other road users. Not only does this smart coordination reduce traffic jams, but it also increases road safety by reducing the chances of accidents due to aggressive driving (Guo et al., 2021).

Urban Mobility Redefined: Beyond alleviating congestion, AVs also facilitate public and shared transportation necessary for sustainable urban growth when aided by IoT. AV adoption is integrated with public transport systems to create a more cohesive transportation network. For instance, autonomous shuttles can work as first and last-mile transports, which, for example, link passengers from their properties or other destinations (i.e., offices) to public transport hubs. This aspect is beneficial in regions that do not have practical public transport or struggling bus services. AVs provide easy access to public transport; thus, more people use public transportation than just relying on private vehicles. In addition, IoT-based shared mobility solutions can help decrease the number of vehicles on the road by a considerable percentage (Nickel et al., 2015).

AVs integrated into ride-sharing platforms enable users to share rides with others in the same direction. Among these is the ability to make cities easier to navigate for people in all circumstances, which IoT-driven AVs do right. Almost everyone can find some barrier to traditional means of getting around, whether it be age, disability, or simply economics. IoT can enable the development of autonomous vehicles that provide services tailored to these underserved demographics. For example, AVs could be developed with equipment that meets the needs of riders with disabilities (e.g., a wheelchair-accessible van and voice-activated controls). Additionally, IoT systems can offer transportation services by personalizing and sensing user choices and travel patterns. This ability allows AVs to provide personalized directions and times around their needs, offering mobility for all. The IoT in AVs can touch the social equity aspects of urban mobility. Shared mobility enables affordable transportation options for low-income residents. When deployed through ride-sharing platforms or community shuttles, it broadens access to services, such as healthcare, education, and employment opportunities. Such improvement in access spreads social recognition and grants independence to those who, otherwise, would have a mobility problem (Nickel et al., 2015).

This facilitates the full benefits of IoT-enabled V2X AVs in urban mobility, but the City must invest in smart infrastructure to support these technologies. This involves the installation of sensors and communication devices in urban habitats, such as smart traffic lights, connected road signs, and active monitoring systems that enable vehicles to interact easily through infrastructure. Smart Infrastructure makes AVs more efficient and ensures that all road/path users reach their destination safely. These technologies, for instance, real-time environmental monitoring systems, can help identify hazardous conditions like icy roads or impaired visibility from fog (Yu et al., 2019). This enables drivers (or autonomous systems) to change their driving behavior based on the information transmitted to them by the connected vehicles. Further, adding IoT technologies to the current public transport systems can ensure high efficiency and commuting enjoyability. With real-time vehicle tracking, passengers can use mobile apps to view digital displays at stations and stop for accurate arrival times. IoT predictive maintenance ensures that a transit vehicle will be continuously monitored for performance issues, which helps with fleet reliability and minimizes downtime (Liu et al., 2023).

Environmental Benefits

Integrating the Internet of Things (IoT) into transportation systems, mainly by deploying autonomous vehicles (AVs), offers significant environmental benefits that contribute to sustainability and improved urban living. One of the most notable advantages is the reduction in emissions. IoT technologies optimize fuel consumption by enabling real-time data analysis and communication between vehicles and infrastructure. For instance, AVs with IoT sensors can adjust their driving patterns based on traffic conditions, reducing unnecessary acceleration and braking and lowering fuel consumption. Additionally, the promotion of electric vehicle (EV) adoption is facilitated by IoT, as smart charging systems can optimize charging times based on grid demand and renewable energy availability. This encourages the use of EVs and helps reduce overall greenhouse gas emissions associated with traditional combustion engine vehicles. Furthermore, IoT enables smart energy utilization by integrating transportation systems with renewable energy sources such as solar and wind. For example, charging stations for electric vehicles can be strategically placed in locations that maximize access to renewable energy, allowing EVs to be charged during peak production times. This integration supports the transition to cleaner energy sources and enhances the resilience of urban energy systems by reducing reliance on fossil fuels. As cities increasingly adopt these technologies, they can create a more sustainable transportation ecosystem that aligns with broader environmental goals.

Improve Urban Planning

In addition to ecological benefits, IoT-driven AVs significantly improve urban planning by facilitating dynamic traffic systems that enhance congestion control. Real-time traffic monitoring allows city planners to analyze traffic patterns and make informed decisions about infrastructure improvements or adjustments to traffic signal timings. This proactive approach helps mitigate congestion and ensures smoother traffic flow throughout urban areas. Moreover, the data collected from IoT-enabled vehicles can inform future urban planning initiatives, enabling cities to design more efficient transportation networks that adapt to changing mobility needs. Optimized land use is another critical aspect of improved urban planning facilitated by IoT technologies. With the rise of shared mobility solutions and autonomous vehicles, the demand for extensive parking infrastructure decreases. Cities can repurpose underutilized parking spaces for green areas, public amenities, or mixed-use developments that enhance community engagement and quality of life. Urban environments can become more pedestrian-friendly and conducive to sustainable living by reducing the need for large parking lots and garages (Tariq, 2025).

Social Equity and Inclusion

are also key considerations in the deployment of IoT-driven AVs. These technologies can potentially enhance mobility for underserved communities and individuals with disabilities, thereby creating more equitable transportation systems. For instance, autonomous shuttles can be deployed in areas with limited public transport options, providing reliable access to essential services such as healthcare and education. Additionally, AVs can be designed with features that accommodate passengers with disabilities, ensuring everyone can access safe and efficient transportation. By leveraging IoT technologies, cities can develop transportation solutions prioritizing equity and inclusivity. This focus on social equity improves individual mobility and fosters community cohesion by ensuring that all residents have equal access to opportunities and resources. As cities continue to evolve in response to demographic changes and environmental challenges, embracing IoT-driven solutions will be essential for creating sustainable urban environments that benefit all members of society. The ecological benefits of IoT-enabled autonomous vehicles extend beyond reduced emissions and smart energy utilization; they also encompass improved urban planning and enhanced social equity. By optimizing fuel consumption and promoting renewable energy integration, these technologies contribute significantly to sustainability goals. Additionally, dynamic traffic systems facilitate better congestion management, while optimized land use allows for more vibrant urban spaces. Most importantly, IoT-driven AVs enhance mobility for underserved communities and individuals with disabilities, paving the way for equitable transportation systems that empower all citizens (Kraft et al., 2023).

CHALLENGES AND BARRIERS

However, using Internet of Things (IoT) technologies in autonomous vehicles (AVs) also presents several challenges and barriers that must be overcome before they can be successfully used at their full potential. These challenges can be divided into four general types: technological problems, security and privacy concerns, regulatory and ethical barriers, and social acceptance obstacles. These domains present unique challenges that stakeholders must overcome to meet or exceed the needs of transportation systems embedded with IoT-enabled AVs. One of the significant technological challenges is handling a massive volume of IoT data as it is generated with minimum latency, which is imperative for making timely decisions and operational efficiency. The high-volume data can flood existing systems, resulting in delays and possibly failure of crucial functions like navigation and obstacle avoidance. To overcome this challenge, it will be essential to develop robust data management frameworks that ensure relevant information is processed promptly and that vehicle performance does not suffer. Seamlessly integrating multiple IoT devices and systems is another critical area with interoperability issues. Given that AVs depend on various sensors and communication technologies, they must operate interoperable as seamless components for safe functioning. Each manufacturer has its protocol and standards, so integration remains complicated. Interoperability is the ability of heterogeneous systems and devices to communicate with one another smoothly, which can be done when standards are set with an agenda that all stakeholders should adhere to (Li et al., 2023).

Security and Privacy Concerns

IoT-powered autonomous vehicles are secure and private. Cyberattacks on vehicle systems and IoT networks represent a significant risk to vehicle safety and users' privacy. The more connected AVs become, the more susceptible they are to hacks used to control their systems or expose sensitive user data. Putting cyber security measures in place is essential to prevent hackers and other unwanted actions at the wheel of these vehicles. Data privacy is an ongoing issue with user information as well. Autonomous vehicles gather and store a wealth of data related to their owners, including travel frequency, destination addresses, personal habits, etc. All this information must be handled carefully to avoid misusing or sharing with third parties without authorization. On the one hand, we need clear policies on collecting, storing, and sharing data to ensure user privacy while maintaining IoT functionality (Schier et al., 2023).

Obstacles of Regulation and Ethics

Another big hurdle comes from regulations, as autonomous vehicles still face many unknown variables. The absence of overarching frameworks regulating the deployment of AVs has caused uncertainty for many manufacturers and operators. Governments must establish clear regulations regarding safety, liability, insurance, and operational parameters for AVs operating on public roadways. These regulations are still catching up to the widespread use of autonomous vehicles; however, without them, we may likely see a major slowdown in their expansive implementation. Another set of ethical dilemmas relates to decision-making in emergencies (Mahon et al., 2020).

Societal Acceptance

Social acceptance of IoT-based autonomous vehicles in transport systems. However, skepticism regarding AV safety and reliability in the population remains a principal hurdle to ubiquitous uptake. Many still fear the dangers of giving up control of machines, especially in complex driving situations. Instilling trust in AV tech will require transparency over safety practices, testing procedures, and real-world performance data. Compounding the problem is an aversion to changing existing transportation systems. Drivers are mostly used to traditional driving and may be resistant to other technologies that upend standard practices; public acceptance of AVs will also depend on effective education campaigns that educate the public on the benefits of AVs, including reductions in traffic congestion, safety records, and mobility expansion opportunities for underserved populations (Xiao et al., 2023).

There is a more profound challenge, though: there are technological hurdles associated with data overload and interoperability that may need bespoke techno-political solutions to optimize the efficiency of these systems without compromising safety. Robust cybersecurity measures and clear policies around data handling should be implemented to ensure security and privacy concerns are paramount. The regulatory frameworks keep pace with technology to clarify boundaries for manufacturers and operators while balancing the ethical problems posed by autonomous driving. Social acceptance will decide whether IoT-powered AVs flourish or dwindle; thus, community outreach efforts framed by education and clear communication are necessary to offset skepticism. Suppose stakeholders can work together to solve these challenges. In that case, IoT-based self-driving cars can turn from a dream into reality, making our urban mobility more sustainable and inclusive for everyone. The road to mass adoption may be paved with hurdles. Still, with united action from the auto industry, policymakers, technology leaders, and society, we can usher in a new era of transportation that uses the power of IoT for safer roads and smarter cities.

STRATEGIC SOLUTIONS

The statutory expansion of Internet of Things (IoT) technologies into autonomous vehicles (AVs) represents a significant opportunity for transformative new urban mobility and improved environmental sustainability, potentially providing an economic stimulus throughout the newly digitized AV IoT landscape. The future of AVs is bright, but challenges and barriers operational, legislative, and societal must be overcome to realize their full benefits; strategic solutions will be required. These solutions include technological development, security and privacy enhancement, policy and governance frameworks, and stakeholder engagement. Furthermore, the evolution of urban IoT and zero-emission mobility outlook will provide guidelines for a sustainable transportation system (Liu et al., 2020).

Advancing Technology

The role of IoT at AVs is to introduce a robust IoT framework that stores data on central cloud servers and implements edge computing. Edge computing enables data to be processed closer to the point where it is generated, in this case, onboard the vehicle itself or other nearby infrastructure rather than implementing a ubiquitous cloud approach. This dramatically minimizes the latency, processes data faster, and makes real-time decisions. For AVs, real-time processing of essential information identifying road conditions, obstacles, and traffic patterns can be nearly instantaneous, enabling rapid responses that improve safety and efficiency. In addition, artificial intelligence (AI) algorithms must be enhanced to function as better decision-making components in AVs. Most current AI systems depend on machine learning algorithms that need massive training data to work correctly. Through retooling these systems to understand both the intricacies of nuanced driving behavior and the need for quick-to-execute ethical decision-making, AVs may be able to traverse challenging driving environments with improved safety. Setting aside cutting-edge research, deep learning techniques are advancing. They will allow for improved object recognition so that cars know when a pedestrian, bicycle, or other road user is nearby (Liu et al., 2020). Also, AI programmers can integrate ethical decision-making frameworks into algorithms, thus allowing AVs to make better decisions in the context of moral dilemmas, such as prioritizing passenger or pedestrian safety.

Enhancing Security and Privacy

AVs have become increasingly connected using IoT technologies, reinforcing security, essential to prevent cyberattacks. Encrypted communication protocols for AVs secure data between vehicles and infrastructure. Such protocols should ensure that all communications are behind a safe wall from interception or tampering by malicious actors. Furthermore, it is imperative to apply regular updates in the software, considering vulnerabilities to find paths through which they enter an organization daily. Another essential aspect of this concern in the realm of IoT-driven AVs is associated with data privacy. Considering the data these vehicles gather on users' travel behavior and preferences, comprehensive data governance frameworks like the General Data Protection Regulation (GDPR) must be adopted. These frameworks should provide guidelines for data collection, storage, and sharing practices while giving users control over their personal information. Transparency in data usage will strengthen people's confidence in AV technology.

Policy and Governance

Deployment of robots on public roads must be done safely, and the principal vehicle for ensuring this will remain effective policy and governance strategies. Creating consistency around how AVs can operate globally will not only broaden innovation but ensure that they do so safely. This could include safety testing protocols, liability frameworks, and other AV deployment standards. The public adoption of AVs can also be incentivized with subsidies and wide awareness campaigns. Financial incentives for purchasing or using autonomous vehicles can be offered by governments/authorities to expedite the penetration into the market. Further, as general awareness moves toward understanding the benefits of AVs (i.e., less traffic congestion, safety record benefits, and more incredible mobility options), this skepticism will wane among potential users (Bonatti et al., 2019).

Promoting Collaboration

However, successfully integrating such IoT-driven autonomous vehicle systems into urban infrastructures requires collaboration among the various stakeholders, including governments, technology companies, and urban planners. Collaboration between the public and private sectors can help cities establish effective integrated transportation plans for new AV technology that builds on existing infrastructure. In parallel with these relationships, urban planners must collaborate with technology providers to construct smart transportation systems responsive to the autonomous fleet while maintaining a steady balance in traffic flow. It allows the implementation of smart traffic control systems by facilitating communication using real-time data from connected vehicles as signals to govern the flows of vehicular behavior and adjust traffic lights automatically depending on current conditions. Also, by including communities as stakeholders in the planning process, we can ensure transportation solutions serve all residents. Public consultations would provide insight into a community's preferred mode of mobility options and establish communication on barriers to adoption (Zheng et al., 2022).

FUTURE PERSPECTIVES

In the future, the IoT-powered evolution of urban systems will dominate transportation systems in cities across the globe. The concept of entirely autonomous smart cities entails interconnected transportation networks that include AVs, energy management systems, and waste management solutions (Werner et al., 2021). Thus, real-time data sharing between systems across sectors would allow for the optimal allocation of shared resources in such cities. For instance, connected systems such as smart transportation and smart energy grids can coordinate electric vehicle (EV) charging by directing them toward charge points at low-demand hours or high renewable generation periods. This coupling of transport and energy management will reduce greenhouse gas emissions while maximizing overall system efficiency. Integrating new technologies like blockchain into the AV ecosystems has a huge potential for secure transactions and mobility services. For example, it can offer a decentralized ledger to track vehicle ownership history or enable secure payment without compromising user privacy for ride-sharing services.

Enabling zero-emission mobility is an essential milestone toward sustainable transportation systems. IoT Enablement. With autonomous vehicle fleets, real-time monitoring of energy consumption patterns can be examined with grid-centric demand forecasts to leverage renewable asset sources optimally. This ability to analyze data permits fleet operators to know the best time for vehicles to charge when renewable energy is available. Similarly, the widespread use of green hydrogen fuel cells and electric propulsion systems will be a key factor in decreasing dependency on fossil fuels in transportation networks. Continued improvements in battery technology, energy density, and fast charging time of EVs will mean that hybrid combination with public transport systems will become more practical (Sun et al., 2023). Secondly, ensuring that new infrastructure accommodates alternative fuel sources, such as hydrogen refueling stations, will complement the advancement of zero-emission mobility solutions throughout cities.

CONCLUSION

The impact of IoT technologies integrated into Autonomous vehicles (AVs) and intelligent mobility can significantly contribute to building sustainable cities and communities. With urban environments grappling with growing needs related to congestion, pollution, and accessibility in the face of a climate crisis, these IoT-enabled AVs provide an integrated solution that mitigates these challenges while fostering environmental sustainability and social equity. Such a transformative approach transforms transport systems and improves the overall quality of urban life. One of the most significant benefits of intelligent mobility is reduced traffic congestion. AVs with IoT capabilities can talk to one another and infrastructure systems to improve route optimization and enhance overall traffic flow. These vehicles use real-time data to make decisions to minimize delays and optimize time. Fewer driver hours reduced fuel consumption and emissions, benefiting air quality and climate objectives. Furthermore, using AVs in public transportation can reduce demand on an already strained system by offering alternatives to private vehicle ownership. Many may overlook the other important part of intelligent mobility: social equity. IoT-driven AVs can help make society more accessible for those who never could travel freely: underserved communities and individuals with disabilities. These technologies can also help fill mobility gaps by providing new forms of dependable transport in places where public transit is unfeasible or unavailable. Reducing the divides between people offers access to efficient transportation for diverse groups in our society, which facilitates the inclusion of all demographics and allows everyone to participate in their communities fully. However, several challenges must be met to make intelligent mobility successful. Technological barriers, data overload, and interoperability challenges must be removed to ensure interaction and integration across diverse IoT devices and systems. Moreover, essential infrastructure is required in the next step to enable robust cybersecurity and strong data governance frameworks to preserve user security and privacy while allowing the platform to operate efficiently. The regulatory frameworks had to be updated as well. Governments must develop clear guidelines for AV deployment, including safety standards, liability questions, and the ethics of emergency decision-making. Establishing trust and ensuring residents' needs are met by transportation solutions that benefit everyone will depend on engaging communities and stakeholders. The global convergence between technology and stakeholders should be brought together for future studies by marrying newer technologies with IoT-based autonomous mobility solution spaces. Also, intelligent mobility is a disruptive opportunity to shape sustainable cities and communities. IoT and automatic vehicles are the brains we need to revamp transportation systems that tackle some of our most pressing urban challenges, such as pollution and social justice, but also demand environmental stewardship and socially responsible action. Technological limitations need to be surmounted, regulatory mechanisms built, and the public educated and engaged with these technologies so that they are accepted. By bringing together the efforts of governments, technology providers, urban planners, and communities, we can harness the full power of intelligent mobility for a better, more equitable future.

REFERENCES

Bonatti, P. A., Decker, S., Polleres, A., & Presutti, V. (2019). AIs: New directions for knowledge representation on the semantic web (Dagstuhl seminar 18371). *Dagstuhl Reports, 8*(9). Schloss Dagstuhl–Leibniz-Zentrum für Informatik.

Dai, S., Liang, Y., Liu, S., Wang, Y., Shao, W., Lin, X., & Feng, X. (2018, March). Learning entity and relation embeddings with entity description for AI completion. In *2018 2nd International Conference on Artificial Intelligence: Technologies and Applications (ICAITA 2018)* (pp. 194–197). Atlantis Press.

Elahi, M. F., Luo, X., & Tian, R. (2020, July). A framework for modeling AIs via processing natural descriptions of vehicle-pedestrian interactions. In *International Conference on Human-Computer Interaction* (pp. 40–50). Springer.

Guo, R., Hong, Z., & Xue, X. (2021). Reward function design via human AI and inverse reinforcement learning for intelligent driving (No. 2021-01-0180). *SAE Technical Paper.*

Halilaj, L., Dindorkar, I., Lüttin, J., & Rothermel, S. (2021). An AI-based approach for situation comprehension in driving scenarios. In *The Semantic Web: 18th International Conference, ESWC 2021, Virtual Event, June 6–10, 2021* [Springer.]. *Proceedings*, *18*, 699–716.

Halilaj, L., Luettin, J., Henson, C., & Monka, S. (2022, September). AIs for automated driving. In *2022 IEEE Fifth International Conference on Artificial Intelligence and Knowledge Engineering (AIKE)* (pp. 98–105). IEEE.

Halilaj, L., Luettin, J., Monka, S., Henson, C., & Schmid, S. (2023). AI-based integration of autonomous driving datasets. *International Journal of Semantic Computing*, *17*(2), 249–271. DOI: 10.1142/S1793351X23600048

Huang, H., Sun, L., Du, B., Liu, C., Lv, W., & Xiong, H. (2021, August). Representation learning on AIs for node importance estimation. In *Proceedings of the 27th ACM SIGKDD Conference on Knowledge Discovery & Data Mining* (pp. 646–655).

Kim, J. E., Henson, C., Huang, K., Tran, T. A., & Lin, W. Y. (2021). Accelerating road sign ground truth construction with AI and machine learning. In *Intelligent Computing:Proceedings of the 2021 Computing Conference,* Volume 2 (pp. 325–340). Springer.

Kraft, M., Bai, J., Mosbach, S., Taylor, C., Karan, D., Lee, K. F., ... & Lapkin, A. (2023). From platform to AI: Distributed self-driving laboratories.

Lecue, F. (2020). On the role of AIs in explainable AI. *Semantic Web, 11*(1), 41–51. DOI: 10.3233/SW-190374

Li, X., Liu, J., Li, J., Yu, W., Cao, Z., Qiu, S., . . . Jiao, X. (2023, August). ML structure-based implicit risk reasoning for long-tail scenarios of automated driving. In *2023 4th International Conference on Big Data, Artificial Intelligence and Internet of Things Engineering (ICBAIE)* (pp. 415–420). IEEE.

Liu, Q., Li, X., Tang, Y., Gao, X., Yang, F., & Li, Z. (2023). ML reinforcement learning-based decision-making technology for connected and autonomous vehicles: Framework, review, and future trends. *Sensors (Basel), 23*(19), 8229. DOI: 10.3390/s23198229 PMID: 37837063

Luettin, J., Monka, S., Henson, C., & Halilaj, L. (2022, November). A survey on AI-based methods for automated driving. In *Iberoamerican AIs and Semantic Web Conference* (pp. 16–31). Springer.

Mahon, L., Giunchiglia, E., Li, B., & Lukasiewicz, T. (2020, December). AI extraction from videos. In *2020 19th IEEE International Conference on Machine Learning and Applications (ICMLA)* (pp. 25–32). IEEE.

Nickel, M., Murphy, K., Tresp, V., & Gabrilovich, E. (2015). A review of relational machine learning for AIs. *Proceedings of the IEEE, 104*(1), 11–33. DOI: 10.1109/JPROC.2015.2483592

Schier, M., Reinders, C., & Rosenhahn, B. (2023). Deep reinforcement learning for autonomous driving using high-level heterogeneous ML representations. In *International Conference on Robotics and Automation (ICRA).*

Sun, R., Lingrand, D., & Precioso, F. (2023). Exploring the road ML in trajectory forecasting for autonomous driving. In *Proceedings of the IEEE/CVF International Conference on Computer Vision* (pp. 71–80).

Tan, Z., Zhao, X., Fang, Y., Ge, B., & Xiao, W. (2018). AI representation via similarity-based embedding. *Scientific Programming*, •••, 2018.

Tang, X., Sun, T., Zhu, R., & Wang, S. (2021, January). CKG: Dynamic representation based on context and AI. In *2020 25th International Conference on Pattern Recognition (ICPR)* (pp. 2889–2895). IEEE.

Tariq, M. U. (2024). Generative AI in Curriculum Development in Higher Education. In Fields, Z. (Ed.), *Impacts of Generative AI on Creativity in Higher Education* (pp. 227–258). IGI Global., DOI: 10.4018/979-8-3693-2418-9.ch009

Tariq, M. U. (2025). Leveraging Artificial Intelligence to Enhance Diversity and Drive Business Innovation. In Margaretha van der Poll, H., Andrew van der Poll, J., & Ngwakwe, C. (Eds.), *Diversity, AI, and Sustainability for Financial Growth* (pp. 79–96). IGI Global Scientific Publishing., DOI: 10.4018/979-8-3693-6011-8.ch004

Tariq, M. U. (2025). *AI-Driven Learning and Engagement in Higher Education*. IGI Global., DOI: 10.4018/979-8-3693-4074-5

Tariq, M. U. (2025). Quantum Algorithms and Predictive Analytics: Revolutionizing Consumer Insight. In Gajbhiye, P., Kannan, H., Rodriguez, R., & Rojas-Méndez, J. (Eds.), *The Quantum AI Era of Neuromarketing* (pp. 173–202). IGI Global Scientific Publishing., DOI: 10.4018/979-8-3693-7673-7.ch008

Tariq, M. U., & Sergio, R. (Eds.). (2025). *Cases on AI-Driven Solutions to Environmental Challenges*. IGI Global, 2025. https://doi.org/DOI: 10.4018/979-8-3693-7483-2

Tariq, M. U., & Sergio, R. (Eds.). (2025). *Convergence of AI, Education, and Business for Sustainability*. IGI Global., DOI: 10.4018/979-8-3373-1917-9

Werner, S., Rettinger, A., Halilaj, L., & Lüttin, J. (2021). RETRA: Recurrent transformers for learning temporally contextualized AI embeddings. In *The Semantic Web: 18th International Conference, ESWC 2021, Virtual Event, June 6–10, 2021* [Springer.]. *Proceedings, 18*, 425–440.

Xiao, D., Dianati, M., Geiger, W. G., & Woodman, R. (2023). Review of ML-based hazardous event detection methods for autonomous driving systems. *IEEE Transactions on Intelligent Transportation Systems, 24*(5), 4697–4715. DOI: 10.1109/TITS.2023.3240104

Yu, C., Wang, X., Xu, X., Zhang, M., Ge, H., Ren, J., Sun, L., Chen, B., & Tan, G. (2019). Distributed multiagent coordinated learning for autonomous driving in highways based on dynamic coordination MLs. *IEEE Transactions on Intelligent Transportation Systems, 21*(2), 735–748. DOI: 10.1109/TITS.2019.2893683

Zheng, Z., Zhou, B., Zhou, D., Soylu, A., & Kharlamov, E. (2022, October). Executable AI for transparent machine learning in welding monitoring at Bosch. In *Proceedings of the 31st ACM International Conference on Information & Knowledge Management* (pp. 5102–5103).

KEY TERMS AND DEFINITIONS

Intelligent Mobility: Technology-driven transportation system which integrates IoT, AI and smart infrastructure to enhance the efficiency, sustainability and accessibility of urban mobility.

Autonomous Vehicles: Self-driving vehicles equipped with AI, machine learning, LiDAR, radar and IoT sensors that operate without human intervention.

Vehicle-to-Everything (V2X) Communication: IoT-enabled system that allows AVs to communicate with other vehicles (V2V), infrastructure (V2I), pedestrians (V2P), and networks (V2N) to ensure safer and more efficient transportation. V2X enhances real-time data exchange, predictive traffic management, and eco-friendly driving practices.

Sustainable Urban Mobility: Transportation model that prioritizes low-carbon, energy-efficient and smart mobility solutions to reduce environmental impact while ensuring accessibility.

Mobility-as-a-Service: Digital platform that combines multiple transportation services, including autonomous ride-sharing, public transit, micro-mobility and bike-sharing, into a single accessible system.

Chapter 10
Real-Time Vehicle Detection Embedded System for Self-Driving Cars in Challenging Rainy Conditions

Hung Kiem Nguyen
https://orcid.org/0000-0003-3417-3447
VNU University of Engineering and Technology, Vietnam

Tuan Huu Luong
https://orcid.org/0009-0009-6264-5289
VNU University of Engineering and Technology, Vietnam

ABSTRACT

In this chapter, we present researching and developing an embedded system for autonomous vehicles that detect objects in rainy conditions. We investigated and evaluated popular deep learning models for object detection to identify the most suitable algorithm for real-time embedded systems. The results indicate that the SSD MobileNet-v2 FPNLite320×320 model, after quantization, is the most appropriate algorithm for deployment on embedded systems. Additionally, we supplemented the dataset with 1,221 rain images collected from real-world dashcam videos, enriching the training dataset. We successfully designed an embedded system capable of detecting traffic objects in rainy conditions. Testing under various rainy weather conditions showed that the system could accurately identify traffic objects within a range of 1 to 15 meters. The system achieves up to 6.78fps, allowing for timely reactions to dangerous situations and reducing the risk of accidents. Furthermore,

DOI: 10.4018/979-8-3373-4332-7.ch010

the system consumes very low memory, with a small size of only 3.57 MB and utilizing 11.9% of the total system RAM.

INTRODUCTION

This chapter describes an embedded system design that ensures safe operations under rainy conditions for autonomous vehicles. This detection system accurately identifies all vehicles traveling on the road, especially those that influence the autonomous vehicle operating within its traffic zone, including motorcycles, as well as cars, trucks, and buses. Autonomous vehicles use traffic object data to determine relative distances between these vehicles as the system transmits this information to control systems. Using acquired information, the control system generates suitable decisions that protect autonomous vehicle safety by forcing speed reductions or requiring emergency stops or warning drivers of potential risks. The system employs an architecture that needs limited resources, together with low power consumption capability and seamless adaptability into autonomous vehicle frameworks. The primary contributions of this work consist of two points. The design of our embedded system used a Raspberry Pi 5 Board and IMX219-83 stereo camera module, which identifies motorcycles, cars, buses, trucks, and related vehicles accurately at distances between 1 to 15 meters in rainy weather conditions. The system implements the quantized SSD MobileNet-v2 FPNLite320×320 model that received training with 4154 images obtained from the BDD100K dataset, besides additional data acquired from dashcams found on the internet. System operation at 6.78 frames per second provides timely responses to dangers, which decreases accident possibilities. The system operates at low memory consumption because its file size measures 3.57 MB while it utilizes 11.9% of the system's total RAM. The implemented system preserves vehicle energy while maintaining the overall performance and operational abilities of modern autonomous vehicles. We believe the proposed system will enhance the efficiency and safety of autonomous vehicles when operating during rainfall. The remainder of this chapter is organized as follows. Section 2 presents research and evaluation of various methods, particularly neural networks, used for detecting vehicles in rainy weather conditions. Based on the evaluation results, we select the most suitable method for developing a system to recognize vehicles under such conditions. After selecting the optimal method, we develop the software and implement it on the Raspberry Pi 5 embedded computer system in Section 3, followed by real-world testing and evaluation of the embedded system in Section 4. Section 5 provides conclusions.

BACKGROUND

Technological advancements make autonomous vehicles an emerging significant pattern in road transportation (Stoma, 2021). The vehicles deploy many advanced sensors, including LiDAR sensors, radars, and cameras, to detect their environment to operate autonomously with minimum human oversight. The different sensors serve purpose-specific roles, from obstacle detection to traffic signal interpretation (Vargas, 2021). The combination of multiple sensor data enables autonomous vehicles to develop comprehensive environmental awareness. Technology enables safe and efficient operation in complex traffic conditions by delivering precise decisions either without or with minimal operator intervention.

What Are Autonomous Vehicles?

Figure 1 Illustrates a modern vehicle's driver assistance systems along with sensor operations that enhance safety together with convenience features. The following list describes the systems and sensors displayed:

1. **Adaptive Cruise Control:** maintains a set speed and distance from the vehicle ahead using long-range radar.
2. **Emergency Braking:** a system of radar and cameras that detects collisions before automatically activating emergency braking in the vehicle.
3. **Pedestrian Detection:** The vehicle uses cameras and radar to detect pedestrians in its path for safe driving actions.
4. **Lane Departure Warning:** performs vehicle monitoring through cameras to warn drivers when their vehicle moves unintentionally out of its path.
5. **Collision Avoidance Assist:** The system utilizes radar and camera sensors to automatically control brakes or steering to prevent moving vehicle collisions.
6. **Park Assist:** The system enables safe parking through ultrasound sensor detection of surrounding obstacles.
7. **Traffic Sign Recognition:** The system uses cameras to detect traffic signs and convey them to drivers within the display.
8. **Cross Traffic Alert:** This system identifies side traffic threats when reversing the vehicle by using radar technology.
9. **Blind Spot Detection:** Through radar technology, the system warns drivers about vehicles situated in their blind spot areas.
10. **Surround View:** creates 360-degree views through multiple cameras installed around the car.
11. **Park Distance Control/Assistance/Warning:** uses ultrasound sensors for obstacle detection to help parking maneuvers.

Different detection sensors used by the car are displayed through colored zones that include LiDAR (Light Detection and Ranging), long-range radar and cameras, as well as short/medium range radar and ultrasound systems. All these systems operate in partnership to build an extensive safety network that improves both safety and convenience while driving.

Figure 1. Sensor system on self-driving cars

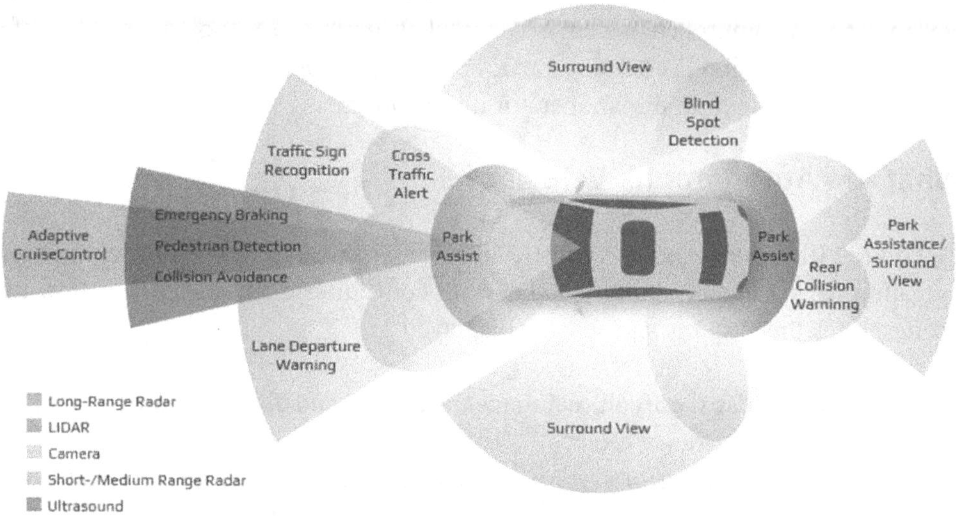

(Vargas, 2021).

What is Vehicle Detection?

Vehicle detection in autonomous driving systems refers to the technology and processes used to identify and track other vehicles on the road. Autonomous vehicles basically use this ability to detect their environment for making their key decisions (Liang, 2024). Automated vehicles implement cameras along with LiDAR (Light Detection and Ranging) systems and extra sensors that include radar and ultrasonic technology for identifying vehicles. Every sensor maintains its advantage because cameras deliver complete visual information, yet LiDAR delivers exact distance detection. Modern algorithms, along with machine learning models, process data obtained from sensors. The algorithms process data to identify and categorize different detected objects. Such systems run in real-time by assessing the environment to spot other vehicles. Such operations allow the autonomous system to act swiftly when traffic changes occur. The system keeps safe distances with other vehicles while forestalling collisions, and it performs well in traffic conditions. Vehicle detection systems represent an essential aspect that protects human passengers and supports the optimal performance of the running vehicle. Vehicle detection usually operates together with self-driving operation functions such as mapping, decision making, and control management. The integration enables the vehicle to base its decisions on detected vehicles in its vicinity.

Vehicle detection serves as a core functionality in self-operated vehicles because it helps them detect obstacles while making decisions for safe pathfinding. However, vehicle detection during rainy conditions raises major challenges because poor visibility and light-reflecting objects, along with view-blocking objects, affect the system's perception capabilities (Chen, 2020).

LIDAR sensors installed in autonomous vehicles serve to identify forward objects while ensuring the prevention of accidents. LiDAR operates as a remote sensing system that implements laser beams to generate accurate and precise environmental maps with high definition. The technology of LiDAR enables precise distance measurements because of its high accuracy while producing detailed 3D maps through high-resolution scans and works effectively across all lighting situations. Besides this, it reveals obscured structures by passing through obstacles and vegetation. However, under harsh weather conditions, particularly during rain, the performance of these sensors is significantly degraded (Liang, 2024). Rain-induced humidity on the surface of observed objects, combined with the effects of Mie scattering, are factors that reduce LIDAR's observation efficiency and can lead to incorrect situational judgments (Ahmed, 2022). Therefore, it is necessary to design a system that compensates for these losses to ensure the safety of autonomous vehicles in traffic. One of the effective supplementary systems for LIDAR sensors is the camera (Ponn, 2020). Cameras can provide visual images to complement LIDAR data. By integrating image processing algorithms and machine learning, cameras can help accurately identify and classify objects even under adverse conditions, such as rain. Here's a concise overview of how they work:

- **Image Capture:** A vehicle can obtain a comprehensive view with image-capturing cameras positioned in front, back, and along the sides of the vehicle. The system operates both high-definition image processing and real-time video transmission capabilities.
- **Image Processing:** The system processes captured images through multiple sequential steps that normalize images after processing them through noise reduction operations and contrast adjustments. The system obtains vehicle contours through the extraction of edges, textures, and shapes.
- **Object Detection** functions get assistance from deep learning Convolutional Neural Networks (CNNs) to detect and classify vehicles effectively by pro-

cessing large amounts of labeled training data. The system marks identified vehicles with placement and dimensional boundary boxes.

- **The tracking process** consists of vehicle monitoring through Kalman filter and optical flow methods that enhance subsequent position detection results. The system examines how things move to predict upcoming vehicle movements.
- **Decision Making:** Using combined camera sensor data with LiDAR and radar sensor data to generate improved reliable environmental information. The directly processed information enables drivers to make spontaneous decisions involving speed adjustments together with changing lanes and braking to prevent accidents.

Real-time vehicle detection through cameras has emerged as an essential system to boost the safety features and reliability and operational performance of autonomous vehicles under difficult weather, such as rainy conditions. Research progress in this field works to enhance the current computer vision along with deep learning capabilities for creating safer and more advanced self-driving features. Detecting vehicles using camera systems encounters several major difficulties when operating during wet weather.

- **Reduced Image Contrast and Blurring:** Noisy image content occurs in rain because total gradient magnitude measures decrease by 49% when comparing clear to rainy conditions (Hasirlioglu, 2019). Visual detection of edges together with corners and other visual features becomes more challenging in such conditions because image clarity declines.
- **Occlusion by Rain Droplets:** When water droplets accumulate on the camera lens, they cause the direct blocking of distant image elements (Hasirlioglu, 2019). The obstructed view during rain creates problems in seeing clearly through to moving vehicles and additional detected objects.
- **Shifting and Distortion of Bounding Boxes:** The positioning and shape of detected vehicle bounding boxes change because of rain appearance compared to dry conditions (Hasirlioglu, 2019). One test showed that bounding box movement reached 13 pixels while its dimensions inflated by 6% under rainy conditions. Errors in bounding box detection lead to misclassifications during object recognition and trackable objects' misidentification.
- **Decreased Detection Scores:** The classification accuracy scores that determine vehicle detection fall drastically in weather with rain. The detection system displayed a score reduction of 21% when operating in rain instead of clear weather, according to the findings presented in Hasirlioglu (2019). A

lower scoring system creates more difficulty in performing trustworthy vehicle detection.

- **Challenges with Histogram-based Features:** Several object detection methods depend on HOG (Histogram of Oriented Gradients) and other histogram-based features, which become challenging under these conditions. Rain lowers gradient intensities and modifies the histogram form, which deteriorates the performance capabilities of these feature-based methods (Hnewa, 2020).

The crucial obstacles to camera-based vehicle detection systems in rainy conditions include weakened image clarity, partial occlusion and deformed bounding boxes, together with deteriorating detection ratings and diminished noticeable visual characteristics, which combine to degrade system performance and increase both incorrect alerts and missed targets while lowering accuracy levels (Abdullah (2021), Xia (2022)). The solution to these challenges depends on both strong deep learning algorithms and fusion techniques between sensors (Hassaballah, 2020; Tahir, 2024; Tariq, 2025).

Deep learning techniques have become the main research focus for developing better systems to detect vehicles during rainy weather conditions. Some key approaches include:

- **Improved CNN Architectures:** YOLOv4 and other standard convolutional neural network (CNN) architectures have been modified to achieve higher detection precision during conditions with haze and rain along with other adverse weather conditions (Wang, 2023).
- **Multi-Modal Sensor Fusion:** A vehicle detection system gains higher reliability through the combination of camera and LiDAR sensor data known as multi-modal sensor fusion.
- **Adversarial Training:** The training of detection models happens through adversarial training by feeding synthetic rainy images to models during their learning process to make them resistant to weather-induced degradations.
- **Attention Mechanisms:** The incorporation of attention modules within CNN architectures enables the network to focus on important features while disregarding unimportant elements like rain streaks and reflections.

Special metrics and datasets designed for rainy environments must be used for evaluating vehicle detection systems in such conditions. Some key datasets include:

- BDD100K provides more than 100,000 diverse driving scene images that incorporate rainy weather elements.

- Raincouver stands out as a collection of vehicle detection data in rainy conditions that comes from Vancouver, Canada.

Precision and recall, together with the F1-score, form the standard metrics for evaluating transportation vehicle detection solution accuracy. The measurement of autonomous driving application speed occurs through both frames per second (FPS) and latency metrics for real-time performance.

Real-time vehicle detection in conditions with rain requires an urgent solution because autonomous driving systems depend on this capability for safety and reliability. The challenge of vehicle detection can be faced by implementing deep learning methods, including enhanced CNN architecture and sensor fusion alongside attention mechanisms, which lead to promising results. Immersive study about vehicle detection must progress to enhance detection accuracy while also improving robustness and achieving real-time detection capabilities despite various weather-induced image degradations (Tariq, 2025).

RELATED WORKS

Deep Learning Models for Object Detection

Object detection is a fundamental and challenging task in the field of computer vision, aimed at identifying objects of a certain class (e.g., people, animals, or cars) within one or more digital images. At the beginning of the field, object detection models analyzed entire images by moving a sliding window at various scales. A feature extraction process occurred at every window position before using the feature data to classify the presence of desired objects. This simple method remains ineffective for detecting multiple-sized or oriented objects across different images since it needs multiple window sizes to enhance detection probabilities. The processing takes a long time because the algorithms demand high computational power. Popular modern algorithms, such as deep learning networks, have improved object detection capabilities to a significant degree. CNNs, together with other advanced models, perform better at handling object variations when it comes to size and orientation and angle-related positions. High accuracy in object detection and classification becomes possible because these models use vast labeled datasets during training. Modern algorithms, when used with image processing methods, enable cameras to identify and categorize objects even through adverse weather, like rainy conditions.

Systems that use object detection benefit from these integrated capabilities, thus improving their operational outcomes.

Based on the search results of Tahir et al. (2024), a few specific deep learning models are effective for vehicle detection in rainy conditions, such as R-CNN, Fast R-CNN, YOLO, and SSD.

R-CNN

The Regional Convolutional Neural Network approach (R-CNN) emerged in 2014 as a regional proposal algorithm developed by Ross Girshick together with his colleagues (2014) for focusing on localized object detection in object identification procedures.

Two principal components serve as the foundation of R-CNN operation: region proposal followed by classification and labeling. An alternative selective search method replaces traditional sliding windows to produce about 2000 probable object regions in the initial phase. These proposals are extracted from the image and resized to a fixed size to be processed by CNN. During the second phase of operation, each region proposal passes through a CNN for important characteristic extraction. A Support Vector Machine (SVM) serves to classify the extracted features, which determine the type of object present in each region. The detected objects receive refinement from a regression model, which enhances the accuracy of localization precision. The main drawback of R-CNN lies in its slow and impractical computational processing. A large amount of hardware and time is required for the execution of 2,000 proposed regions during image processing.

Fast R-CNN

During 2015, researchers developed the Fast R-CNN (Fast Region-based Convolutional Network) model to resolve problems of high computation costs and sluggish operational speed, which were encountered by the previous R-CNN method (Girshick, 2015). Instead of R-CNN, the authors designed Fast R-CNN by operating a complete VGG-16 (Simonyan, 2014) pre-trained convolutional neural network on each entire input image. This feature extraction method shortens computation times while enhancing efficiency because it removes the requirement for individual feature extraction per region, which R-CNN needed. The RoI (Region of Interest) pooling layer in Fast R-CNN generates uniform-sized regions of interest by processing the proposed regions selected by the algorithm. The proposal regions enter a fully connected layer for both classification and bounding box regression operations that detect the objects. A selective search algorithm serves fast R-CNN to produce region proposals from the original image, yet this operation takes place only during the first stage and avoids multiple repetitions across extracted features. Fast R-CNN shortens processing time and lowers computational needs to offer enhanced object detection results than R-CNN.

YOLO

A convolutional neural network named YOLO (You Look Only Once (YOLO) emerged in 2016 through the work of Joseph Redmon and his collaborators for object detection together with recognition and classification (Redmon, 2016). YOLO currently exists in its ninth version, according to Kim (2023) and (Wang, 2024). YOLO transforms object detection into one unified regression problem instead of following traditional methods that operate with separate models for regional proposal and classification. The single direct evaluation of whole images enables the prediction of bounding boxes and class probabilities by this approach. The input is divided into an S×S pattern, as shown in Figure 2. Each grid cell is responsible for predicting a fixed number of bounding boxes, their confidence scores, and class probabilities. The confidence score integrates two components, which demonstrate both box object presence probability and box coordinate precision. The model generates likelihood predictions for each defined box, along with probability ratings for every class type.

Figure 2. YoLo model

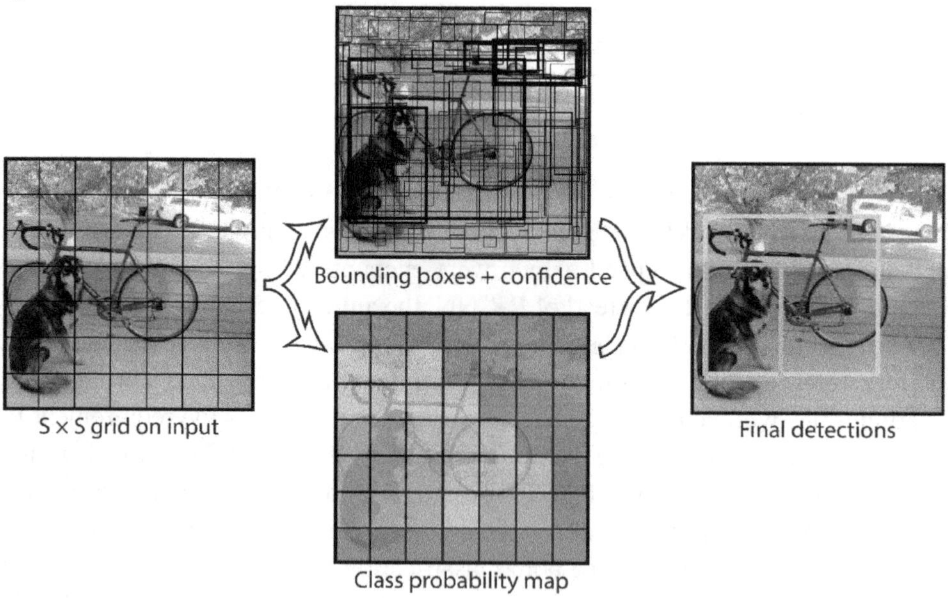

(Redmon, 2016).

The success of the YOLO model arises from significant advantages that make it suitable for various object detection operations. This model achieves exceptional speed performance because of its main benefit. YOLO's image processing capabilities lead to its fast execution speed of 45 frames per second, and its successor, Fast YOLO, reaches 155 frames per second. Real-time detection applications, including autonomous driving and surveillance, get excellent results from YOLO because of its inherent speed capabilities. Another advantage is simplicity. YOLO defines object detection as a single predictive task that directly identifies boxes for individual objects and class assignments throughout complete image evaluation. The unified detection model operates without needing a sequence of separate models that handle both regional proposals and classification tasks, thus simplifying detection operations and deployment protocol. YOLO also benefits from a global context. YOLO makes predictions based on the global context by examining the full image in both training and testing operations, resulting in fewer wrong positive predictions. A comprehensive understanding achieved by YOLO enables it to interpret object-environment relationships effectively, which results in superior detection precision. High accuracy improvements in YOLO versions result in complicated network designs and a large number of parameters, which restrict real-time object detection to GPU (Graphics Processing Unit) computation. The mobile and embedded devices featuring restricted computational and memory resources cannot effectively use these particular detection models. The problem requires resolution through lightweight architecture implementations of the full YOLO network design system (Jiang, 2020; Tariq, 2025).

SSD

The SSD (Single Shot MultiBox Detector) model, introduced by Wei Liu and colleagues in 2016, is a convolutional neural network designed for object detection, recognition, and classification (Liu, 2016). Like YOLO, SSD aims to identify objects from a single view of the image, but it introduces several unique features and advantages.

The general architecture of an SSD model consists of two main components: a base network and multi-scale feature layers connected in sequences. Figure *3* Show an SSD model with a VGG-16 network as a base.

Figure 3. VGG-166-based SSD model

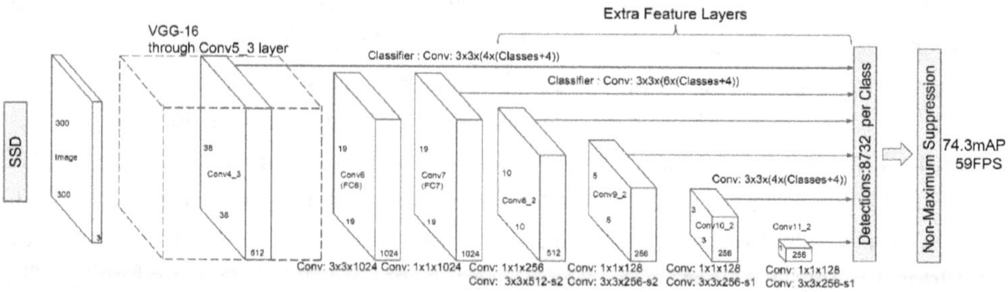

(Liu, 2026).

The base network serves as the first step because it extracts image features using deep convolutional neural network architecture. The image goes through sequential convolution, pooling, and activation steps that create a sophisticated feature map that contains critical data about image objects. The base network plays an essential role in SSD models since it determines the quality of the multi-scale feature layers' accuracy. SDS permits multiple network types as its base architecture with specific performance qualities that determine different trade-offs between speed and accuracy. Commonly used base networks include:

- The model VGG-16 represents an effective solution for extracting features because of its straightforward design (Simonyan, 2014).
- The MobileNet-v2 architecture works best for real-time applications due to its optimized design for mobile embedded devices. The network applies depth-wise separable convolutions, which decrease both parameter count and computational expenses (Sandler, 2018).
- EfficientDet belongs to the EfficientNet family and achieves high-speed performance along with accuracy levels. Through compound scaling methods, this network maintains a balance between network depth, width, and resolution, which leads to exceptional performance for detecting objects (Tan, 2020).

Base network and multi-scale feature layers are arranged to detect objects at multiple scales. Multiple additional convolutional layers with successive decreasing sizes make up the feature layers that enable the model to process information at varied resolutions. Every layer has separate responsibilities to detect objects of particular sizes by generating both bounding boxes and class probabilities. SSD implements multiple feature layers to detect both large and small objects and objects with different proportions inside a single image. The multi-scale approach of the model detects small objects that single-scale detectors would normally miss. The

SSD model stands out as a preferred option for object detection because it provides several useful operational advantages. The main benefit of the SSD model stems from its fast operational speed and improved efficiency. SSD operates differently from two-stage detectors, including Faster R-CNN, because it performs object localization and classification simultaneously during a single network forward operation. The single-stage operation makes SSD deliver real-time performance while reducing computational needs, thus making it ideal for systems that need speedy detection like automatic driving and video surveillance applications. Feature maps in SSD operate at various scales, which generates another advantage for the system. The SSD detection system works through its usage of different resolution maps, which enable the detection of objects at different scales. The multi-scale feature map structure in the model enables the detection of objects from small to large dimensions inside the same image boundary. Using feature maps from diverse network layers allows SSD to obtain detailed object information, which leads to better detection accuracy for both small and large objects. The integration of an advanced base network design and multi-scale feature layers allows SSD to gain exceptional efficiency alongside precision in its object detection operations across multiple applications. In this chapter, we utilize base networks from the "TensorFlow 2 Detection Model Zoo" collection, which was created by Vighnesh Birodkar and his colleagues (2021) for the COCO 2017 dataset. The SSD model relies on these fine-tuned base networks for high performance.

Discussion

The research results indicate some statements as follows.

- Real-time object detection becomes possible with YOLO (You Only Look Once) models, including YOLOv3 and YOLOv4, because they provide a good balance between accuracy and processing speed. These detection systems operate fast while keeping strong accuracy standards, which allows efficient operation in unpredictable wet environments.
- We receive balanced detection speed and precision through the Single Shot MultiBox Detector (SSD). With a single direct pass in operation, this system achieves fast object recognition as well as competitive detection accuracy.
- The object detection capability of Faster R-CNN remains accurate because it uses region proposal networks despite its slower processing rate when compared to YOLO and SSD. The model works appropriately in situations that require precise results over speed, even though its operation speed can be enhanced through optimization efforts.

YOLOv3 and YOLOv4 stand as the preferred options for their optimized results between speed and accuracy when identifying vehicles in real time during unfavorable weather conditions like rain. SSD also offers a good alternative, while Faster R-CNN can be considered depending on the specific requirements for accuracy and processing speed. The specific requirements will determine whether Faster R-CNN or SSD serves as an appropriate solution.

Methods to Improve the Accuracy of Object Detection Models

How autonomous vehicles determine image pixel values depends on their camera system capabilities, which use illumination from the environment. Multiple negative factors that include falling raindrops along with fog combined with water droplets on object surfaces and wind speed together with dirt adversely affect the ability to record meaningful information about environmental objects during rain. These environmental factors damage critical object information, which subsequently causes autonomous vehicles to miss objects during observation. Researchers have introduced different approaches to increase vehicle observation capacities for autonomous operations under rainy weather settings as part of present-day research initiatives. Widespread interest has focused on two main strategies among researchers:

- The training data gets improved through data augmentation methods. Data augmentation techniques function as key elements in improving object detection model accuracy under rainy conditions as they develop robust generalization abilities for these models. The most efficient approach involves the generation of synthetic images containing rain effects by implementing techniques based on Generative Adversarial Networks (GANs) and style-transfer networks (Gupta, 2024). The methods generate authentic rain effects on established images, thus allowing the model to understand objects in rainy environments. The rephrasing of synthetic rainy images enables the model to perform outstandingly in actual rainy situations. Another dataset construction method involves collecting images that were recorded under different weather situations, including rain, to build a complete weather dataset. The diverse weather conditions present in BDD100K help the model achieve better generalization abilities. However, the collected datasets have an unbalanced image distribution according to different weather conditions. The addition of augmented rain images will help achieve a more balanced training set (Gupta, 2024).
- Deep learning-based image preprocessing techniques eliminate rain from images to achieve better object detection precision. The model's input is substantially better images after denoising, and contrast enhancement procedures

are applied to the data. The denoising algorithms remove both rain streaks and other image noise elements, whereas contrast enhancement improves the visibility of objects in dark environments (Hao, 2024). The combination of preprocessing methods added to data augmentation helps the model process better-quality information.

Deep learning techniques work to transform images with raindrops into clear scenes, along with the removal of moisture traces. However, research by Mazin Hnewa and Hayder Radha [8] demonstrates that these sophisticated techniques fail to enhance the quality of detected objects in images effectively. Furthermore, embedded systems that deploy deep learning techniques together with object detection models need major computational resources, which eventually leads to reduced processing speed. The approaches demand substantial processing speed along with significant memory allocation, whereas embedded systems usually operate without such availability. Both technologies combine to increase latency and diminish real-time operation. These advanced algorithms consume elevated amounts of energy, which becomes a critical issue when operating battery-powered products. Consequently, the deployment of deep learning-based raindrop removal methods in resource-constrained embedded systems faces restrictions because it decreases system efficiency and performance (Xia, 2022; Tariq, 2025).

A better solution compared to deep learning-based rain removal techniques exists in improving the model training dataset. The model acquires improved detection accuracy by learning from different rainy conditions when more real-world images are added to its training set. Various images with rainfall at different intensities, raindrop sizes, and light conditions improve model generalization, thus preventing the model from overfitting to specific rainy conditions. Real-world images with diverse content in training datasets help models perform better in various contexts, which minimizes performance variations during real-world applications, thus boosting their resistance. The model acquires knowledge about complex features of rainy conditions that can detect and classify objects more accurately in demanding situations because of the larger training dataset. This method demonstrates efficiency when deployed on embedded systems with limited resources because it needs lower computational power and memory use. The method demonstrates scalability because adding new rainy condition data sets allows a gradual enhancement of performance levels. The strategy of obtaining additional data and annotations proves more affordable than building sophisticated rain removal systems, which makes it a useful solution primarily suited for limited-budget applications. The model obtains better performance in rainy situations by using training data enhancement that eliminates the need for complicated preprocesses while delivering increased efficiency for real-world deployment. The approach achieves the best fit for design specifications described in this chapter, thanks to its simplicity and operational efficiency.

MODEL ANALYSIS

As discussed in the previous section, there are currently several machine learning-based object detection methods, such as R-CNN, Faster R-CNN, YOLO, and SSD (Tahir, 2024). Generally, R-CNN methods involve multiple complex processing steps, resulting in slower computational speeds. Therefore, for embedded systems that require low energy consumption and resources while ensuring real-time performance, these algorithms do not meet the requirements. Modern methods like YOLO and SSD, which are designed to detect objects in a single pass of the image, are the most promising algorithms for this purpose. In this section, we focus on analyzing and evaluating the YOLO and SSD models to identify a model that achieves a balance between frame processing speed, resource utilization, and algorithm accuracy, making it the most suitable for deployment on embedded systems.

Evaluation Criteria

Mean Average Precision (mAP): In object detection tasks, the mAP (Everingham, 2010) is a commonly used metric to evaluate the accuracy of object detection models such as YOLO, Fast R-CNN, and SSD. It provides an overview of the precision of object detection across a given dataset. mAP provides a comprehensive measure of a model's performance across different classes and varying confidence thresholds, summarizing both precision and recall into a single metric.

The formula of the mean Average Precision is given below:

$$\text{mAP} = \frac{1}{n}\sum_{k=1}^{k=n} AP_k \tag{1}$$

where AP_k is the Average Precision of class k, and n is the number of classes.

The study employs mAP@0.5:0.95 as the evaluation metric, which derives its value from calculating average mAP across the specified IoU (Intersection over Union) thresholds from 0.5 to 0.95 in increments of 0.05. The mAP@0.5:0.95 metric demonstrates exceptional importance because it reveals how well a model functions across different performance conditions. The model evaluates object localization and detection through multiple IoU thresholds because this method delivers nuanced performance measurements that handle different degrees of detection strictness. The performance calculation using mAP@0.5:0.95 provides better representation than a single-IoU measurement because the system forces models to operate with excellent precision regardless of the degree of object overlap. Both academic researchers and practitioners now use mAP@0.5:0.95 as their standard evaluation criterion because this metric enables the comparison of different detection and object localization methods. This performance measurement tool enables experts to provide detailed assessments about individual class performance and support the development of improved detection solutions for complex multi-class detection applications.

Frame Processing Speed indicates the number of picture frames processed per second (FPS). Real-time image processing speed plays an essential role in vehicle detection applications for autonomous vehicles because it determines the time available for making accurate decisions. Such systems equipped with a high FPS rate check more scenarios at once, thus enabling fast delivery of solution options. At extremely low Frame Per Second measurements, a detection delay occurs that may create unsafe circumstances because reaction times become slower than required. The system's ability to sustain high FPS depends on three key elements, which are hardware computational strength, the optimization of processing algorithms, and the quality standards of input data. Deep learning techniques applied to advanced machine learning models provide substantial improvements to vehicle detection precision levels. The computational requirements for these models usually reach high levels. Several methods, including model pruning with quantization and specialized hardware like GPUs (Graphics Processing Units) along with FPGAs (Field Programmable Gate Arrays), help systems achieve accuracy-speed balance in their operations. These optimizations allow for faster processing times while maintaining a high level of detection accuracy, which is essential for the safe operation of autonomous vehicles.

Used Resource: Resource availability in embedded systems is often limited. A system design requires an aim to achieve maximum accuracy with minimal resource consumption. The application's resource utilization rates consist of CPU and RAM capacity usage metrics.

Dataset Building

Training Dataset

The training image dataset comprises 2,933 images of vehicles in rainy weather sourced from the labelled BDD Dataset 100K, which focuses on highway driving scenes, along with 1,221 self-collected images from dashcam videos on YouTube.

The BDD Dataset 100K (Yu, 2020) is a large image dataset collected by sampling frames at the 10th second from videos intended for autonomous vehicle research. This dataset contains 100,000 images that vehicle-mounted cameras captured throughout clear daylight and various weather conditions, including rain and snow, as well as cloud cover and under urban and highway driving conditions. A total of 2,933 images fell within the category of rainy weather and highway driving scenarios for this investigation (Figure 4).

Figure 4. Sample images in the BDD dataset 100K.

The image dataset used to augment this model comprises 1,221 images of driving in real-world rainy conditions, collected from 20 long-distance driving videos shared on YouTube channels. The research team chose study-specific video sequences from YouTube channels because they displayed excellent visuals under three rain intensity levels (light rain, moderate rain, and heavy rain) with multiple vehicle types (cars, motorcycles, trucks, and buses). Each video was segmented into individual frames using a Python-based video cutting technique. Subsequently, the frames were manually reviewed to discard those of insufficient quality. The remaining filtered images were then input into the *Make Sense* tool, an image labeling tool. The resulting dataset consists of 1,221 accurately labeled images of four main vehicle types (cars, motorcycles, trucks, buses) across different rainy weather conditions. Some samples are shown in Figure 5.

Figure 5. Augmented dataset.

The labels for these two datasets are stored in .txt and .xml files to accommodate the respective training methods for the YOLO and SSD models. The dataset is further divided into two subsets: 90% of the videos are allocated for training, and 10% are designated for validation.

Test Dataset

To evaluate the effectiveness of the vehicle recognition model in rainy weather conditions, an independent test dataset must be constructed, separate from the training dataset. The test dataset needs to contain images depicting true weather situations that will occur when the model is deployed to gain reliable assessment results. The DAWN dataset can serve as an appropriate choice to establish the test dataset (Kenk, 2020). A set of 1,000 images obtained in real traffic areas forms this dataset, which splits its content between the weather types of fog, snow, rain, and sandstorm. The evaluation process is simplified by using the images that contain bounding boxes for both YOLO and SSD models. The dataset includes 100 driving images captured in rainy conditions among different traffic environments for performing model assessment professionally. Some of the provided images are displayed in Figure 6.

Figure 6. Sample images in the DAWN dataset.

Evaluation

The YOLO-Tiny and SSD models, including various versions, were trained on the Google Colab platform and quantized. These models were then evaluated based on the criteria of mean Average Precision (mAP), frames per second (FPS), and resource usage when deployed on the Raspberry Pi 5 embedded computing platform.

YOLOv3-Tiny

Table 1. mAP accuracy (at IoU threshold from 0.5 to 0.95) of the YOLOv3-Tiny model.

Vehicle	mAP @ 0.5:0.95
Motorcycle	15%
Car	60.62%
Bus	75%
Truck	54.27%

Table 1 Provides a detailed summary of the mean Average Precision (mAP) accuracy (at IoU thresholds from 0.5 to 0.95) for different vehicles using the YOLOv3-Tiny model. The average mAP accuracy of the YOLOv3-Tiny model is approximately 51.22%. The YOLOv3-Tiny model achieves high accuracy with an

mAP of 54.27% or higher for vehicles such as cars, buses, and trucks. However, this model is less effective at detecting motorcycles, as shown in Figure 7.

The frame rate achieved when running the YOLOv3-Tiny algorithm on Raspberry Pi 5 is 0.64 fps. During operation, the YOLOv3-Tiny model utilizes only 19,6% of the system's RAM resources.

Figure 7. A screenshot of the experimental results of the YOLOv3 – tiny.

YOLOv4-Tny

Table 2. mAP accuracy (at IoU threshold from 0.5 to 0.95) of the YOLOv4-Tiny model.

Vehicle	mAP @ 0.5:0.95
Motorcycle	32.5%
Car	70.62%
Bus	75%
Truck	76.46%

Table 2 Details the mAP accuracy (at an IoU threshold from 0.5 to 0.95) for various vehicles using the YOLOv4-Tiny model. The average mAP accuracy of the YOLOv4-Tiny model is approximately 63.65%. The YOLOv4-Tiny model demonstrates high accuracy, reaching up to 70.62% or higher for vehicles such as cars, buses, and trucks. All vehicles are accurately detected, with motorcycles achieving

an accuracy above 32.5%. The ability to recognize specific vehicles is illustrated in Figure 8.

Figure 8. This is a screenshot of the experimental results of the YOLOv4-Tiny model.

The frame rate achieved when running the YOLOv4-Tiny algorithm on a Raspberry Pi 5 is 0.81 fps. During operation, the YOLOv4-Tiny model only uses 22% of the system's RAM resources.

SSD-MobileNet-v2

Table 3. mAP accuracy (at IoU threshold from 0.5 to 0.95) of the SSD-MobileNet-v2 model.

Vehicle	**mAP @ 0.5:0.95**
Motorcycle	20.00%
Car	44.92%
Bus	65.00%
Truck	51.7%

Table *3* Details the mAP accuracy (at IoU thresholds from 0.5 to 0.95) for vehicles using the SSD model with MobileNet-v2 as the base network. The model achieves an average mAP accuracy of approximately 45.64%. Vehicles are relatively accurately identified, with mAP exceeding 44.92% for cars, buses, and trucks and over 20% for motorcycles. However, running this model encounters several object misidentification errors, as illustrated in Figure 9.

Figure 9. This is a screenshot of the experimental results of the SSD-MobileNetv2 model.

The frame rate achieved when running the SSD-MobileNet-v2 algorithm on Raspberry Pi 5 is 5.26 fps. During operation, the SSD-MobileNet-v2 model only utilizes 12.9% of the system's RAM resources.

Quantized SSD-MobileNet-v2

Table 4. mAP accuracy (at IoU threshold from 0.5 to 0.95) of the quantized SSD-MobileNet-v2 model.

Vehicle	**mAP @ 0.5:0.95**
Motorcycle	0.62%
Car	40.17%
Bus	42.50%
Truck	38.41%

Table 4 Details the mAP accuracy (at an IoU threshold from 0.5 to 0.95) for vehicles using the SSD model with the MobileNet-v2 backbone after quantization. The model experiences a significant drop in mAP, with the highest accuracy only reaching 42.50% and an average of just 30.43%. Vehicle recognition is no longer accurate, with many instances of misclassification, as shown in Figure 10.

Figure 10. This is a screenshot of the experimental results of the quantized SSD-MobileNet-v2 model.

The frame rate achieved when running the quantized SSD-MobileNet-v2 algorithm on a Raspberry Pi 5 is 7.96 fps. During operation, the model utilizes only 11.9% of the system's RAM.

SSD-MobileNetv2-FPNLite320×320

Table 5. mAP accuracy (at IOU threshold from 0.5 to 0.95) of the SSD-MobileNetv2-FPNLite320×320 model.

Vehicle	mAP @ 0.5:0.95
Motorcycle	16.61%
Car	51.51%
Bus	70%
Truck	58.36%

Table *5* Provides a detailed breakdown of the mAP at an IoU threshold of 0.5 to 0.95 for various vehicles using the SSD model with the MobileNet-v2 FPNLite 320×320 backbone. The highest achieved mAP accuracy for this model is approximately 70%, with an average mAP accuracy of about 49.12%. Vehicles such as cars, buses, and trucks are accurately detected, as indicated in Figure 11. However, this model demonstrates reduced effectiveness in detecting motorcycles.

Figure 11. This is a screenshot of the experimental results of the SSD-MobileNetv2-FPNLite320×320 model.

The frame rate achieved when running the SSD-MobileNetv2-FPNLite320×320 algorithm on the Raspberry Pi 5 is 3.88 fps. During operation, this model utilizes only 12.7% of the system's RAM resources.

Quantized SSD-MobileNetv2-FPNLite320×320

Table 6. mAP accuracy (at IOU threshold from 0.5 to 0.95) of the Quantized SSD-MobileNetv2-FPNLite320×320 model.

Vehicle	**mAP @ 0.5:0.95**
Motorcycle	11.25%
Car	46.59%
Bus	65.00%
Truck	39.66%

Table *6* Provides a detailed breakdown of the mAP at IoU thresholds ranging from 0.5 to 0.95 for vehicles using the SSD model with the MobileNet-v2 FPNLite 320×320 backbone after quantization. The model achieves a maximum mAP of approximately 65%, with an average mAP of around 40.63%. The model demonstrates high accuracy in identifying vehicles such as cars, buses, and trucks, as illustrated in Figure 12. However, it shows limited effectiveness in detecting motorcycles.

Figure 12. This is a screenshot of the experimental results of the quantized SSD-MobileNetv2-FPNLite320×320 model.

The frame rate achieved when running the quantized SSD-MobileNetv2-FPNLite320×320 algorithm on the Raspberry Pi 5 is 6.78 fps. During operation, this model utilizes only 11.9% of the system's RAM resources.

Discussion

The results from mAP accuracy evaluation and frame rate measurements, as well as resource utilization comparison, lead to a comparative summary in Table 7.

Table 7. Models' evaluation summary.

Model	FPS	mAP @ 0.5:0.95	Resource	
			ROM	RAM
YOLOv4-tiny	0.81	63.65%	33.1 Mb	22%
YOLOv3-tiny	0.64	51.22%	22.4 Mb	19.6%
SSD MobileNet v2	5.26	45.64%	17.9 Mb	12.9%
MobileNet v2 (8-bit)	7.96	30.43%	5.04 Mb	11.9%
MobileNet v2 FPNLite 320×320	3.88	49.12%	10.9 Mb	12.7%
MobileNet v2 FPNLite 320×320 (8-bit)	6.78	40.63%	3.57 Mb	11.9%

The compilation of evaluated parameters presented in Table 7 Generates these observations:

- Both YOLOv4-Tiny and YOLOv3-Tiny deliver poor performance because they have a very low frame rate combined with high resource requirements. Consequently, these models are not suitable for the objectives of this study.
- The quantized SSD-MobileNet-v2 achieves the best results in terms of frame rate combined with resource usage. However, this model returns results with poor accuracy rates, which renders it non-applicable for the intended purpose.
- Among the tested models, the quantized SSD-MobileNet-v2-FPNLite320×320 achieves optimal performance by maintaining a good balance between frame processing speed, accuracy levels, and resource usage efficiency.

Under rainy conditions, the proposed system chooses the quantized SSD-MobileNet-v2-FPNLite320×320 model to perform vehicle detection.

SYSTEM INTEGRATION

Hardware System

Figure 13 Shows the diagram of how embedded system's hardware elements connect for detecting vehicles during rainfall. The system includes these technical parts:

1. **9-Inch IPS Screen:** The system features a 9-Inch IPS Screen, which uses capacitive IPS technology to deliver 2K resolution (2560x1600) for high-quality image display. The screen and the Raspberry Pi 5 kit are connected via a micro-HDMI port.

2. **Raspberry Pi 5 Kit (Figure 14**The Raspberry Pi 5 features a 64-bit quad-core ARM Cortex-A64 processor operating at 2.4 GHz. This unit also supports Wi-Fi 5 (802.11ac) and Bluetooth 5.0 and Low Energy protocols, as well as Gigabit Ethernet services. The device incorporates two micro-HDMI ports together with two MIPI camera/display ports. Serving as the central processing unit of the system, it receives data from the camera, executes object detection algorithms, calculates the distance from the camera to objects, and outputs the result on the display. The Raspberry Pi 5 is connected to the display via HDMI and to the camera via CSI (Camera Serial Interface) ports. Additionally, it has USB ports for connecting other peripheral devices if necessary.
3. **Stereo Camera IMX219-83 module (Figure 15)** contains two IMX219 cameras in one board at an 8-megapixel resolution for each camera. The image and video data captured by the camera are used not only for object detection but also for measuring the distance to objects. This camera connects to the Raspberry Pi 5 through two CSI ports, providing image data for processing and detecting vehicles.

The connected system combines all components to create a complete detection and alert capability for vehicles in rainy weather conditions.

Figure 13. Block diagram of the system.

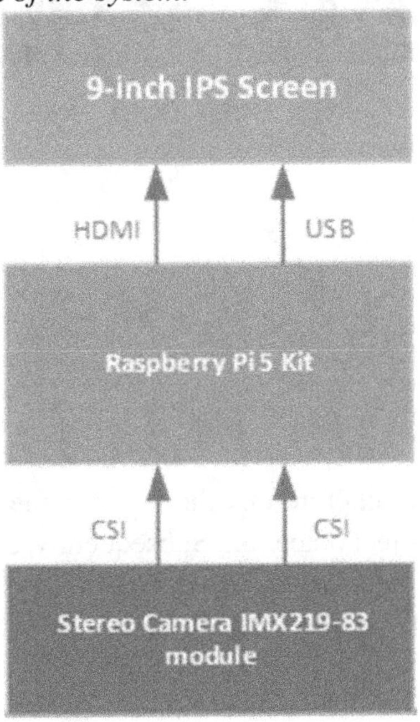

Figure 14. Raspberry Pi 5 kit.

Figure 15. Stereo Camera IMX219-83 module.

Software System

Overall Function

The system uses images captured by the stereo camera module and employs a deep learning model for object detection. Subsequently, a regression model is used to measure distance.

Figure 16. The flowchart of the software program.

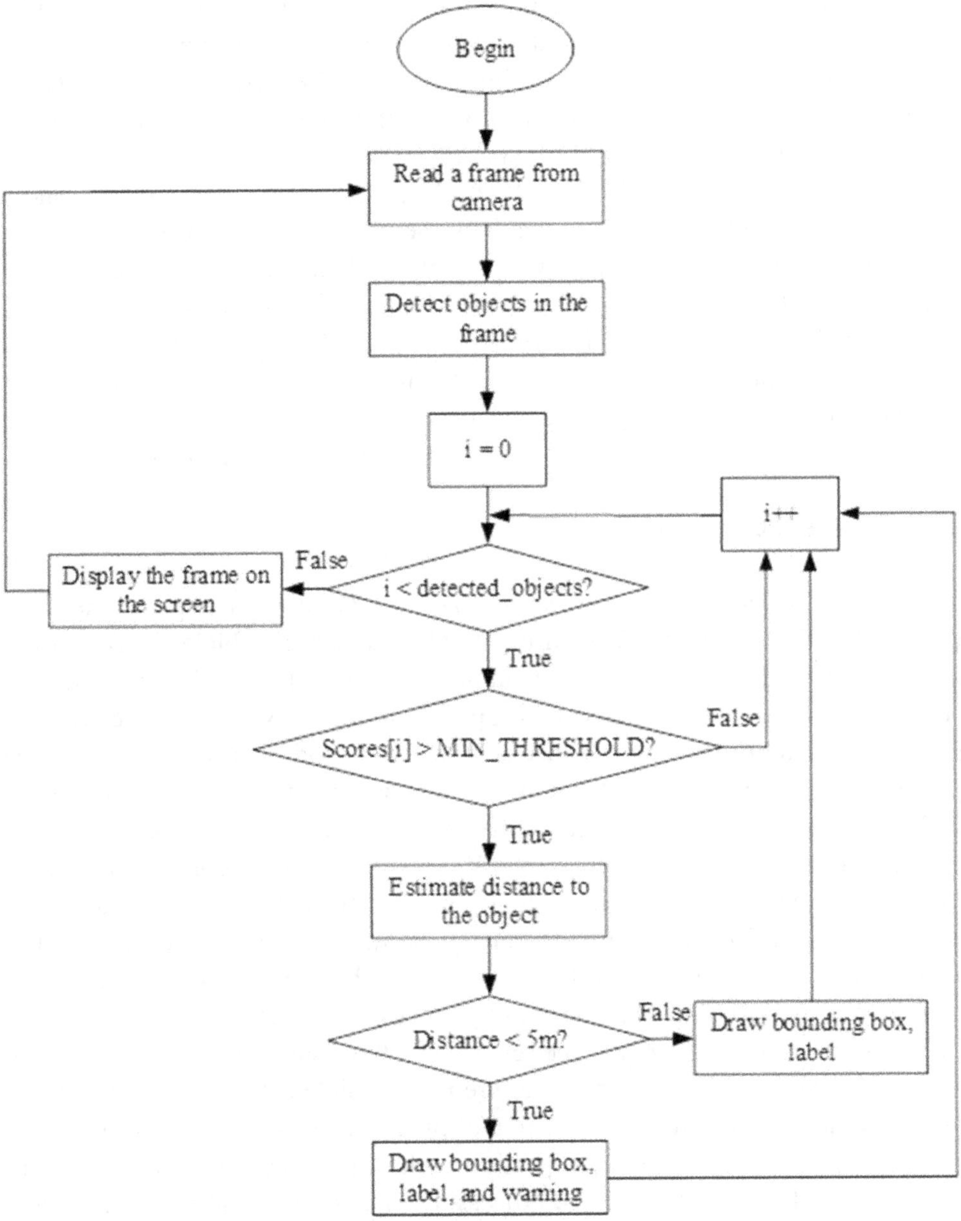

The flowchart in Figure 16 Describes the operation of the software program for detecting and measuring the distance of vehicles in rainy weather conditions. The process begins with reading video frames from the camera. Next, the system performs the task of object detection within the frame using a deep neural network. A counter variable *(i)* is used to count objects in the frame starting from the initial value 0.

The system evaluates each object detected by running the counter variable *(i)* starting from the first detected object up to the last (*i<detected_objects*). If the current value of *i* is less than the number of detected objects in the frame, then the system examines the detection score (*scores[i]*) of the current object against the minimum threshold (*MIN_THRESHOLD*). When the score exceeds the threshold value, the system establishes an estimate for the distance to the object.

An object detection alert, along with a bounding box, will appear when the distance between the object and the system is less than 5 meters. The system performs this task while omitting the alert function when the condition is not met. The process runs until it completes the examination of all objects through repeated increases of counter variable *i* by 1. The system concludes the process by transferring the frame to the display screen before resuming the camera image reading process.

Object Detection Task

Based on the analyses in Section 3, the system adopts the pre-trained and quantized SSD-MobileNet-v2-FPNLite320×320 model for vehicle detection from camera-captured frames under rainy conditions. The quantized SSD-MobileNet-v2-FPNLite320×320 model consists of three parts: MobileNet-v2 for the base network, SSD (Single Shot Detector) as the detection network, and FPNLite320×320 as the feature extractor. The SSD object detection system acquires its high-level features from the base network. Detecting objects at different scales, particularly small objects, is challenging. The Feature Pyramid Network (FPN) functions as an extractor through its feature pyramid design structure to achieve better processing speed alongside higher precision. The system performance is boosted by quantizing the model to the int8 one through TensorFlow Lite conversion. The int8 quantization also helps to shrink the model's size while decreasing processing expenses.

The combination of these advanced elements in the system enables efficient video frame processing, which results in more reliable vehicle detection during adverse weather situations. The int8 quantization of models results in smaller memory requirements and, at the same time, speeds up inference time, which is essential for embedded platforms with restricted processing capabilities. Real-time processing, which is enabled through this optimization of the model, is necessary for autonomous driving applications together with traffic monitoring functions.

Distance Estimation Task

Distance measurement of objects from a camera can be achieved with stereo camera-based vision systems. Research publications by Adi (2017), Salman (2017), Dandil (2019), Zaarane (2020), and Adil (2022) demonstrate various techniques for measuring object distances through stereo camera systems. The distance measurement system requires camera calibration followed by image rectification together with object detection on stereo images and disparity calculation and 3D reconstruction (Adil, 2022). The distance of an object is determined using the central point of the object's segmented bounding box and the Euclidean distance measurement method to find the shortest distance from the camera. The difference between the images captured by the left and right cameras enables calculations for detecting the 3D positions and measuring distances of detected objects. The stereo vision-distancing methods are widely implemented across domains, including autonomous vehicle technology and smart contact lens development, as well as plant surveillance systems.

This algorithm demonstrates how to use stereo camera IMX219-83 for distance measurement of objects:

Step 1: Camera Calibration

A stereo camera system calibration process produces the intrinsic and extrinsic parameters that belong to the left and right cameras. Correct 3D reconstruction alongside distance estimation depends on this step.

Step 2: Image Rectification

A process must align the epipolar lines horizontally throughout both left and right camera images. The image rectification process reduces how complicated stereo matching operations become and how disparities are computed.

Step 3: Object Detection

The object detection algorithm should identify objects in stereo images that have been rectified. The detection process yields boxes that bound the identified objects.

Step 4: Disparity Calculation

A disparity map gets generated through pixel correspondence discovery between rectified left and right camera images. A stereo matching algorithm, such as block matching or semi-global matching algorithm, should be used to produce disparity values.

Step 5: 3D Reconstruction

The 3D positions of detected objects can be computed from the disparity map combined with calibration information. The 3D position of objects results from trigonometry that combines disparity data with camera-setting parameters.

Step 6: Distance Estimation

The distance between the camera and the bounding box center can be measured through calculation based on reconstructed 3D positions. The distance measurement expresses the vector length connecting the camera origin to the 3D position that represents the object center.

Step 7: Output Distance

The algorithm should return the distance evaluation results for recognized objects through its output.

This algorithm leverages the principles of stereo vision to estimate the distance of objects from a calibrated stereo camera system. The method enables measuring object distances accurately through a pair of cameras, which include object detection and disparity calculation along with 3D position reconstruction and Euclidean distance computation.

EXPERIMENT AND EVALUATION

This section implements testing of the integrated system in actual rainy weather to conduct objective assessments. Four different weather conditions, including light rain and moderate rain as well as heavy rain and nighttime rain, will be used to evaluate the system. The system evaluation will analyze its operational performance through three key criteria, which include visibility along with accuracy and processing speed in different conditions. Experts will document the evaluation process by recording system performance as well as all detected issues and its response to altering weather conditions.

Testing in Light Rain Conditions

Figure 17. System screenshot tested in light rain weather scenario.

System performance tests under light rain conditions indicate that the system performs well during such weather conditions. The detection system accurately identifies vehicles ranging from 1 to 15 meters at a frame rate of 6.5 FPS. The experimental evaluation of the system is shown through visual evidence in Figure 17.

Testing in Moderate Rain Conditions

Figure 18. System screenshot tested in moderate rain weather scenario.

The system maintains its performance capabilities in experimental conditions of moderate rainfall. The system achieves accurate car and truck detection between 1-13.5 meters when running at 6.35 FPS frame rate. The experimental evaluation of the system is shown through visual evidence in Figure 18.

Testing in Heavy Rain Conditions

Figure 19. System screenshot tested in heavy rain weather scenario.

The system proves its capabilities when tested through experiments conducted in heavy rainfall conditions. The detection system successfully identifies cars throughout 1-14.59 meters at an average frame speed of 6.46 FPS. The experimental evaluation of the system is shown through visual evidence in Figure 19.

Testing in Nighttime Rain Conditions

Figure 20. System screenshot tested in nighttime rain weather scenario.

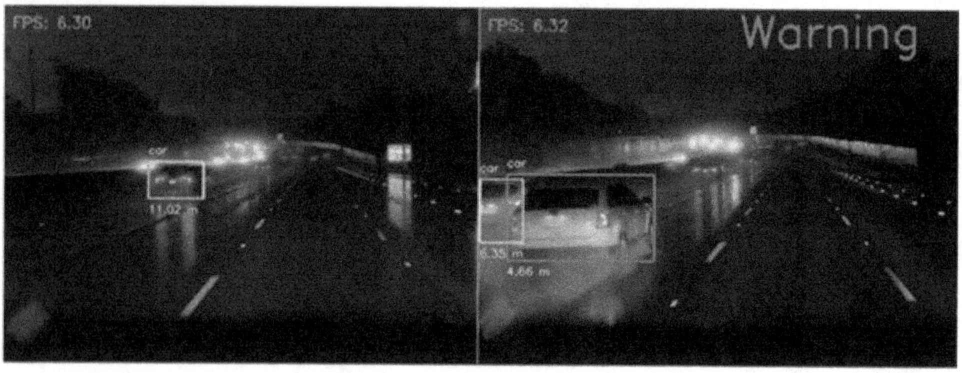

Practical tests examining the system under rainy nighttime conditions prove its reliability in such conditions. The system provides accurate detection of cars between 1 and 14 meters at a frame rate that approximates 6.33 FPS. The experimental evaluation of the system is shown through visual evidence in Figure 20.

Discussion

The experimental findings under different test conditions produced significant conclusions.

The detection system reliably identifies traffic vehicles regardless of rain or changing light situations. The detection system works for traffic vehicles within a range of 1 to 15 meters while delivering approximately 6.5 frames per second to ensure the safety of the vehicle. This can be illustrated as follows. With a frame rate of 6.5 FPS, the average time to process one frame is approximately 0.15 seconds. The available processing time fulfills requirements to handle possible unwanted situations. For instance, when a car is detected ahead at a distance s = 5 meters, the system will provide the next result after $\Delta t = 0.15$ seconds. Let v be the speed of the vehicle, and we have:

distance covered =

$$v.\Delta t < s \qquad (2)$$

Therefore,

$$v < \frac{s}{\Delta t} = \frac{5}{0.15} = 33.3 m/s \approx 120 km/h \qquad (3)$$

The processing speed of the system indicates it meets the requirements regarding time-sensitive tasks, including detection and situation assessment and system notifications. The system can deliver immediate responses to potential situations.

CONCLUSIONS

The road transportation industry sees autonomous vehicle technology developing into a major industry trend in current times. However, the performance capabilities of autonomous vehicles experience decline during times of poor weather, especially when rainfall occurs. The sensors on the vehicle lose their visibility during rainfall, which results in impaired performance as well as accuracy of autonomous driving functions. Our designed embedded system functions to precisely identify motorcycles and cars, along with buses and trucks, from a distance of 1-15 meters in rainy conditions for autonomous vehicles. The system uses a quantized SSD MobileNet-v2 FPNLite320×320 model, which received training through 4154 pictures from the BDD 100K dataset, along with various images from dashcam videos. This system delivers real-time processing speed at which it accomplishes up to 6.78 frames per second, therefore enabling quick responses to potential risks that prevent accidents. The system requires minimal system memory because its size amounts to just 3.57 MB, which represents 11.9% of the total available RAM. Due to its small energy consumption level, the system can be seamlessly incorporated into current autonomous vehicles while maintaining their standard operational capabilities. We assert that this system will substantially boost the effectiveness and safety of autonomous vehicles when driving in harsh weather conditions. The system still exhibits some constraints because it fails to recognize every automotive vehicle that operates on the road. The identification capability for motorcycles operates at a low rate due to inadequate training data. The detection system makes occasional mistakes by recognizing objects such as trees or billboards as vehicles, but these wrong identifications happen infrequently. We will improve and enhance the system in the future to overcome existing problems for better performance. The system will receive additional recognition functions to detect adverse situations when weather conditions become dark from fog or being covered by snow. We will enhance the detection capabilities to monitor pedestrians together with animals and traffic lanes while also scanning traffic signals to construct a complete system that ensures driving safety. The anticipated application improvements will eventually develop into a safer driving system that reduces accidents and ensures everyone experiences safe and comfortable driving conditions. The system needs to adopt advanced machine learning algorithms since they improve how it learns in real-world driving environments. Driving data collected on a vast scale will enable the system to improve its performance continually so it responds successfully to new challenges and environments.

REFERENCES

Abdullah, M. N., & Ali, Y. H. (2021). Vehicle detection system at different weather conditions. *Iraqi Journal of Science*, *62*(3), 2040–2052. DOI: 10.24996/ijs.2021.62.6.30

Adi, K., & Widodo, C. E. (2017). Distance measurement with a stereo camera. *International Journal of Innovative Research in Advanced Engineering*, *4*(11), 24–27.

Adil, E., Mikou, M., & Mouhsen, A. (2022). A novel algorithm for distance measurement using stereo camera. *CAAI Transactions on Intelligence Technology*, *7*(2), 177–186. DOI: 10.1049/cit2.12098

Ahmed, M., Hegazy, M., Klimchik, A. S., & Boby, R. A. (2022). Lidar and camera data fusion in self-driving cars. *Компьютерные исследования и моделирование [Computer research and modeling]*, *14*(6), 1239–1253.

Chen, X. Z., Chang, C. M., Yu, C. W., & Chen, Y. L. (2020). A real-time vehicle detection system under various bad weather conditions based on a deep learning model without retraining. *Sensors (Basel)*, *20*(20), 5731. DOI: 10.3390/s20205731 PMID: 33050173

Dandil, E., & Çevik, K. K. (2019). Computer vision-based distance measurement system using stereo camera view. In *2019 3rd International Symposium on Multidisciplinary Studies and Innovative Technologies (ISMSIT)* (pp. 1–4). IEEE. DOI: 10.1109/ISMSIT.2019.8932817

Everingham, M., Van Gool, L., Williams, C. K., Winn, J., & Zisserman, A. (2010). The PASCAL visual object classes (VOC) challenge. *International Journal of Computer Vision*, *88*(2), 303–338. DOI: 10.1007/s11263-009-0275-4

Girshick, R. (2015). Fast R-CNN. In *Proceedings of the IEEE International Conference on Computer Vision* (pp. 1440–1448).

Girshick, R., Donahue, J., Darrell, T., & Malik, J. (2014). Rich feature hierarchies for accurate object detection and semantic segmentation. In *Proceedings of the IEEE Conference on Computer Vision and Pattern Recognition* (pp. 580–587). DOI: 10.1109/CVPR.2014.81

Gupta, H., Kotlyar, O., Andreasson, H., & Lilienthal, A. J. (2024). Robust object detection in challenging weather conditions. In *Proceedings of the IEEE/CVF Winter Conference on Applications of Computer Vision* (pp. 7523–7532). DOI: 10.1109/WACV57701.2024.00735

Hao, C. Y., Chen, Y. C., Chen, T. T., Lai, T. H., Chou, T. Y., Ning, F. S., & Chen, M. H. (2024). Synthetic data-driven real-time detection transformer object detection in raining weather conditions. *Applied Sciences (Basel, Switzerland)*, *14*(11), 4910. DOI: 10.3390/app14114910

Hasirlioglu, S., & Riener, A. (2019). Challenges in object detection under rainy weather conditions. In *Intelligent Transport Systems, From Research and Development to the Market Uptake: Second EAI International Conference, INTSYS 2018, Guimarães, Portugal, November 21–23, 2018* [Springer International Publishing.]. *Proceedings*, *2*, 53–65.

Hassaballah, M., Kenk, M. A., Muhammad, K., & Minaee, S. (2020). Vehicle detection and tracking in adverse weather using a deep learning framework. *IEEE Transactions on Intelligent Transportation Systems*, *22*(7), 4230–4242. DOI: 10.1109/TITS.2020.3014013

Hnewa, M., & Radha, H. (2020). Object detection under rainy conditions for autonomous vehicles: A review of state-of-the-art and emerging techniques. *IEEE Signal Processing Magazine*, *38*(1), 53–67. DOI: 10.1109/MSP.2020.2984801

Huang, J., Rathod, V., Birodkar, V., Myers, A., Lu, Z., Votel, R., Chen, Y., & Chow, D. (2021). TensorFlow 2 detection model zoo. https://github.com/tensorflow/models/blob/master/research/object_detection/g3doc/tf2_detection_zoo.md

Jiang, Z., Zhao, L., Li, S., & Jia, Y. (2020). Real-time object detection method based on improved YOLOv4-tiny. *arXiv preprint* arXiv:2011.04244.

Kenk, M. A., & Hassaballah, M. (2020). DAWN: Vehicle detection in adverse weather nature dataset. *arXiv preprint* arXiv:2008.05402.

Kim, T., Jeon, H., & Lim, Y. (2023). Challenges of YOLO series for object detection in extremely heavy rain: CALRA simulator based synthetic evaluation dataset. *arXiv preprint* arXiv:2312.07976.

Liang, L., Ma, H., Zhao, L., Xie, X., Hua, C., Zhang, M., & Zhang, Y. (2024). Vehicle detection algorithms for autonomous driving: A review. *Sensors (Basel)*, *24*(10), 3088. DOI: 10.3390/s24103088 PMID: 38793942

Liu, W., Anguelov, D., Erhan, D., Szegedy, C., Reed, S., Fu, C. Y., & Berg, A. C. (2016). SSD: Single shot multibox detector. In *Computer Vision–ECCV 2016: 14th European Conference, Amsterdam, The Netherlands, October 11–14, 2016* [Springer International Publishing.]. *Proceedings*, *14*(Part I), 21–37.

Ponn, T., Kröger, T., & Diermeyer, F. (2020). Identification and explanation of challenging conditions for camera-based object detection of automated vehicles. *Sensors (Basel), 20*(13), 3699. DOI: 10.3390/s20133699 PMID: 32630350

Redmon, J., Divvala, S., Girshick, R., & Farhadi, A. (2016). You only look once: Unified, real-time object detection. In *Proceedings of the IEEE Conference on Computer Vision and Pattern Recognition* (pp. 779–788). DOI: 10.1109/CVPR.2016.91

Salman, Y. D., Ku-Mahamud, K. R., & Kamioka, E. (2017, April). Distance measurement for self-driving cars using stereo camera. In *International Conference on Computing and Informatics* (Vol. 1, No. 105, pp. 235–242).

Sandler, M., Howard, A., Zhu, M., Zhmoginov, A., & Chen, L. C. (2018). MobileNetV2: Inverted residuals and linear bottlenecks. In *Proceedings of the IEEE Conference on Computer Vision and Pattern Recognition* (pp. 4510–4520).

Simonyan, K., & Zisserman, A. (2014). Very deep convolutional networks for large-scale image recognition. *arXiv preprint* arXiv:1409.1556.

Stoma, M., Dudziak, A., Caban, J., & Droździel, P. (2021). The future of autonomous vehicles in the opinion of automotive market users. *Energies, 14*(16), 4777. DOI: 10.3390/en14164777

Tahir, N. U. A., Zhang, Z., Asim, M., Chen, J., & ELAffendi, M. (2024). Object detection in autonomous vehicles under adverse weather: A review of traditional and deep learning approaches. *Algorithms, 17*(3), 103. DOI: 10.3390/a17030103

Tan, M., Pang, R., & Le, Q. V. (2020). EfficientDet: Scalable and efficient object detection. In *Proceedings of the IEEE/CVF Conference on Computer Vision and Pattern Recognition* (pp. 10781–10790).

Tariq, M. U. (2025). Harnessing Technology to Combat Bias: Innovative Tools and Techniques in HR. In Kanwal, S. (Ed.), *Navigating Bias in Human Resource Management* (pp. 155–184). IGI Global Scientific Publishing., DOI: 10.4018/979-8-3693-7132-9.ch006

Tariq, M. U. (2025). Smart Forecasting With AI: Revolutionizing Energy Consumption and Storage. In Derbali, A. (Ed.), *Integrating Artificial Intelligence Into the Energy Sector* (pp. 165–184). IGI Global Scientific Publishing., DOI: 10.4018/979-8-3693-7112-1.ch009

Tariq, M. U. (2025). Innovative Business Models: Integrating Social, Economic, and Environmental Goals in Rural Manufacturing. In Prabhakar, A., Yu, P., & Erokhin, V. (Eds.), *Rural Social Entrepreneurship Development: Network-Based Manufacturing System Model* (pp. 147–174). IGI Global Scientific Publishing., DOI: 10.4018/979-8-3693-7515-0.ch005

Tariq, M. U. (2025). Revolutionizing Communication: EEG-Based Brain-Computer Interface for Speech and Mood Detection. In Prabhakar, A., Yu, P., & Erokhin, V. (Eds.), *Rural Social Entrepreneurship Development: Network-Based Manufacturing System Model* (pp. 237–264). IGI Global Scientific Publishing., DOI: 10.4018/979-8-3693-7515-0.ch008

Vargas, J., Alsweiss, S., Toker, O., Razdan, R., & Santos, J. (2021). An overview of autonomous vehicles sensors and their vulnerability to weather conditions. *Sensors (Basel), 21*(16), 5397. DOI: 10.3390/s21165397 PMID: 34450839

Wang, C. Y., Yeh, I. H., & Mark Liao, H. Y. (2024, September). YOLOv9: Learning what you want to learn using programmable gradient information. In *European Conference on Computer Vision* (pp. 1–21). Cham: Springer Nature Switzerland.

Wang, R., Zhao, H., Xu, Z., Ding, Y., Li, G., Zhang, Y., & Li, H. (2023). Real-time vehicle target detection in inclement weather conditions based on YOLOv4. *Frontiers in Neurorobotics, 17*, 1058723. DOI: 10.3389/fnbot.2023.1058723 PMID: 36968300

Xia, H., Lan, Y., Song, S., & Li, H. (2022). Raindrop removal from a single image using a two-step generative adversarial network. *Signal, Image and Video Processing, 16*(3), 677–684. DOI: 10.1007/s11760-021-02007-z

Yu, F., Chen, H., Wang, X., Xian, W., Chen, Y., Liu, F., & Darrell, T. (2020). BDD100K: A diverse driving dataset for heterogeneous multitask learning. In *Proceedings of the IEEE/CVF Conference on Computer Vision and Pattern Recognition* (pp. 2636–2645). DOI: 10.1109/CVPR42600.2020.00271

Zaarane, A., Slimani, I., Al Okaishi, W., Atouf, I., & Hamdoun, A. (2020). Distance measurement system for autonomous vehicles using stereo camera. *Array (New York, N.Y.), 5*, 100016. DOI: 10.1016/j.array.2020.100016

Chapter 11
Legal Challenges in the Integration AIoT (Artificial Intelligence and the Internet of Things) in AVs (Autonomous Vehicles):
National Lessons for Global Governance

Aliakbar Shams
https://orcid.org/0000-0001-9154-5679
Independent Researcher, Austria

ABSTRACT

This paper explores the legal challenges surrounding the integration of Artificial Intelligence and the Internet of Things (AIoT) in Autonomous Vehicles (AVs), drawing national lessons from key countries—the United States, China, the United Kingdom, Germany, and Singapore—to inform frameworks for global governance. AVs, characterized by their ability to operate autonomously, present transformative opportunities across sectors, including transportation, logistics, and mobility services. However, these advancements introduce complex regulatory, legal, and ethical dilemmas, such as liability, accountability, cybersecurity, and privacy. The convergence of AI and IoT enables AVs to enhance safety, fuel efficiency, and environmental sustainability but also necessitates comprehensive legal frameworks to address uncertainties and societal impacts. By analyzing national approaches to AV integration, this paper identifies critical insights to balance innovation with safety, equity, and accountability.

DOI: 10.4018/979-8-3373-4332-7.ch011

INTRODUCTION

This introductory section offers a comprehensive overview of Artificial Intelligence, and the Internet of Things (AIoT) as applied to autonomous vehicles (AVs). It highlights critical aspects of their current state and addresses widespread misconceptions. By dispelling these myths, the chapter presents a detailed perspective on the legal challenges and opportunities associated with integrating AIoT into AVs, drawing from national experiences to inform global governance frameworks. This discussion is a foundation for a multifaceted analysis in the following chapters. Autonomous vehicles (AVs) represent a groundbreaking innovation in the automotive industry. They are characterized by their ability to operate independently of direct human intervention. These vehicles can perceive their surroundings, make decisions, and act autonomously. This section aims to provide a thorough overview of AVs, exploring their technological foundations, societal implications, economic impacts, ethical challenges, environmental considerations, and, most importantly, their regulatory and legal challenges. AVs rely on an advanced integration of state-of-the-art technologies, demonstrating the transformative potential of artificial intelligence (AI), machine learning (ML), the Internet of Things (IoT), sensors, actuators, and communication systems. At their core, AVs are designed to sense and interpret their environment, process data, and navigate complex scenarios. Leveraging AI and ML, these systems continually learn and adapt, improving their functionality and responsiveness. Connectivity enhances these capabilities by enabling seamless interaction with other vehicles, infrastructure, and cloud platforms. A comprehensive understanding of these technological underpinnings is essential for appreciating AVs' potential and limitations. It provides a foundation for addressing their broader technical, social, and regulatory challenges.

Autonomous vehicles (AVs) are classified into six levels of automation, as defined by the Society of Automotive Engineers (SAE), ranging from Level 0 (no automation) to Level 5 (full automation). Each level represents a specific stage of vehicle autonomy, defined by the extent of human intervention required:

- Level 0 (No Automation): The human driver performs all tasks without assistance from automated systems.
- Level 1 (Driver Assistance): The vehicle can assist with specific functions, such as steering or acceleration, but the human driver always retains full control.
- Level 2 (Partial Automation): Under certain conditions, the vehicle can control both steering and acceleration simultaneously, but the driver must remain engaged and ready to intervene.

- Level 3 (Conditional Automation): The vehicle can perform all driving tasks under certain conditions, but the driver must be prepared to take control if needed.
- Level 4 (High Automation): The vehicle can handle all driving tasks within specific environments or conditions without human intervention, although limitations may exist in certain situations (e.g., severe weather).
- Level 5 (Full Automation): The vehicle is fully autonomous and capable of operating without human input in all conditions and environments.

ROUTE: NATIONAL LESSONS

New technologies are reshaping industries and society as the world continues to automate various aspects of everyday life, including processes such as manufacturing and logistics (Dunn, 2017). As automation continues to advance across various aspects of everyday life, it is only natural that the concept of autonomous cars is at the forefront of many people's minds. This prospect has generated excitement and apprehension as the possibility of autonomous vehicles (AVs) on the road soon becomes increasingly tangible. Vehicle manufacturers are making significant progress in this area, with prototypes of fully autonomous vehicles emerging globally (Palmer, 2016). Legislators are increasingly questioning how these innovations should be regulated. This chapter section prioritizes analyzing various countries' legal and policy frameworks relevant to the research.

- *THE UNITED STATES*

The United States needs a unified federal regulatory framework for overseeing the deployment and monitoring of autonomous vehicles (AVs). Since 1966, the National Traffic and Motor Safety Act has granted the U.S. Department of Transportation (DOT), through the National Highway Traffic Safety Administration (NHTSA), the authority to establish federal motor vehicle safety standards for "motor vehicles" and "motor vehicle equipment" (Light, 2017). NHTSA has utilized this authority to address gaps in Federal Motor Vehicle Safety Standards (FMVSS) (NHTSA, n.d.). These federal standards remain in place; however, they are inadequate for regulating AVs because they "were created with human drivers in mind" (Jansma, 2016). Rather than implementing comprehensive regulations for AVs, NHTSA has granted manufacturers limited exemptions to the FMVSS, which are valid for only two to three years (Roth, 2020). A manufacturer can exempt only 2,500 vehicles per year from these regulations. The DOT and NHTSA have published several voluntary standards and guidelines to support the development of AV regulations. The most recent report, "Ensuring American Leadership in Automated Vehicles 4.0," was released in January 2020 and builds upon reports that began in 2016. These reports outline various safety standards and regulations, including a "15-point 'Safety Assessment,'" and emphasize the operation of AVs at levels three and higher. The latest report also introduces a requirement for reporting the types of accidents and other incidents to NHTSA and DOT (National Science & Technology Council & U.S. Department of Transportation, 2020). Under the American system of federalism, laws can be enacted at both the state and federal levels, with state laws applying only to individuals within that state. Laws enacted at the federal level are considered the "supreme Law of the Land" and have the authority to preempt any state law that conflicts with them. However, states retain general police powers (or reserved powers), as the Constitution grants them all powers not explicitly delegated to the federal government. Driving laws are typically established at the state level, with states generally regulating the drivers of vehicles. At the same time, the National Highway Traffic Safety Administration (NHTSA) is responsible for regulating the vehicles themselves. Since state legislatures have full authority to regulate drivers, the United States needs uniform regulations nationwide. For instance, in New York, a driver is prohibited from passing a stopped school bus, even if the bus is on the opposite side of a median. However, in New Jersey, a driver can continue driving as long as they do not exceed ten miles per hour, while in Indiana, the driver is not required to stop at all. (Werner, 2020) This is just one example of how laws can vary significantly from state to state, highlighting the challenges of implementing AVs nationwide under the current regulatory structure. Manufacturers must navigate a complex landscape of differing state regulations to ensure their AVs can operate nationwide. Since states have jurisdiction over driving laws, AV-specific legislation would likely stem from existing laws. Each state would likely develop regulations based on current definitions for standard terms for operating a motor vehicle. However, using existing definitions for AVs presents challenges due to the inconsistent interpretations of key terms, such as "driver" and "operator" (Smith, 2014). For instance, Alaska's statutes define a driver as a "person who drives or is in actual physical control of a vehicle." In contrast, Maryland's statute defines a driver as "any individual who drives a vehicle." The distinction between these definitions is notable: Maryland's law requires a human to be inside the vehicle, whereas Alaska's definition could allow a vehicle to operate without anyone physically present. These differences create challenges when designing an AV that complies with regulations nationwide, as manufacturers need help determining how to accommodate human drivers. Currently, the regulation of AVs has primarily been delegated to individual states. As previously mentioned, states are not obligated to enforce the guidance standards issued by the DOT. While some states have begun developing regulatory frameworks, such as introducing testing procedures, state and local laws remain uneven, resulting in an ineffective "patchwork" of regulations. As of September 2022, forty-one states and the District of Columbia have implemented some form of regulation for autonomous vehicles, resulting in forty-two standards that manufacturers must adhere to if they wish to deploy AVs in the U.S. State regulators have established these regulations through executive orders signed by the state's governor, legislation passed by the state legislature, or a combination of both. This has resulted in inconsistency and challenges for manufacturers and drivers navigating this emerging legal framework (National Conference of State Legislators, 2020).

- *CHINA*

In China, the Ministry of Public Security issued the Draft Proposed Amendments to the Road Traffic Safety Law (the "MPS Proposed Amendments") on March 24, 2021. These amendments clarify road testing requirements and vehicle access to vehicles equipped with automated driving functions and outline how liability for traffic violations and accidents will be assigned (Ministry of Public Security, 2021). This represents the first instance in which China has proposed specific legislation for autonomous vehicles under the Road Traffic Safety Act, marking a significant milestone. Additionally, on March 23, 2021, the Standing Committee of the Shenzhen Municipal People's Congress published the Draft for Comments of the Regulations of Shenzhen Special Economic Zone on the Administration of Intelligent and Connected Vehicles (the "Shenzhen Draft Regulations") on its website for public consultation. The Shenzhen Draft Regulations cover the entire lifecycle of autonomous vehicle development, including road testing, access registration, usage management, road transportation, traffic accidents, handling of violations, and legal liability. Shenzhen is positioning itself as the first region in China to enable the commercialization of autonomous vehicles. It is anticipated that other regions in China will follow suit and rise to meet the challenge set by Shenzhen. The MPS Proposed Amendments and the Shenzhen Draft Regulations mark the beginning of China's specific legislative efforts concerning autonomous vehicles. The MPS Proposed Amendments outline specific requirements for road testing and access to autonomous vehicles. These vehicles with automated driving functions must undergo testing at designated times, areas, and routes, as stipulated by law. Initially, test vehicles must pass trials on closed roads or venues and obtain temporary license plates before road testing. Once they complete road tests, these vehicles can be manufactured, imported, and sold in compliance with relevant laws and regulations. Vehicles requiring public road access must apply for motor vehicle registration plates. Although the MPS Proposed Amendments only briefly mention the requirements for testing and access to autonomous vehicles, they mark a significant step in integrating autonomous cars into China's transportation regulatory framework.

If the provisions are passed, this will be the first time China has enacted specific legislation regarding autonomous vehicles (MPS Draft Proposed Amendments, 2021). China has an access management system for motor vehicle products wishing to enter the market, with the Ministry of Industry and Information Technology (MIIT) overseeing the management of both motor vehicle products and companies. However, autonomous vehicles currently do not have access standards, meaning they cannot be included in the motor vehicle products catalog. As a result, these vehicles cannot be sold, registered, or driven on the roads in the same manner as traditional vehicles. In 2017, the MIIT and the Standardization Administration of the People's Republic of China (PRC) jointly issued the "Guide for the Construction of the National Internet of Vehicles Industry Standards and System." This guide proposed developing around 269 standards for autonomous vehicles and related products. Fifty of these standards have either been issued or are in the public comment phase. Most of these standards pertain to intelligent vehicles, while those specifically related to automated driving are still in the early stages of research. Consequently, it is unlikely that any national or industry standards for autonomous vehicle products will be issued shortly. In response, the Shenzhen Draft Regulations allow for local and group standards and an exemption from the usual access requirements. Under these regulations, autonomous vehicles that meet local or group standards will be included in the Shenzhen autonomous vehicle products catalog and will be eligible for sale, registration, and licensing within the Shenzhen Special Administrative Region. The regulations also state that an exemption system can be implemented if autonomous vehicles fail to meet the access conditions due to new technologies, techniques, or materials. This exemption system is already in place through the MIIT's "Administrative Measures for Access of Road Motor Vehicle Manufacturers and Products," which came into effect on June 1, 2019. According to Article 24 of these measures, road motor vehicle manufacturers are encouraged to innovate technologically. Companies can apply for an exemption if the conditions in the measures cannot be met due to the adoption of new technologies or materials. The Shenzhen Draft Regulations also provide restrictive measures for autonomous vehicle access. Relevant authorities can impose restrictions on access validity, automation level, and specific application scenarios for autonomous vehicle products by the provisions of the MIIT Access Administrative Measures. Furthermore, the Shenzhen Draft Regulations allow autonomous vehicles to be driven on the roads within the Shenzhen Special Administrative Region once registered with the Public Security Traffic Administrative Bureau and obtain registration certificates, special number plates, and driving licenses (Ministry of Industry and Information Technology, 2021). The Shenzhen Draft Regulations include a specific chapter that addresses cybersecurity and data protection for autonomous vehicles. Cyberattacks pose a persistent threat in an increasingly connected world. The interconnectivity of autonomous cars with external systems heightens the challenge of maintaining robust cybersecurity. Furthermore, the widespread use of autonomous vehicles will collect significant amounts of personal data, including information about the driver or passengers, driving patterns, location, route history, speed, and mileage. This raises serious privacy concerns for drivers, passengers, and even other road users, such as pedestrians. The Shenzhen Draft Regulations impose clear responsibilities on companies involved with autonomous vehicles regarding cybersecurity and data protection. First, these companies must establish cybersecurity evaluation and management systems to safeguard internet data from leakage, theft, or tampering, ensuring the data's accuracy, integrity, and usability. Second, companies must implement measures to prevent the leakage, loss, or destruction of personal information in compliance with relevant laws and create data security and privacy protection systems. Finally, the unauthorized collection, processing, or use of personal data is strictly prohibited. This prohibition also extends to the unlawful collection of data related to national security. A significant feature of the Shenzhen Draft Regulations is the precise allocation of liability for traffic violations and accidents involving autonomous vehicles. The regulations make a distinction between autonomous vehicles with a driver and those without. Autonomous vehicles with a driver: In this case, the liability for traffic violations or accidents will be handled according to existing road traffic laws. The driver will be held responsible for any violations or accidents. However, if a defect in the autonomous vehicle causes an accident, the driver has the right to seek compensation from the vehicle's manufacturer or distributor. Autonomous vehicles without a driver: The regulations specify that the vehicle's controller or owner is responsible for traffic violations or accidents. However, if a defect in the vehicle causes the accident, the controller or owner may seek compensation from the manufacturer or distributor. The regulations further clarify the liability of the controller or owner of autonomous vehicles without a driver: The controller is responsible for any traffic violations or accidents during the execution of all dynamic driving tasks by the automated driving system, including when taking over the dynamic driving task. In other situations, the vehicle owner will be held responsible for traffic violations or accidents caused by improper management, use, or vehicle maintenance (Tariq, 2025).

China has recently introduced national safety guidelines for autonomous vehicles (AVs), aiming to accelerate their deployment nationwide. These guidelines cover AVs used in public transportation, including buses, self-driving taxis, and cargo vehicles. They are intended to form the foundation of a nationwide regulatory framework for AVs, addressing the need for consistency in their operation. Until now, the regulation of AVs has primarily been left to local authorities, with cities like Shenzhen leading the way in experimentation. Given China's rapid progress in the field, the new guidelines indicate a more structured approach to China's AV deployment than expected. While the country is advancing faster than the United States in the rollout of AVs, the regulatory environment is not a free-for-all. The guidelines stipulate specific operational scenarios: Buses and trams are allowed only on "physically closed, relatively closed, or fixed routes with simple road conditions." Similarly, autonomous cargo transport is restricted to point-to-point routes on trunk roads and urban streets where traffic safety can be controlled and the transportation of dangerous goods is prohibited. The guidelines also address the role of human safety operators. A safety operator must be on board for buses, trams, and cargo vehicles. While the requirement for safety operators on autonomous cargo vehicles is stated as "in principle," it remains a key part of the regulatory framework. Depending on their level of autonomy, self-driving taxis will also need a human operator on board for conditionally or highly autonomous vehicles. In contrast, staff can monitor fully autonomous taxis remotely, though each operator can oversee a maximum of three taxis simultaneously. The guidelines further specify how AVs should handle accidents and incidents. AVs must transmit operating status information in real time to operators and local authorities, including 360-degree exterior video monitoring, in-car audio, and remote-control instructions. In a crash, the system must record and store information for at least 90 seconds before the incident. Finally, safety measures also require AVs to feature distinctive patterns, text, or colors to signal their status to other road users. These new guidelines have been issued to all provinces and municipalities in China, with a directive to implement them accordingly (Tariq, 2025).

- *UNITED KINGDOM*

The Automated and Electric Vehicles Act 2018 (UK) provides a potential solution for addressing liability issues related to autonomous, automated, or driverless vehicles. The Act extends the scope of compulsory motor vehicle insurance to include the operation of automated vehicles in automated mode. Under this framework, victims of accidents—including the "driver" who has legitimately handed over control to the automated vehicle—are covered by the vehicle's compulsory insurance. According to the Act, the insurer is initially responsible for compensating all victims of an accident caused by a defect in the automated vehicle. This ensures that individuals receive timely compensation, regardless of whether the fault lies with the vehicle's technology. After settling claims, the insurer retains the right to recover costs from the party ultimately responsible for the defect by existing common law and product liability law. This mechanism provides a clear and practical approach to addressing liability while ensuring the protection of accident victims (Legislation.gov.uk, 2018). A compelling rationale for the Automated and Electric Vehicles Act 2018 (UK) lies in the United Kingdom's long-standing focus on ensuring that victims of road traffic accidents are compensated swiftly and equitably. Unlike traditional road collisions, accidents involving automated vehicles operating in automated mode may not result from human error but from failures within the vehicle itself. Without the provisions of the Act, victims in such cases might be forced to seek recourse through litigation against the vehicle manufacturer, a process likely to be expensive, complex, and protracted. The Act mandates compulsory insurance for automated vehicles under a "single insurer" model to address this challenge. This framework ensures that, following an accident, the insurer of the automated vehicle is liable for damages caused to the insured or any third party while the vehicle is driving itself. However, the insurer or vehicle owner is not held liable when the person responsible for the vehicle negligently allowed it to operate in automated mode under inappropriate conditions. Moreover, the Act allows insurers to limit or exclude liability for damages in specific circumstances, such as when an accident results from unauthorized software modifications or the insured's failure to install critical safety updates that they were aware of or reasonably should have been aware of. Notably, the Act also preserves the insurer's or vehicle owner's right to pursue claims against any third party responsible for the accident, ensuring a balanced approach to liability distribution (Automated and Electric Vehicles Act, 2018). Consequently, if the manufacturer of the automated vehicle, or another third party, is found responsible for the damage or injury caused, the insurer or vehicle owner remains liable to compensate the injured party. However, the insurer or vehicle owner retains the right to recover the compensation paid from the manufacturer or third party responsible under applicable laws, including existing product liability laws. This approach ensures that victims receive prompt compensation while enabling insurers or vehicle owners to seek reimbursement from those ultimately at fault, thereby maintaining a fair and efficient liability framework (Hand & Appt, 2020). The insurance industry has expressed strong support for the approach taken in the Automated and Electric Vehicles Act 2018 (UK), which extends compulsory motor vehicle insurance to cover automated vehicles. Notably, this legislation provides a clear framework for assigning liability, ensuring that victims of accidents involving automated vehicles are compensated efficiently. Car owners must purchase insurance policies that comply with existing road traffic regulations, maintaining consistency with traditional vehicle insurance requirements while addressing the unique challenges of automated driving technology (Kennedy, 2019). With many automated cars capable of switching between automated and driver-controlled modes, insurers have expressed concerns about the complexity of requiring two distinct insurance policies to cover each scenario. Such a system could create confusion for vehicle owners and complicate claims processes. A unified approach, like the one under the Automated and Electric Vehicles Act 2018 (UK), which integrates coverage for automated and driver-controlled modes under a single insurance policy, is preferred. This approach simplifies liability management and ensures seamless protection for all parties involved, regardless of the vehicle's mode of operation during an accident. Moreover, this legislation prioritizes the recovery of the driver or insured party by ensuring there are "no gaps" in coverage. It places strict obligations on insurers to provide compensation in the event of an incident involving the insured vehicle, regardless of fault. Only after compensating the victim does the insurer have the right to pursue recovery as a secondary action. This approach ensures that victims are compensated quickly and fairly without delays caused by disputes over fault while still allowing insurers to recover costs from the responsible party later (Williams, as cited in Carey et al., 2021).

Even with this early and broadly accepted legislative initiative to address autonomous vehicle liability issues, the UK government recognized that a comprehensive overarching regulatory framework review was necessary to manage this transport revolution. Accordingly, in 2018, the government asked the Law Commission of England and Wales and the Scottish Law Commission (the Law Commission) to review legislation to introduce self-driving vehicles on Great Britain's roads safely (Law Commission, 2022). Even with the early and broadly accepted legislative initiative to address autonomous vehicle (AV) liability issues, the UK government recognized that a comprehensive, overarching review of the regulatory framework was necessary to manage the transport revolution brought about by self-driving vehicles. To this end, in 2018, the government tasked the Law Commission of England and Wales and the Scottish Law Commission (the Law Commission) with reviewing existing legislation to prepare for the safe introduction of self-driving vehicles on the roads of Great Britain. The review concluded in January 2022, publishing a report with 75 recommendations for the government. These recommendations, taken together, outlined a new regulatory framework for self-driving vehicles. The review was comprehensive, involving extensive consultation across the self-driving vehicle landscape, including manufacturers, insurers, academics, and civil society. The review generated a wealth of evidence and provided international thought leadership on how self-driving vehicles should be regulated. In August 2022, the UK government published a detailed response to the Law Commission's report, committing to a new legislative framework for self-driving vehicles in Great Britain. The government adopted the Law Commission's recommendations with only a few refinements and adjustments. The government's response outlined a consultation period, which concluded on October 12, 2022, and set an implementation timeline of three years to conclude the legislative process by 2025.

- *GERMANY*

Germany has been researching to advance autonomous vehicles (AVs) and bring them to fruition since the 1980s. Considering the automotive industry's pivotal role in the German economy, it is unsurprising that automated driving technology is emerging as an increasingly significant focus across the entire value chain of Germany's automotive sector (Norton Rose Fulbright, 2016). In response, the German legislature has enacted significant amendments to the Road Traffic Act, incorporating specific provisions for automated and autonomous driving. 2017, the Road Traffic Act was revised to accommodate motor vehicles with highly or fully automated driving functions (i.e., SAE Level 3 vehicles). Then, in 2021, the Autonomous Driving Act was introduced to establish a legal framework for vehicles with autonomous driving capabilities (i.e., SAE Level 4 vehicles) within designated operating areas. Fabian Pütz et al. characterize the 2017 amendments as primarily maintaining the existing liability and insurance framework, marked by a blend of strict liability for the vehicle owner and fault-based liability for the (human) driver (Pütz et al., 2019). The Road Traffic Act stipulates that the individual operating a highly or fully automated vehicle is still considered the vehicle's driver. The Road Traffic Act also allows the driver to disengage from the traffic situation and control the vehicle while driving. However, the driver must remain alert enough to resume control of the vehicle immediately—either when prompted by the system or when they recognize or should recognize, based on precise circumstances, that the conditions for the intended use are no longer met. If the driver fails to comply with these obligations, the fault is presumed, and they are liable for damages to the injured party per the Act. This liability framework is complemented by mandatory insurance that the vehicle owner must maintain, covering the liability costs of the vehicle's owner, driver, and titleholder.

Original equipment manufacturers (OEMs) are not directly included in the legal framework. The legislature expects motor insurance companies to investigate and pursue subrogation claims under existing product liability law if a defect in the automated driving system causes an accident. The driver and registered owner of the vehicle are primarily and directly liable rather than establishing a direct pathway to the manufacturer in cases of accidents occurring while the vehicle is in automated mode. In contrast, the Autonomous Driving Act of 2021, which legalized the use of vehicles with autonomous driving functions within defined operating areas (SAE Level 4), exempts the user from any special responsibilities (Appt, 2017). However, since Article 8, paragraph 5bis of the Vienna Convention on Road Traffic stipulates that autonomous vehicles are only permitted if a person, either inside or outside the vehicle, can at least deactivate them, Germany's new law introduced the role of a "technical supervisor," who is responsible, among other duties, for deactivating the system instead of the driver (Vienna Convention on Road Traffic, 1968). A natural person fills this new role with the authority to deactivate the motor vehicle during its autonomous operation and initiate driving maneuvers for the vehicle. The registered keeper of the vehicle serves as the technical supervisor and is responsible for ensuring that passengers comply with traffic regulations unrelated to vehicle control. The technical supervisor is not required to be present in the vehicle or to monitor the autonomous system constantly. Instead, they receive emergency alerts from the vehicle's autonomous system and decide whether to deactivate the vehicle or initiate an alternative driving maneuver. Martin Ebers raises a critical question regarding the compatibility of this arrangement with the Vienna Convention, as the Convention assigns the responsibility of deactivation or overriding to a "driver," who, according to the conventional interpretation of the statutory text and the overall framework of the Convention, must be present in the vehicle. The new section outlines technical requirements for autonomous vehicles, which must be capable of independently adhering to traffic regulations, autonomously bringing the vehicle into a minimal-risk state if continuing the journey would require violating road traffic laws, and having an accident-avoidance system. Additionally, the system must immediately notify the technical supervisor of any functional impairments and be capable of being deactivated by the supervisor at any time. The accident-avoidance system must be designed to prevent and mitigate damage, prioritizing the protection of human life in the event of an unavoidable accident. The responsibility for complying with traffic regulations unrelated to vehicle control, such as wearing a seatbelt—tasks that a machine cannot perform—remains with the human occupants of the vehicle. (Ebers, 2022). The Autonomous Driving Act 2021 does not alter the liability provisions for the technical supervisor under the Road Traffic Act. As a result, the technical supervisor can only be held liable according to the general civil liability rules outlined in the German Civil Code ("Bürgerliches Gesetzbuch"). The key provision in this area of tort law is Section 823(1) of the Civil Code, which generally requires fault on the part of the tortfeasor. Notably, the legislature rejected a proposal to impose liability for presumed fault on the technical supervisor, akin to a driver's liability under Section 18(1)(1) of the Road Traffic Act.

- *SINGAPORE*

Singapore has consistently ranked among the most prepared nations for integrating autonomous vehicles (AVs) into society. According to KPMG's Autonomous Vehicles Readiness Index, Singapore was ranked as the most prepared country for AV implementation, rising from second place in 2019. The index assessed "policy and legislation, technology and innovation, infrastructure, and consumer acceptance." The Road Traffic Act regulates the use of automobiles in Singapore. The country has been considering legislation for AVs since 2013, with the Land Transport Authority beginning to actively study the feasibility of AVs as a deployable and scalable mode of transportation by 2014 (Tan & Taeihagh, 2021). The Committee on Autonomous Road Transport for Singapore (CARTS) was established in 2014 to transform land transport by deploying autonomous vehicles (AVs). On August 24, 2017, the Minister for Transport amended the Road Transport Act under Chapter 276 to accommodate AVs. The 2017 amendments included several key measures: (1) expanding trial testing areas to cover the entire western region of the country; (2) setting standards for the design and use of AV equipment; (3) requiring AVs to purchase liability insurance; (4) implementing regulatory sandboxes to offer flexibility to regulators; and (5) mandating data sharing for all AVs, including reports on malfunctions or accidents. In 2019, the national standards body, Enterprise Singapore (ESP, formerly SPRING Singapore and IE Singapore), released national guidelines under the authority of the Road Traffic Act to support the development of AVs, including level four and level five AVs (Land Transport Authority [LTA] et al., 2019). The objective of these guidelines, outlined in Technical Reference (TR) 68, was to "provide AV companies with a comprehensive set of guidelines for the safe deployment of AV solutions in Singapore." However, TR 68 is a provisional standard and is not legally binding (Land Transport Authority [LTA] & Singapore Standards Council [SSC], 2021). Enterprise Singapore (ESP) has established ongoing working groups to assess the effectiveness of the guidelines, updating them in 2021 to "keep pace with recent technological advancements and further facilitate the transition from development to the operationalization of autonomous vehicles." As a result, these guidelines will continue to evolve as additional testing takes place. Ultimately, they may become formal standards or be withdrawn if they are ineffective (Mohan, 2021). In parallel with the development of these regulations, the Committee for Autonomous Road Transport Singapore (CARTS) was established to address the "broader potential impacts of AVs beyond transportation." CARTS comprises representatives from "various ministries, agencies, and industrial stakeholders" (Razdan, 2020). The Insurance Council of Australia has expressed its support for this approach, deeming it the most appropriate model for implementing reform. It believes the existing schemes provide the framework to ensure individuals have timely access to treatment, care, and financial support (Fernyhough, 2019).

In June 2020, the National Transport Commission (NTC) drafted a decision Regulation Impact Statement (RIS) regarding the in-service safety framework for automated vehicles. This led to decisions by infrastructure and transport ministers on key elements of the framework, including the creation of a new national law, the Automated Vehicle Safety Law (AVSL), and the establishment of a regulator. The AVSL will oversee Automated Driving System Entities (ADSEs) and ensure the safe operation of their automated driving systems throughout their lifecycle. Further consultations on the law's detailed content and potential variations in legislative implementation resulted in several proposals, which were reviewed by infrastructure and transport ministers in May 2021. A subsequent round of targeted consultations focused on how the AVSL framework would interact with existing state, territory, and Commonwealth transport laws. In February 2022, ministers agreed on a Commonwealth law approach for automated vehicles. In its document "The Regulatory Framework for Automated Vehicles in Australia," the NTC outlines an end-to-end regulatory framework for automated vehicles, emphasizing the safety assurance framework.

Three key regulators for automated vehicle safety have been proposed. The first is a first-supply regulator, responsible for approving the entry of new automated vehicles into the market. This regulator will assess the quality of evidence applicants provide regarding their automated driving system (ADS) safety and certify Automated Driving System Entities (ADSEs) at first supply. An ADSE is an entity that will self-certify the safety of the ADS and take responsibility for its safety throughout its lifecycle. The second regulator is the in-service regulator, whose primary role will be to ensure that regulated parties maintain the safety of an ADS over its lifecycle. This regulator will have a range of functions and powers to ensure that safety risks are comprehensively managed. The third regulator consists of state and territory road transport regulators, who will retain responsibility for an automated vehicle's access to the road network, vehicle registration, road management, regulation of human drivers and other road users, and human driver licensing under existing state and territory laws. Access to the public road network for automated vehicles will be regulated through a registration process similar to that of conventional vehicles. A person can register an automated vehicle in the same manner as any other vehicle, and registered owners will be required to hold compulsory third-party insurance for their vehicles to be registered. The Automated Driving System Entity (ADSE) will be subject to a general safety duty, ensuring the safe operation of its automated vehicles as far as reasonably practicable. This duty will apply to the entire design life of the automated driving system (ADS) and will be detailed in the Automated Vehicle Safety Law (AVSL). The general safety duty places the responsibility on the ADSE to identify risks and manage them accordingly. The ADSE is permitted to mitigate risks as it deems appropriate. In addition to this general duty, the ADSE must comply with specific prescriptive duties outlined in the AVSL. These duties clarify minimum safety requirements while allowing the ADSE flexibility to fulfill its general safety duty. For example, the ADSE must ensure, as far as reasonably practicable, that systems are developed, used, and maintained to uphold the general safety duty; that system upgrades to the ADS are safely installed and do not compromise the vehicle's safety; and that efforts are made to prevent third-party interference with the ADS, as much as possible. Furthermore, the ADSE must ensure that the vehicle complies with relevant road rules when the ADS is engaged, accounting for jurisdictional differences and any future amendments. The ADSE is also responsible for fulfilling dynamic driving task (DDT) obligations when the ADS is active. All jurisdictional Motor Accident Injury Insurance (MAII) schemes, including compulsory third-party and national injury insurance schemes, must cover injuries and deaths caused when the Automated Driving System (ADS) is engaged. The guiding principle behind this approach is to "ensure no person is better or worse off, financially or procedurally, in the relevant jurisdiction, if they are injured by a vehicle whose ADS was engaged, compared to being injured by a vehicle controlled by a human driver." As part of their first-supply corporate obligations, Automated Driving System Entities (ADSEs) must specify their ongoing data recording and sharing capabilities, detailing the ADS data they will record and how it will be made available to relevant parties. The operational aspects of data access by motor accident injury insurers—specifically, how they can access data to assess liability (including the necessary legal frameworks)- will be addressed later in the in-service safety work. This will occur after states and territories assess whether existing legal frameworks support insurers' access to data for liability assessment in the event of crashes. Regarding non-motor Accident Injury Insurance (MAII) injury, damage, and loss, as part of the first-supply minimum financial requirements, Automated Driving System Entities (ADSEs) must maintain an appropriate level of insurance to cover personal injury, death, and property damage caused by the Automated Driving System (ADS). Individuals affected by non-MAII injury, damage, or loss must pursue private legal action. This could involve individuals taking legal action for negligence related to injury, damage, or loss not covered by the MAII scheme in their jurisdiction or seeking compensation for property damage. National legislation governing autonomous vehicles and establishing a new safety regulator is being developed. The aim is to have national laws in place by 2026.

NAVIGATION: CORE INSIGHTS

National lessons for global governance offer the following suggestions for non-legally binding international documents, such as a Code of Conduct: Governments should prioritize the development of systematic and standardized legal regulations for autonomous vehicles (AVs). Establishing clear legal frameworks, particularly concerning licensing, security (cybersecurity), safety, liability, and data privacy, as these regulations directly impact AVs' operational effectiveness and compliance. Additionally, governments should ensure a stable political environment and offer support to promote AVs within the shared mobility market. Lastly, governments should encourage the integration of AVs with public transportation, though this should be considered a lower priority. Specifically, AVs could be deployed during times and in locations outside of traditional public transport services, contributing to a more sustainable transportation model that enhances accessibility and reduces costs. The following section clarifies the key terms for regulating autonomous vehicles (AVs).

- *Safety*

It is estimated that at least 90% of vehicle accidents are caused by human error (Sun et al., 2016). Adopting autonomous vehicles (AVs) can potentially reduce or eliminate the leading cause of car accidents—human error—while surpassing human drivers in perception, decision-making, and execution. However, introducing AVs also brings new safety challenges that must be addressed (Collingwood, 2017). It is important to highlight that adopting autonomous vehicles (AVs) may reduce seatbelt usage by vehicle occupants and decrease caution among pedestrians, as both groups may feel safer due to the technology's perceived reliability. However, the elimination of human error does not equate to the elimination of machine error. As AV technology becomes more complex, the likelihood of technical malfunctions or errors that could compromise vehicle safety also increases. The fatal crash involving Tesla's autopilot in 2016 highlights the uncertainties surrounding machine perception. This incident underscores the challenges in ensuring autonomous systems can accurately interpret complex, real-world environments and make reliable decisions under all conditions (Banks, Plant, & Stanton, 2018). Concerns also arise regarding how autonomous vehicles (AVs) should be programmed with "crash algorithms" to respond in the event of unavoidable accidents. These algorithms raise ethical questions about how AVs should prioritize decisions, such as determining whom to harm or protect in emergencies, further complicating their deployment's safety and moral implications. Due to the "lack of blame" associated with autonomous vehicles (AVs), the damage caused by accidents cannot be assessed subjectively. Therefore, it is essential to establish rules that govern AVs' responses to moral dilemmas. These regulations are crucial to ensure that AVs make ethically sound decisions when harm is unavoidable and to provide clarity regarding accountability in such incidents. Algorithms may be programmed to prioritize the safety of the AV's occupants "above all else," ensuring the economic viability of AV development. However, justifying harm to others based on the individual self-interest of AV occupants challenges the very principles of law, as it undermines the broader societal responsibility and ethical considerations that legal frameworks are designed to uphold. In contrast, algorithms may be programmed to make decisions that maximize social benefits, considering a range of factors. However, determining which factors to prioritize and how to weigh them remains to be seen, posing significant challenges in creating ethical and legally sound decision-making frameworks for autonomous vehicles (Coca-Vila, 2018).

Additionally, regulators have yet to reach a consensus on an acceptable level of safety or establish legitimate methods for determining the safety of autonomous vehicles (AVs). While AV performance could improve over time through real-world driving experience, this progress is contingent upon public acceptance of the technology. With widespread trust and adoption, the potential benefits of AVs may be improved (Kalra, 2017). In the United States, the federal government traditionally establishes "national safety standards," while state governments are responsible for issuing licenses and regulating driver behavior. However, the federal government is reluctant to impose strict regulations on autonomous vehicles (AVs). As the Transportation Secretary stated, the federal government is "not in the business... to pick the best technology," favoring a market-oriented approach that allows innovation to drive the development of AV technologies (Halsey, 2018). The National Highway Traffic Safety Administration (NHTSA) has outlined a Vehicle Performance Guidance for all entities involved in manufacturing, designing, supplying, testing, selling, operating, or deploying autonomous vehicles (AVs) in the United States. Although NHTSA intends to enforce these recommendations in the future, it requests that these entities submit a Voluntary Safety Assessment. This assessment outlines compliance with the guidance, including detailed specifications on system safety, such as descriptions of safety strategies and design redundancies to address potential AV malfunctions (NHTSA, 2017). The responsibilities of federal and state governments regarding autonomous vehicles (AVs) were clarified in the "Self Drive Act" passed in late 2017, which designates the National Highway Traffic Safety Administration (NHTSA) as the "preeminent regulating body" for AV oversight and regulation (Stone, 2018). The act permits states to enforce new standards on autonomous vehicles (AVs) only if those standards are "identical" to those prescribed by federal law, reinforcing federal authority over AV regulations and ensuring uniformity across jurisdictions (H.R.3388, 2017). With the advent of autonomous vehicles (AVs), the federal government's legal competence is likely to expand. In contrast, state governments' competence may diminish as their traditional role in regulating driver behavior becomes increasingly redundant in the context of self-driving technology.

Similarly, the UK's Department for Transport (DfT) published a Code of Practice in 2015 for testing autonomous vehicles (AVs) aimed at ensuring AV safety in various situations throughout their service life. However, this code needs legal status. It permits and encourages testing on any public road in the UK without requiring prior approval from authorities or the provision of a surety bond (CCAV, 2016). Despite this flexibility, frameworks for minimizing risks during public testing have yet to be established. This relatively relaxed approach aligns with the UK's broader Industrial Strategy, which aims to create a national "cluster of excellence" in AV testing. The strategy seeks to build human capital, attract foreign investment, and develop "high-skill, well-paying jobs" to boost the economy, achieve greener economic growth, improve mobility, and address the needs of an aging society (DBEIS, 2017). The US and the UK have adopted cautious stances, avoiding overly stringent regulations while refraining from being excessively lenient regarding AV safety. This balance is intended to provide sufficient room for innovation while setting and aligning expectations around safety standards. These efforts reflect a light control-oriented strategy, fostering innovation without imposing restrictive barriers (Kang, 2016). Likewise, Australia's National Transport Commission (NTC) has published non-mandatory guidelines for safe AV testing, which align with a light control-oriented strategy. In 2016, the Transport and Infrastructure Council approved the NTC's proposal to develop a national safety assurance system to evaluate the safety levels of autonomous vehicles (NTC, 2017). This approach emphasizes controlling access to AVs while supporting their commercial deployment as a long-term objective. Although regulations for AV deployment have yet to be established, individual cases are evaluated as they arise (NTC, 2016). The NTC has outlined four regulatory options for safety and actively seeks stakeholder feedback (NTC, 2017). This effort represents a step toward consensus-building and public participation, suggesting a shift toward an adaptation-oriented strategy. Similarly, China has adopted a light control-oriented strategy to address safety risks while implementing preventive measures to limit AV exposure to real-world road conditions. Regulations require human drivers to remain in vehicles with their hands on the steering wheel during testing. AVs can only be tested under actual road conditions once the government formulates a framework for granting road test exemptions (KPMG, 2018). While draft rules for AV testing on public roads have been developed, testing has progressed slowly due to the need to update existing laws (The Straits Times, 2018). China's National Technical Committee of Auto Standardization reviewed vehicle standards and regulations in 2016 to identify necessary regulatory adjustments. In 2017, the China-New Car Assessment Program was launched to incorporate safety measures into the assessment framework. Concurrently, research on AV industry policy and stakeholder engagement is underway to assist regulatory authorities (ERTRAC, 2017). AVs have been prioritized as a key sector in China's broader plan to become a global leader in artificial intelligence by 2025 and to compete with the United States' AI industries. China aims to establish a "friendly policy environment" that fosters rapid AV development (Cadell & Jourdan, 2017).

- *Liability*

The driver typically retains control in traditional vehicle accidents, assuming primary liability. However, with autonomous vehicles (AVs), liability dynamics shift significantly as drivers are no longer directly in control (Collingwood, 2017).

Accidents involving AVs increasingly resemble product safety or efficacy issues, exposing third parties such as manufacturers, software providers, and system designers to more significant liability risks (Pinsent Masons, 2016). This shift introduces uncertainties about how responsibility is divided between the human driver and the AV's autonomous system. For instance, would the human driver bear partial responsibility if a manual override feature exists but is not used (Collingwood, 2017)? Event data recorders (EDRs) or "black boxes" may be utilized to address liability accurately, although this raises privacy concerns (Dhar, 2016). Furthermore, the lack of a clear legal framework complicates the allocation of liability among manufacturers, suppliers, and software operators when determining the cause of malfunctions (Collingwood, 2017). Manufacturers also face reputational risks from accidents, attributed to design flaws, which could deter innovation (Tien, 2017). The absence of defined legal and moral responsibilities for programmers developing "crash algorithms" further complicates matters. These algorithms, which determine decisions during unavoidable accidents, raise ethical concerns about life-and-death choices, such as prioritizing between injury severity, the number of people involved, or other factors (Fleetwood, 2017). Governments have not yet standardized the criteria for such algorithms, leaving a gap in both regulation and public trust. While the UK has amended its legal framework to address these complexities, most countries have yet to follow suit. In the U.S., liability rules are primarily left to state governments, with the federal government urging states to consider liability allocation, insurance requirements, and tort liability rules (NHTSA, 2017). Many states have taken initial steps by revising AV definitions to address liability concerns. The UK stands out for adopting a toleration-oriented strategy, as reflected in the passage of the Automated and Electric Vehicles Act (Bill HC 143, 2017). This legislation clarifies insurer and AV owner liability under various circumstances. Insurers are held automatically liable for damages caused by insured AVs, with certain exceptions where the owner is at fault. This framework ensures that victims receive faster compensation under the existing motor vehicle insurance scheme (DfT, 2017). It also offers manufacturers protection under the Consumer Protection Act if they can prove that vehicle defects were undetectable during production due to scientific advancements (Coates, 2017). These developments demonstrate a move toward balancing consumer protection with incentives for innovation, though much work remains to create globally harmonized AV liability and insurance regulations.

In Singapore and Australia, governments have actively acknowledged the need to address the liability and insurance challenges posed by autonomous vehicles (AVs). In Singapore, the Road Traffic Act (RTA) was amended in 2017 to exempt AVs, their operators, and participants in AV trials from existing provisions of the RTA, which typically assign liability to human drivers (CNA, 2017). This adjustment recognizes the shift in control from humans to the AV system, challenging the traditional notion of human responsibility at the core of road and criminal laws (MOT, 2017). Similarly, in Australia, the government has committed to a phased update of liability and insurance laws. By November 2017, the National Transport Commission (NTC) aimed to clarify "control" definitions for AVs. Following this, the NTC planned to review existing driving laws, establish specific obligations for AV driving entities, and address potential barriers to eligibility for occupants and accident victims under compulsory injury systems by 2018 (NTC, 2017). The European Union (EU) has also explored AV-related liability and insurance solutions but has yet to make formal regulatory amendments.

One of the major initiatives in this regard is GEAR 2030, which focuses on addressing AV-related issues, including liability and insurance challenges. In addition, the European Parliament has recommended creating a mandatory insurance scheme, an accompanying compensation fund for victims of AV accidents, and establishing a legal status for robots to clarify liability responsibilities (EPCLA, 2016). However, the EU has yet to finalize legislation to address these issues comprehensively. Germany, in contrast, has enacted permanent legislation addressing AV liability risks. In June 2017, Germany passed a law requiring AVs to install black boxes to record the entire journey, which will be used to determine liability during accidents (JDSUPRA, 2017). The law also doubles the maximum liability limits imposed by the existing Road Traffic Act (RTA) and establishes clear rules for the apportionment of liability: manufacturers are held responsible for accidents that occur when the AV system is in control, and a system failure is the leading cause (Wacket et al., 2017). However, the law lacks clarity on the "adequate time reserve" drivers are permitted to have before taking control of the vehicle and who owns the data collected by the black boxes (JDSUPRA, 2017). In addition, Germany's Ethics Commission has published the world's first ethical guidelines for AVs, recommending that there must be clarity regarding who is considered the driver and prohibiting the use of personal data (such as age or gender) for decision-making in unavoidable accident scenarios (FMTDI, 2017). While not mandatory, these guidelines are a significant step toward addressing the ethical concerns surrounding AVs. Comparing these strategies, countries like the UK and Germany have already enacted comprehensive frameworks addressing AV-related liability and insurance, offering clarity for all stakeholders involved. Singapore and Australia are following a more gradual path toward regulatory reform, balancing the need for innovation with safety concerns. The EU is still in the exploration phase, while China and South Korea have yet to make significant moves to establish liability frameworks, citing the need for international standardization. Furthermore, Germany has also taken the lead in addressing ethical concerns with AV algorithms, setting an important precedent for other nations. These countries are adopting diverse approaches to managing AV-related liability, balancing technological innovation, public safety, and ethical considerations.

- *Privacy*

Autonomous vehicles (AVs) rely heavily on sensors, high-definition maps, and other instruments to collect and optimize data to ensure their safe operation (West, 2016). However, concerns about the control and use of this information have emerged, with numerous issues surrounding informational privacy still needing to be solved. For instance, it needs to clarify why specific data is being collected, what types of information are gathered, who has access to it, and how long it can be stored. Vehicle-to-vehicle (V2V) and vehicle-to-infrastructure (V2I) communications, which allow AVs to exchange data for safety reasons, also pose risks. These technologies can expose an AV's movements and geographical location to external networks, potentially allowing individuals to track AV users (Glancy, 2012). Another challenge highlighted by Schoonmaker is the need for more protection of location-based data through customer consent. Users must often fully understand the implications of accepting terms and conditions. Additionally, using event data recorders (EDRs) to determine the causes of accidents raises concerns. EDR data could be sold to third parties, such as insurance companies, and used against drivers in ways they may not anticipate (Schoonmaker, 2016). Further risks to privacy involve the potential misuse of the data collected by AVs. Information could be exploited to target users with unwanted marketing and advertising, steal identities, create detailed profiles of individuals, or predict their future actions. This concentration of data could lead to an over-concentration of power and control over large populations (Glancy, 2012). While the anonymization of collected data is possible, it is not foolproof. Deanonymization algorithms can reverse the anonymization process, re-identifying individuals with high certainty. This is particularly concerning for location-based data, as unique human traces enable adversaries to track movements even with minimal additional information (Gambs, Killijian, & del Prado Cortez, 2014). The interconnected nature of AVs' wireless networks also raises concerns about remote surveillance by public and private agencies, which could undermine individual autonomy and lead to psychological manipulation or intimidation (Glancy, 2012). Another emerging issue is video surveillance in AVs, particularly those used as transportation services, like autonomous taxis. Since users do not own these vehicles, it becomes unclear whether they are considered "public spaces" where surveillance might be deemed acceptable (Schoonmaker, 2016). This blurring of boundaries between public and private spaces complicates the debate over the permissibility of surveillance in AVs. In response to these concerns, the US and South Korean governments have enacted new legislation on data privacy that applies to all vehicles, including conventional vehicles and AVs. In the US, the SPY Car Act gives the National Highway Traffic Safety Administration (NHTSA) the authority to regulate the use and access to driving data in all vehicles manufactured for sale. This law mandates that vehicle owners or lessees must be allowed to stop data collection, except for data critical for safety or post-incident investigations. Manufacturers are prohibited from using collected data for marketing or advertising without explicit consent from the vehicle's owners or lessees (SCA, 2017).

China and Japan have also implemented legislation to control privacy and cybersecurity risks, reflecting a control-oriented strategy. Japan amended its Privacy Protection Law in 2017 to address the growing concerns over personal data security in a digital era (The Japan Times, 2017). Similarly, China enacted a new Cybersecurity Law that mandates the anonymization of all personal data. The law emphasizes customer consent and requires network operators to be transparent about data collection and usage's purpose, method, and scope (KPMG, 2017). While the law establishes many controls on data collection, sharing, and use, it does not contain AV-specific provisions. Singapore's government has adopted a control-oriented strategy to manage privacy risks in general and public sectors. The government is currently amending the Personal Data Protection Act (PDPA). A public consultation in July 2017 proposed amendments to increase transparency regarding the collection and use of personal data and allow individuals to withdraw consent for data collection activities (PDPC, 2018). Additionally, the government passed the Public Sector (Governance) Bill, which prohibits the unauthorized use and sharing of data between public sector agencies. This bill aims to improve public service delivery, especially efficiency and program management (PSGB, 2017). Germany and Australia must still amend their legislation to address AV-specific privacy risks. Germany's new AV bill does not address data privacy but focuses on safety and liability issues. However, the German government has indicated that privacy concerns will be considered when the bill is revised in two years (Wacket et al., 2017). Australia's National Transport Commission (NTC) has issued privacy recommendations, such as adopting a "privacy by design" approach and minimizing personal information generation whenever possible. These recommendations may be more rhetorical than practical, suggesting a formal commitment to managing risks rather than detailing specific actions to control AV privacy risks. The NTC has also suggested that the upcoming national safety assurance system include the highest level of privacy protection (NTC, 2017). More recently, the House of Representatives Standing Committee on Industry, Innovation, Science, and Resources (SCIISR) called for public participation in an inquiry into the social implications of AVs and recommended further exploration of the data rights of consumers, insurers, government agencies, and manufacturers. This reflects an adaptation-oriented strategy, engaging the public to build consensus around privacy risks and their management (NTC, 2017).

In the UK, the Department for Transport (DfT), in collaboration with the Centre for the Protection of National Infrastructure (CPNI), developed key principles for privacy and cybersecurity in the context of autonomous vehicles (AVs). These guidelines suggest that manufacturers adhere to ISO standards, such as the Privacy Architecture framework outlined by ISO 29101, indicating a light control-oriented strategy. The principles recommend that personal information be "properly managed" regarding what is stored, transmitted, and how it is used. They also emphasize the importance of giving AV users control over the data, including the ability to delete "sensitive data." However, the guidelines do not define what constitutes "proper" management of personal information or what qualifies as "sensitive" data. This indicates the government's awareness of the unique privacy risks associated with AVs but also reflects a non-binding approach, aiming to avoid impeding the growth of AV research and development in the UK. The guidelines suggest that the UK government is cautious about privacy while fostering innovation in the AV sector (DfT, 2017). In Germany, 13 voluntary recommendations have been released for managing privacy and cybersecurity risks in AVs. These recommendations notably advocate for specific rules clarifying the data businesses can process without AV users' "explicit consent" (FMTDI, 2017). Similar to the EU's GDPR, these recommendations emphasize transparency and the need for drivers to have complete control over how their data collected by AVs is used. Germany's data protection laws are stringent when defining personal data, applying this definition even to information that is only loosely connected to an individual. Consequently, most data generated by connected AVs will likely be considered personal unless the data-generating components have been specifically designed to anonymize the information (Pinsent Masons, 2016). This approach highlights the German government's commitment to ensuring strong privacy protections while allowing AV technologies to be developed (Tariq, 2025).

- *Cybersecurity*

Cybersecurity threats to conventional vehicles with automated features are already a significant concern. A survey of 5,000 respondents across 109 countries found that software hacking and misuse were the most pressing concerns for people with vehicles at all levels of automation (Kyriakidis, Happee, & de Winter, 2015). With the increasing reliance on wireless networks, such as Bluetooth, keyless entry systems, and cellular connections, hackers can take control of a vehicle by exploiting these connections as the car interacts with its environment (Lee, 2017). Autonomous vehicles (AVs), in particular, present an attractive target for hackers because they store and transmit sensitive transaction and lifestyle data, which can be sold for financial gain or used to inflict harm by extremists or criminals (König & Neumayr, 2017). For example, Miller and Valasek demonstrated the vulnerability of AVs when they successfully hacked a Chrysler Jeep via its internet connection, gaining control of its engines and brakes. AVs rely more on computer systems to control their movements, so they are more vulnerable to hacking than conventional vehicles, with the driver less able to intervene during an attack (Hern, 2016). If AVs' vehicle-to-vehicle (V2V) and vehicle-to-infrastructure (V2I) communication channels lack sufficient security, these could be hacked, leading to serious accidents (Pinsent Masons, 2016). Cybersecurity threats also include spoofing of global navigation satellite systems (GNSS) to manipulate critical safety functions and the use of sensor manipulation, such as bright lights to blind cameras or interference with ultrasound and radar systems to obstruct the vehicle's view of obstacles (Page & Krayem, 2017). While some systems may be implemented to detect such malfunctions, they require continuous software updates and changes to existing security architectures (Bagloee et al., 2016). Many governments have developed non-mandatory guidelines on cybersecurity best practices for AVs in response to these concerns. The US, China, the EU, and Singapore governments have adopted a control-oriented strategy and have introduced or enacted new legislation to address cybersecurity risks. In the US, the National Highway Traffic Safety Administration (NHTSA) has issued voluntary guidelines recommending that manufacturers and software companies design AV systems by existing international standards from bodies like the National Institute for Standards and Technology, NHTSA, SAE, and the Alliance of Automobile Manufacturers (NHTSA, 2017). Furthermore, NHTSA has set up an electronics systems safety research department to monitor and assess potential cyber vulnerabilities, and the Electronics Council has been established to enhance collaboration on cybersecurity research (NHTSA, 2018). The SPY Car Act was also introduced to enhance cybersecurity and privacy controls for all vehicles. This law mandates the separation of critical and noncritical software systems in every vehicle, ensuring that vehicles are evaluated according to best practices. The law also specifies that vehicles must secure collected information in electronic systems, whether the data is stored in the vehicle, transmitted, or stored offboard. It further requires vehicles to have the ability to detect, stop, and report attempts to capture driving data or take control of the vehicle and to disclose the level of privacy and cybersecurity protection they offer to consumers (SCA, 2017).

These measures represent a proactive approach by governments to address cybersecurity risks as AV technology evolves. Cybersecurity has been a longstanding concern in the EU. Over the past few years, the region has made incremental steps to manage cybersecurity risks, although these measures are not explicitly tailored to autonomous vehicles (AVs). The EU introduced its Cybersecurity Strategy in 2013, followed by the Security of Network and Information Systems Directive in

2016, marking the first EU-wide legislation on cybersecurity (EC, 2017). This Directive laid the groundwork for further actions by various EU organizations to raise awareness and offer recommendations on addressing cybersecurity issues. For instance, in 2016, the EU's independent advisory body on data protection and privacy, the Data Protection Working Party, published its views on the developments in the Internet of Things (IoT) and its associated security concerns (Pillath, 2016). Similarly, China's latest Cybersecurity Law adopts a control-oriented strategy. Key provisions of this law include personal information protection, critical information infrastructure protection, network operators' responsibilities to ensure security, and the preservation of sensitive information within the country. The law requires critical information infrastructure operators to store personal data within China, and companies must obtain approval and pass national reviews before transferring data. Additionally, the law mandates that critical cyber equipment and cybersecurity products only be sold after receiving security certifications (KPMG, 2017). Singapore has also taken steps to address cybersecurity risks through legislative amendments. The country's Computer Misuse and Cybersecurity Act was amended in April 2017 to enhance businesses' response to computer-related offenses (Kwang, 2017). Singapore's government has further prioritized cybersecurity awareness, fostering partnerships between academia and the private sector to position the country as a leading cybersecurity service provider. This reflects an adaptation-oriented strategy and plans to establish a national Defense Cyber Organisation (Srikanthan, 2017). In the UK, while the government has not yet enacted legal control over cybersecurity risks in AVs, it has taken proactive steps to strengthen awareness and resilience against such risks. The National Cybersecurity Strategy 2016–2021 aims to promote further research into cybersecurity for all systems and position the UK as a world leader in cybersecurity by 2021 (Cabinet Office, 2016). As part of this strategy, the National Cyber Security Centre (NCSC) was established in 2016 to analyze and detect cyber threats. The strategy also includes autonomous systems as a target for research funding, demonstrating the country's efforts to enhance its adaptive capacity. Additionally, the Department for Transport (DfT) and the Centre for the Protection of National Infrastructure (CPNI) have developed principles for privacy and cybersecurity, recommending that AV systems be designed to be resilient to attacks and capable of responding appropriately when their defenses or sensors fail (DfT, 2017).

CONCLUSION

The findings from this study highlight the importance of adopting a comprehensive approach to autonomous vehicle (AV) policy that incorporates both public and private sector perspectives. Governments must continue refining regulations prioritizing safety, data security, and public trust in AV technologies. Meanwhile, the private sector must collaborate closely with policymakers to ensure that innovations align with regulatory frameworks, especially as AV technologies move toward commercialization. This collaboration will be crucial in addressing challenges such as liability, safety standards, and infrastructure readiness, which are essential for the widespread adoption of autonomous vehicles. By combining government-led regulatory measures with private sector-driven technological advancements, a robust policy framework can be developed to guide the future of AVs. The coordinated efforts of both sectors will bridge the gap between technological innovation and societal readiness, ensuring that AVs are deployed in a way that benefits both consumers and industry stakeholders. Ultimately, three levels of documentation are necessary: national, regional, and international. Given the low global acceptance of binding documents (Hard Law), the author proposes an international code of conduct (Soft Law). At the national level, however, mandatory laws and regulations are essential. Additionally, due to the movement of vehicles across different regions, such as within the European Union, regional agreements are also needed to ensure consistency and cooperation.

REFERENCES

Appt, S. (2017, July 14). *Germany introduces law permitting automated vehicles.* Pinsent Masons. https://www.pinsentmasons.com/out-law/news/germany-introduces-law-permitting-automated-vehicles

Automated and Electric Vehicles Act. (2018). *UK Public General Acts. Section 2(1).* https://www.legislation.gov.uk/ukpga/2018/18/contents/enacted

Bagloee, S. A., Tavana, M., Asadi, M., & Oliver, T. (2016). Autonomous vehicles: Challenges, opportunities, and future implications for transportation policies. *Journal of Modern Transportation, 24*(4), 284–303. DOI: 10.1007/s40534-016-0117-3

Banks, V. A., Plant, K. L., & Stanton, N. A. (2018). Driver error or designer error: Using the perceptual cycle model to explore the circumstances surrounding the fatal Tesla crash on 7th May 2016. *Safety Science, 108*, 278–285. DOI: 10.1016/j.ssci.2017.12.023

Cabinet Office. National Security and Intelligence, HM Treasury, and The Rt Hon Philip Hammond MP. (2016). *National Cyber Security Strategy 2016–2021.* https://www.gov.uk/government/publications/national-cyber-security-strategy-2016-to-2021

Cadell, C., & Jourdan, A. (2017, April 12). China aims to become world leader in AI, challenges U.S. dominance. *Reuters.* https://www.reuters.com/article/us-china-ai/china-aims-to-become-world-leader-in-ai-challenges-u-s-dominance-idUSKBN1A5103?il=0

Carey, N., Lienert, P., & Bellon, T. (2021, April 21). Britain's driverless car ambitions hit speed bump with insurers. *Reuters.* https://www.reuters.com/article/us-autos-autonomous-insurance-insight-idUSKBN2C814K

CCAV. (2016). *Pathway to driverless cars: Proposals to support advanced driver assistance systems and automated vehicle technologies.* Crown. https://www.gov.uk/government/consultations/advanced-driver-assistance-systems-and-automated-vehicle-technologies-supporting-their-use-in-the-uk

Coates, C. (2017). *UK bill to set a precedent for autonomous car legislation.* https://www.dwf.law/news-events/legal-updates/2017/03/uk-bill-to-set-a-precedent-for-autonomous-car-legislation/

Coca-Vila, I. (2018). Self-driving cars in dilemmatic situations: An approach based on the theory of justification in criminal law. *Criminal Law and Philosophy, 12*(1), 59–82. DOI: 10.1007/s11572-017-9411-3

Collingwood, L. (2017). Privacy implications and liability issues of autonomous vehicles. *Information & Communications Technology Law*, *26*(1), 32–45. DOI: 10.1080/13600834.2017.1269871

DBEIS. (2017). *Industrial strategy: The grand challenges*. Department for Business, Energy & Industrial Strategy, UK. Crown. https://www.gov.uk/government/publications/industrial-strategy-the-grand-challenges/industrial-strategy-the-grand-challenges

Department of Infrastructure. Transport, Cities and Regional Development. (n.d.). *Automated vehicles in Australia*. Australian Government. https://www.infrastructure.gov.au/transport/automatedvehicles/index.aspx

Dhar, V. (2016). Equity, safety, and privacy in the autonomous vehicle era. *Computer*, *49*(11), 80–83. DOI: 10.1109/MC.2016.326

Draft Proposed Amendments, M. P. S. (2021, March 24). *Road Traffic Safety Law (Draft amendments)*. https://www.mps.gov.cn/n2254536/n4904355/c7787881/content.html

Dunn, J. (2017, March 14). 17 ridiculous 'smart' gadgets that really exist. *Business Insider*. https://www.businessinsider.com/weirdest-smart-gadgets-internet-ofthings-smart-home-2017-3

EPCLA. (2016). *Draft report with recommendations to the Commission on Civil Law Rules on Robotics (2015/2103(INL))*. European Parliament Committee on Legal Affairs. http://www.europarl.europa.eu/sides/getDoc.do?pubRef=-//EP//NONSGML+COMPARL+PE-582.443+01+DOC+PDF+V0//EN

European Commission (EC). (2017). *The directive on security of network and information systems (NIS directive)*. https://ec.europa.eu/digital-single-market/en/network-and-information-security-nis-directive

Fernyhough, J. (2019, January 17). Motorists, states set to pay for self-driving car crashes. *Financial Review*. https://www.afr.com/companies/financial-services/motorists-states-set-to-pay-for-self-driving-car-crashes-20190117-h1a5ru

Fleetwood, J. (2017). Public health, ethics, and autonomous vehicles. *American Journal of Public Health*, *107*(4), 532–537. DOI: 10.2105/AJPH.2016.303628 PMID: 28207327

FMTDI. (2017). *Data protection law recommendations of the Federal Commissioner for Data Protection and Freedom of Information for automated and connected driving*. Federal Ministry of Transport and Digital Infrastructure. https://www.bfdi.bund.de/SharedDocs/Publikationen/Allgemein/DatenschutzrechtlicheEmpfehlungenVernetztesAuto.pdf

Gambs, S., Killijian, M. O., & del Prado Cortez, M. N. (2014). De-anonymization attack on geolocated data. *Journal of Computer and System Sciences*, *80*(8), 1597–1614. DOI: 10.1016/j.jcss.2014.04.024

Glancy, D. J. (2012). Privacy in autonomous vehicles. *Santa Clara Law Review*, *52*(4), 1171–1239.

Halsey, A. (2018, March 1). "We're listening," Department of Transportation says on the future of driverless cars. *The Washington Post*. https://www.washingtonpost.com/local/trafficandcommuting/were-listening-department-of-transportation-says-on-the-future-of-driverless-cars/2018/03/01/8992682a-1d72-11e8-b2d9-08e748f892c0_story.html?noredirect=on

Hand, C., & Appt, S. (2020, November 26). Car insurance must evolve for autonomous vehicles. *Pinsent Masons*. https://www.pinsentmasons.com/out-law/analysis/car-insurance-evolve-autonomous-vehicles

HC 143. (2017). *Vehicle technology and aviation bill HC 143*. UK Parliament. https://www.publications.parliament.uk/pa/bills/cbill/2016-2017/0143/cbill_2016-20170143_en_2.htm#pt1-l1g2

Hern, A. (2016). Car hacking is the future – and sooner or later, you will be hit. *The Guardian*. https://www.theguardian.com/technology/2016/aug/28/car-hacking-future-self-driving-security

Hern, A. (2016). Car hacking is the future – and sooner or later, you will be hit. *The Guardian*. https://www.theguardian.com/technology/2016/aug/28/car-hacking-future-self-driving-security

Jansma, S. D. (2016, August 11). Autonomous vehicles: The legal landscape in the US. *Norton Rose Fulbright*. https://www.nortonrosefulbright.com/en/knowledge/publications/2951f5ce/autonomous-vehicles-the-legal-landscape-in-the-us

Kalra, N. (2017). *Challenges and approaches to realising autonomous vehicle safety. Testimony submitted to the House Energy and Commerce Committee*. RAND.

Kang, C. (2016, September 20). Self-driving cars gain powerful ally: The government. *The New York Times*. https://www.nytimes.com/2016/09/20/technology/self-driving-cars-guidelines.html

Kennedy, E. (2019, March 21). How do you insure a driverless car? *CNN Business*. https://edition.cnn.com/2019/03/21/cars/driverless-cars-insurance-liability-explained/index.html

König, M., & Neumayr, L. (2017). Users' resistance towards radical innovations: The case of the self-driving car. *Transportation Research Part F: Traffic Psychology and Behaviour*, *44*, 42–52. DOI: 10.1016/j.trf.2016.10.013

KPMG. (2017). *Overview of China's cybersecurity law*. KPMG Advisory Limited. https://assets.kpmg.com/content/dam/kpmg/cn/pdf/en/2017/02/overview-of-cybersecurity-law.pdf

KPMG. (2018). *Autonomous vehicles readiness index*. KPMG International. https://assets.kpmg.com/content/dam/kpmg/xx/pdf/2018/01/avri.pdf

Kyriakidis, M., Happee, R., & de Winter, J. C. F. (2015). Public opinion on automated driving: Results of an international questionnaire among 5000 respondents. *Transportation Research Part F: Traffic Psychology and Behaviour*, *32*, 127–140. DOI: 10.1016/j.trf.2015.04.014

Land Transport Authority (LTA). Enterprise Singapore, Standards Development Organisation & Singapore Standards Council. (2019, January 31). *Singapore develops provisional national standards to guide the development of fully autonomous vehicles*. Land Transport Authority. https://www.lta.gov.sg/content/ltagov/en/newsroom/2019/1/2/joint-media-release-by-theland-transport-authority-lta-enterprise-singapore-standards-development-organisationsingapo.html

LTA. Land Transport Authority (LTA) & Singapore Standards Council (SSC). (2021, September 3). *Enhanced national standards for the safe deployment of autonomous vehicles in Singapore*. Land Transport Authority. https://www.lta.gov.sg/content/ltagov/en/newsroom/2021/9/newsreleases/enhanced-national-standards-for-the-safe-deployment-of-autonomou.html

Law Commission. (2022). *Automated vehicles joint report (CVR-03-02-22)*. https://s3-eu-west-2.amazonaws.com/cloud-platform-e218f50a4812967ba1215eaecede923f/uploads/sites/30/2022/01/Automated-vehicles-joint-report-cvr-03-02-22.pdf

Lee, C. (2017). Grabbing the wheel early: Moving forward on cybersecurity and privacy protections for driverless cars. *Federal Communications Law*, *69*, 25–52.

Legislation.gov.uk. (2018). *Commentary on provisions of Bill/Act, Automated and Electric Vehicles Act 2018*. https://www.legislation.gov.uk/ukpga/2018/18/notes/division/6/index.htm

Light, S. E. (2017). Autonomous Vehicle Bill leaves safety gaps. *The Regulatory Review*. https://www.theregreview.org/2017/09/25/light-autonomous-vehicle-billgaps/

Ministry of Industry and Information Technology. (2021). *Access Administrative Measures (Article 24, Clause 2)*. https://www.szrd.gov.cn

Ministry of Public Security. (2021, March 24). *Road Traffic Safety Law (draft for public consultation)*. https://www.mps.gov.cn/n2254536/n4904355/c7787881/content.html

Mohan, M. (2021, September 3). *Enhanced national standards for the safe deployment of autonomous vehicles in Singapore*. Channel News Asia. https://www.channelnewsasia.com/singapore/autonomous-vehicles-tr68-enhancednational-standards-safe-2153946

MOT. (2017). *Opening speech by second minister for transport Ng Chee Meng for the road traffic (amendment) bill second reading*. Ministry of Transport, Singapore. https://www.mot.gov.sg/News-Centre/News/2017/Opening-Speech-by-Second-Minister-for-Transport-Ng-Chee-Meng-for-the-Road-Traffic-Amendment-Bill-Second-Reading/

National Conference of State Legislators. (2020, February 18). *Autonomous vehicles | Self-driving vehicles enacted legislation*. https://www.ncsl.org/research/transportation/autonomous-vehicles-self-driving-vehicles-enacted-legislation.aspx

National Highway Traffic Safety Administration (NHTSA). (2018). *NHTSA and vehicle cybersecurity*. National Highway Traffic Safety Administration.

National Science & Technology Council & U.S. Department of Transportation. (2020). *Ensuring American leadership in automated vehicle technologies: Automated vehicles 4.0*. https://www.transportation.gov/sites/dot.gov/files/2020-02/EnsuringAmericanLeadershipAVTech4.pdf

NHTSA. (2017). *Automated driving systems 2.0: A vision for safety*. National Highway Traffic Safety Administration, U.S. Department of Transportation.

NHTSA. (n.d.). *Understanding NHTSA's regulatory tools*. https://www.nhtsa.gov/sites/nhtsa.gov/files/documents/understanding_nhtsas_current_regulatory_tools-tag.pdf

Norton Rose Fulbright. (2016, August 11). *Autonomous vehicles: The legal landscape in Germany*. https://www.nortonrosefulbright.com/en/knowledge/publications/0e91a75d/autonomous-vehicles-the-legal-landscape-in-germany#

NTC. (2016). *Regulatory reforms for automated vehicles: Policy paper*. National Transport Commission.

Office of Impact Analysis. (2022, April 4). *The regulatory framework for automated vehicles in Australia*. Australian Government. https://oia.pmc.gov.au/published-impact-analyses-and-reports/regulatory-framework-automated-vehicles-australia

Page, F. D., & Krayem, N. M. (2017). Are you ready for self-driving vehicles? *Intellectual Property & Technology Law Journal, 29*(4), 14.

Palmer, D. (2016, June 26). *Photos: The amazing prototypes in the race to build self-driving cars*. ZD Net. https://www.zdnet.com/pictures/photos-the-amazing-prototypes-in-the-race-to-build-self-driving-cars/

PDPC. (2018). *Response to feedback on the public consultation on approaches to managing personal data in the digital economy*. Personal Data Protection Commission Singapore. https://www.pdpc.gov.sg/-/media/Files/PDPC/PDF-Files/Legislation-and-Guidelines/PDPCResponse-to-Feedback-for-Public-Consultation-on-Approaches-to-Managing-Personal-Data-inthe-Dig.pdf

PSGB. (2017). *The public sector (governance) bill*. https://sso.agc.gov.sg/Bills-Supp/45-2017/Published/20171106?DocDate=20171106

Pütz, F., Murphy, F., Mullin, M., & O'Malley, L. (2019). Connected automated vehicles and insurance: Analyzing future market structure from a business ecosystem perspective. *Technology in Society, 59*, 101182. DOI: 10.1016/j.techsoc.2019.101182

Razdan, R. (2020, March 1). *Singapore and autonomous vehicles – Interesting lessons in governance, planning, and safety. Forbes*. https://www.forbes.com/sites/rahulrazdan/2020/03/01/singapore-and-autonomousvehicles---interesting-lessons-in-governance-planning-and-safety/?sh=7066d3975fe2

Roth, M. (2020). Regulating the future: Autonomous vehicles and the role of government. *Iowa Law Review, 105*, 1411–1429.

SCA. (2017). *S. 680, SPY Car Act of 2017, 115th United States Congress*. https://www.congress.gov/bill/115th-congress/senate-bill/680

Schoonmaker, J. (2016). Proactive privacy for a driverless age. *Information & Communications Technology Law, 25*(2), 96–128. DOI: 10.1080/13600834.2016.1184456

SCIISR. (2017). *Social issues relating to land-based automated vehicles in Australia*. Parliament of the Commonwealth of Australia. https://www.aph.gov.au/Parliamentary_Business/Committees/House/Industry_Innovation_Science_and_Resources/Driverless_vehicles/Report

Smith, B. W. (2014). Automated vehicles are probably legal in the United States. *Texas A&M. Law Review*, *1*, 411–466. DOI: 10.37419/talr.v1i3.307

Srikanthan, T. (2017). *Commentary: Cybersecurity is the next economic battleground.* Channel News Asia. https://www.channelnewsasia.com/news/singapore/commentary-cybersecurity-is-the-next-economic-battleground-8591642

Stone, B. (2018, March 26). *Regulators are asleep at the wheel on self-driving cars. Bloomberg*. https://www.bloomberg.com/news/articles/2018-03-26/regulators-are-asleep-at-the-wheel-on-self-driving-cars

Sun, Y., Olaru, D., Smith, B., Greaves, S., & Collins, A. (2016). Road to autonomous vehicles in Australia. *Australasian Transport Research Forum 2016*, Melbourne, 16–18 November.

Tan, S. Y., & Taeihagh, A. (2021). Adaptive governance of autonomous vehicles: Accelerating the adoption of disruptive technologies in Singapore. *Government Information Quarterly*, *38*(1), 5. DOI: 10.1016/j.giq.2020.101546

Tariq, M. U. (2025). AI-Driven Personalized Learning – Enhancing Student Engagement and Outcomes. In H. Mamede & A. Santos (Eds.), *AI and Learning Analytics in Distance Learning* (pp. 151–178). IGI Global Scientific Publishing. https://doi.org/DOI: 10.4018/979-8-3693-7195-4.ch007

Tariq, M. U. (2025). Bridging Cultures and Languages: Innovative Literacy Strategies for Inclusive and Engaged Classrooms. In Tussey, J., & Haas, L. (Eds.), *Supporting Cultural Differences Through Literacy Education* (pp. 257–282). IGI Global Scientific Publishing., DOI: 10.4018/979-8-3693-9271-3.ch010

Tariq, M. U. (2025). Co-Living With Augmented Digital Beings – Navigating the Intersection of AI and Everyday Life. In Moutinho, L., & Martins, L. (Eds.), *Impacts of Sensetech on Society* (pp. 187–206). IGI Global Scientific Publishing., DOI: 10.4018/979-8-3693-7147-3.ch009

Tariq, M. U. (2025). Harnessing Technology to Combat Bias: Innovative Tools and Techniques in HR. In Kanwal, S. (Ed.), *Navigating Bias in Human Resource Management* (pp. 155–184). IGI Global Scientific Publishing., DOI: 10.4018/979-8-3693-7132-9.ch006

Tariq, M. U. (2025). Revitalizing Pedagogy: Integrating Indigenous Knowledge and Practices Into Higher Education Disciplines. In Kgari-Masondo, M. (Ed.), *Indigenous Teaching Disciplines and Perspectives for Higher Education* (pp. 379–400). IGI Global Scientific Publishing., DOI: 10.4018/979-8-3693-9296-6.ch018

The Japan Times. (2017). *Amended privacy protection law. The Japan Times.* https://www.japantimes.co.jp/opinion/2017/06/01/editorials/amended-privacy-protection-law/#.WrHxmZ7H-Z

The Straits Times. (2018, April 25). *China completes first draft of national rules to allow road tests for driverless vehicles. The Straits Times.* https://www.straitstimes.com/asia/east-asia/china-completes-first-draft-of-national-rules-to-allow-road-tests-for-driverless

Tien, J. M. (2017). The Sputnik of servgoods: Autonomous vehicles. *Journal of Systems Science and Systems Engineering*, *26*(2), 133–162. DOI: 10.1007/s11518-016-5325-1

Vienna Convention on Road Traffic, 1968, BGBl. 1977 II, 809 (811), last amended by agreement of 24 March 2014, BGBl. 2016 II, 1306.

Wacket, M., Escritt, T., & Davis, T. (2017). Germany adopts self-driving vehicles law. *Reuters.* https://www.reuters.com/article/us-germany-autos-self-driving-idUSKBN1881HY

Werner, D. (2020, December 7). Vehicle and traffic laws should be standard nationwide. *Adirondack Daily Enterprise: Opinion.* https://www.adirondackdailyenterprise.com/

KEY TERMS AND DEFINITION

Autonomous Vehicles (AVs): Vehicles capable of sensing their environment and operating without direct human control.

Artificial Intelligence of Things (AIoT): The convergence of AI and IoT technologies to enable intelligent, interconnected systems, such as AVs.

SAE Levels of Automation: A classification system from Level 0 to Level 5 that defines the extent of driving automation in vehicles.

Federal Motor Vehicle Safety Standards (FMVSS): U.S. federal regulations governing vehicle safety requirements, traditionally designed for human-driven vehicles.

Patchwork Regulation: A fragmented legal landscape where different jurisdictions adopt varying rules, leading to inconsistency and complexity in compliance.

Compilation of References

Abduljalil, J. M., & Abduljalil, B. M. (2020). *Epidemiology, genome, and clinical features of the pandemic SARS-CoV-2: a recent view*. New Microb. New Infect., DOI: 10.1016/j.nmni.2020.100672

Abdullah, M. N., & Ali, Y. H. (2021). Vehicle detection system at different weather conditions. *Iraqi Journal of Science*, *62*(3), 2040–2052. DOI: 10.24996/ijs.2021.62.6.30

Abdulmalek, S., Nasir, A., Jabbar, W. A., Almuhaya, M. A., Bairagi, A. K., Khan, M. A.-M., & Kee, S.-H. (2022). IoT-based healthcare-monitoring system towards improving quality of life: A review. *Health Care*. PMID: 36292441

Abouelmehdi, K., Beni-Hessane, A., & Khaloufi, H. (2018). Big healthcare data: Preserving security and privacy. *Journal of Big Data*, *5*(1), 1–18. DOI: 10.1186/s40537-017-0110-7

Aceto, G., Persico, V., & Pescapé, A. (2020). Industry 4.0 and health: Internet of things, big data, and cloud computing for Healthcare 4.0. *Journal of Industrial Information Integration*, *18*, 100129. DOI: 10.1016/j.jii.2020.100129

Act, A. (1996). Health Insurance Portability and Accountability Act of 1996, Pub. L. No. 104-191, 110 Stat. 1936. https://www.govinfo.gov/content/pkg/PLAW-104publ191/pdf/PLAW-104publ191.pdf

Adi, K., & Widodo, C. E. (2017). Distance measurement with a stereo camera. *International Journal of Innovative Research in Advanced Engineering*, *4*(11), 24–27.

Adil, E., Mikou, M., & Mouhsen, A. (2022). A novel algorithm for distance measurement using stereo camera. *CAAI Transactions on Intelligence Technology*, *7*(2), 177–186. DOI: 10.1049/cit2.12098

Agyemang, J. O., Yu, D., & Kponyo, J. J. (2021, September). Autonomic IoT: towards smart system components with cognitive IoT. In *Pan-African Artificial Intelligence and Smart Systems Conference* (pp. 248-265). Cham: Springer International Publishing.

Ahmed, M., Hegazy, M., Klimchik, A. S., & Boby, R. A. (2022). Lidar and camera data fusion in self-driving cars. *Компьютерные исследования и моделирование [Computer research and modeling], 14*(6), 1239–1253.

Ahmid, M., Kazar, O., & Barka, E. (2024). Internet of Things Overview: Architecture, Technologies, Application, and Challenges. In Boulila, W., Ahmad, J., Koubaa, A., Driss, M., & Farah, I. R. (Eds.), *Decision Making and Security Risk Management for IoT Environments. Advances in Information Security* (Vol. 106). Springer., DOI: 10.1007/978-3-031-47590-0_1

Akkaş, M. A., Sokullu, R., & Çetin, H. E. (2020). Healthcare and patient monitoring using IoT. *Internet of Things : Engineering Cyber Physical Human Systems, 11*, 100173. DOI: 10.1016/j.iot.2020.100173

Al Ridhawi, I., Aloqaily, M., & Karray, F. (2022). Intelligent blockchain-enabled communication and services: Solutions for moving internet of things devices. *IEEE Robotics & Automation Magazine, 29*(2), 10–20.

Alabdulatif, A., Khalil, I., Yi, X., & Guizani, M. (2019). Secure Edge of Things for Smart Healthcare Surveillance Framework. *IEEE Access : Practical Innovations, Open Solutions, 7*, 31010–31021. DOI: 10.1109/ACCESS.2019.2899323

Alam, F., Almaghthawi, A., Katib, I., Albeshri, A., & Mehmood, R. (2021). iResponse: An AI and IoT-Enabled Framework for Autonomous COVID-19 Pandemic Management. *Sustainability (Basel), 13*(7), 3797. https://www.mdpi.com/2071-1050/13/7/3797. DOI: 10.3390/su13073797

Al-Atawi, A. A., Khan, F., & Kim, C. G. (2022). Application and Challenges of IoT Healthcare System in COVID-19. *Sensors (Basel), 22*(19), 7304. DOI: 10.3390/s22197304 PMID: 36236404

Al-Humairi, S. N. S., Yih, A. E., & Daud, R. J. (2022). Design and Development of an IoT Virtual Doctor Robot (IVDR) for the Hospital Station. In *2022 IEEE 10th Conference on Systems, Process & Control (ICSPC)*

Al-Humairi, S. N. S., & Hajamydeen, A. I. (2022). IoT-based healthcare monitoring practices during Covid-19: Prospects and approaches. In *Healthcare Systems and Health Informatics* (pp. 163–185). CRC Press.

Ali, Z., Mahmood, S., Mansoor Ul Hassan, K., Daud, A., Alharbey, R., & Bukhari, A. (2024). A lightweight and secure authentication scheme for remote monitoring of patients in IoMT. *IEEE Access : Practical Innovations, Open Solutions*, *12*, 73004–73020. DOI: 10.1109/ACCESS.2024.3400400

Alotaibi, S., Mehmood, R., & Katib, I. (2020). The role of big data and twitter data analytics in healthcare supply chain management. *Smart Infrastructure and Applications: Foundations for Smarter Cities and Societies*, 267-279.

Alsaeed, N., Nadeem, F., & Albalwy, F. (2024). A scalable and lightweight group authentication framework for Internet of Medical Things using integrated blockchain and fog computing, *Future Generation Computer Systems*, vol. 151, Pages 162–181, ISSN 0167-739X, DOI: DOI: 10.1016/j.future.2023.09.032

Alsaeed, N., & Nadeem, F. (2022). Authentication in the Internet of Medical Things: Taxonomy, Review, and Open Issues. *Applied Sciences (Basel, Switzerland)*, *12*(15), 7487–7487. DOI: 10.3390/app12157487

Aminizadeh, S., Heidari, A., Toumaj, S., Darbandi, M., Navimipour, N. J., Rezaei, M., Talebi, S., Azad, P., & Unal, M. (2023). The applications of machine learning techniques in medical data processing based on distributed computing and the Internet of Things. *Computer Methods and Programs in Biomedicine*, *241*, 107745. DOI: 10.1016/j.cmpb.2023.107745 PMID: 37579550

Anjos, J. C. S. D., Matteussi, K. J., Orlandi, F. C., Barbosa, J. L. V., Silva, J. S., Bittencourt, L. F., & Geyer, C. F. R. (2023). A Survey on Collaborative Learning for Intelligent Autonomous Systems. *ACM Computing Surveys*, *56*(4), 98. Advance online publication. DOI: 10.1145/3625544

Appt, S. (2017, July 14). *Germany introduces law permitting automated vehicles*. Pinsent Masons. https://www.pinsentmasons.com/out-law/news/germany-introduces-law-permitting-automated-vehicles

Arifiani, L., & Arifiani, L. (2019). The effect of disruption technology, opportunities and challenges of telecommunication industry 4.0 in Indonesia. *Int. J. Recent Technol. Eng*, *7*(6), 808–819.

Arinze, C. A., Ajala, O. A., Okoye, C. C., Ofodile, O. C., & Daraojimba, A. I. (2024). *Evaluating the integration of advanced IT solutions for emission reduction in the oil and gas sector*. Engineering Science and Technology Journal., DOI: 10.51594/estj/v5i3.862

Arisdakessian, S., Wahab, O. A., Mourad, A., Otrok, H., & Kara, N. (2020). FoGMatch: An intelligent multi-criteria IoT-fog scheduling approach using game theory. *IEEE/ACM Transactions on Networking*, *28*(4), 1779–1789. DOI: 10.1109/TNET.2020.2994015

Arowoogun, J. O., Babawarun, O., Chidi, R., Adeniyi, A. O., & Okolo, C. A. (2024). A comprehensive review of data analytics in healthcare management: Leveraging big data for decision-making. *World Journal of Advanced Research and Reviews*, *21*(2), 1810–1821. DOI: 10.30574/wjarr.2024.21.2.0590

Arute, F., Arya, K., Babbush, R., Bacon, D., Bardin, J. C., Barends, R., Biswas, R., Boixo, S., Brandao, F. G. S. L., Buell, D. A., Burkett, B., Chen, Y., Chen, Z., Chiaro, B., Collins, R., Courtney, W., Dunsworth, A., Farhi, E., Foxen, B., & Martinis, J. M. (2019). Quantum supremacy using a programmable superconducting processor. *Nature*, *574*(7779), 505–510. DOI: 10.1038/s41586-019-1666-5 PMID: 31645734

Ashisha, G. R., Anitha Mary, X., George, T., Martin Sagayam, K., & Unai Fernandez-Gamiz Hatıra Günerhan, M. N. (2023). Uddin and S. Pramanik, Analysis of Diabetes disease using Machine Learning Techniques: A Review. *Journal of Information Technology Management*.

Asiedu, B., Failler, P., & Beygens, Y. (2018). Ensuring food security: An analysis of the industrial smoking fishery sector of Ghana. *Agriculture & Food Security*, *7*(1), 38. DOI: 10.1186/s40066-018-0187-z

Aslam, M., Khan, T. M., Naqvi, S. S., Holmes, G., & Naffa, R. (2019). On the application of automated machine vision for leather defect inspection and grading: A survey. *IEEE Access : Practical Innovations, Open Solutions*, *7*, 176065–176086. DOI: 10.1109/ACCESS.2019.2957427

Aslam, M., Khan, T. M., Naqvi, S. S., Holmes, G., & Naffa, R. (2020). Ensemble convolutional neural networks with knowledge transfer for leather defect classification in industrial settings. *IEEE Access : Practical Innovations, Open Solutions*, *8*, 198600–198614. DOI: 10.1109/ACCESS.2020.3034731

Ataç, H. O., Kayabaşı, A., & Aslan, M. F. (2024). The study on multi-defect detection for leather using object detection techniques. *Collagen and Leather*, *6*(1), 1–12. DOI: 10.1186/s42825-024-00186-2

Atefeh Hemmati, A. M. R. (2022). *The Internet of Autonomous Things applications: A taxonomy, technologies, and future directions, Internet of Things* (Vol. 20) https://doi.org/https://doi.org/10.1016/j.iot.2022.100635

Automated and Electric Vehicles Act. (2018). *UK Public General Acts. Section 2(1)*. https://www.legislation.gov.uk/ukpga/2018/18/contents/enacted

Awonuga, K. F., Nwankwo, E. E., Oladapo, J. O., Okoye, C. C., Odunaiya, O. G., & Uzondu, C. S. (2024). *Driving sustainable growth in SME manufacturing: The role of digital transformation, project, and capture management. International Journal of Science and Research Archives*. IJSRA., DOI: 10.30574/ijsra.2024.11.1.0270

Azad, M. A., Arshad, J., Mahmoud, S., & Imran, M. (2019). A Privacy-preserving Framework for Smart Context-aware Healthcare Applications. *Transactions on Emerging Telecommunications Technologies*, *33*(8), e3634. Advance online publication. DOI: 10.1002/ett.3634

Baeza-Yates, R. (2018). Bias on the Web. *Communications of the ACM*, *61*(6), 54–61. https://dl.acm.org/doi/10.1145/3209581. DOI: 10.1145/3209581

Bagdadee, A. H., Hoque, M. Z., & Zhang, L. (2020). IoT based wireless sensor network for power quality control in smart grid. *Procedia Computer Science*, *167*, 1148–1160. DOI: 10.1016/j.procs.2020.03.417

Bagloee, S. A., Tavana, M., Asadi, M., & Oliver, T. (2016). Autonomous vehicles: Challenges, opportunities, and future implications for transportation policies. *Journal of Modern Transportation*, *24*(4), 284–303. DOI: 10.1007/s40534-016-0117-3

Balch, J. A., Ruppert, M. M., Loftus, T. J., Guan, Z., Ren, Y., Upchurch, G. R., Ozrazgat-Baslanti, T., Rashidi, P., & Bihorac, A. (2023). Machine learning–enabled clinical information systems using fast healthcare interoperability resources data standards: Scoping review. *JMIR Medical Informatics*, *11*, e48297. DOI: 10.2196/48297 PMID: 37646309

Balsamo, D., Merrett, G. V., Zaghari, B., Wei, Y., Ramchurn, S., Stein, S., . . . Beeby, S. (2017, September). Wearable and autonomous computing for future smart cities: Open challenges. In 2017 25th International Conference on Software, Telecommunications and Computer Networks (SoftCOM) (pp. 1-5). IEEE.

Banduka, N., Tomić, K., Živadinović, J., & Mladineo, M. (2024). Automated Dual-Side Leather Defect Detection and Classification Using YOLOv11: A Case Study in the Finished Leather Industry. *Processes (Basel, Switzerland)*, *12*(12), 2892. DOI: 10.3390/pr12122892

Banerjee, A., Chakraborty, C., Kumar, A., & Biswas, D. (2020). Emerging trends in IoT and big data analytics for biomedical and health care technologies. *Handbook of data science approaches for biomedical engineering*, 121-152.

Banerjee, M., Lee, J., & Choo, K.-K. R. (2018). A blockchain future for internet of things security: a position paper, *Digital Communications and Networks*, vol. 4, no. 3, pp. 149–160, ISSN 2352-8648, DOI: DOI: 10.1016/j.dcan.2017.10.006

Banks, V. A., Plant, K. L., & Stanton, N. A. (2018). Driver error or designer error: Using the perceptual cycle model to explore the circumstances surrounding the fatal Tesla crash on 7th May 2016. *Safety Science*, *108*, 278–285. DOI: 10.1016/j.ssci.2017.12.023

Barbosa, S., Cosley, D., Sharma, A., & Cesar-Jr, R. M. (2016). Averaging gone wrong: Using time-aware analyses to better understand behavior. *Proceedings of the ... Conference on Human Factors in Computing Systems,* April 2016, 829–841. https://dl.acm.org/doi/10.1145/2872427.2883083

Benson, T., & Grieve, G. (2021). *Principles of health interoperability*. Springer International. DOI: 10.1007/978-3-030-56883-2

Berdigaliyev, N., & Aljofan, M. (2020). An overview of drug discovery and development. *Future Medicinal Chemistry*, *12*(10), 939–947. DOI: 10.4155/fmc-2019-0307 PMID: 32270704

Bergmann, P., Batzner, K., Fauser, M., Sattlegger, D., & Steger, C. (2021). The MVTec anomaly detection dataset: A comprehensive real-world dataset for unsupervised anomaly detection. *International Journal of Computer Vision*, *129*(4), 1038–1059. DOI: 10.1007/s11263-020-01400-4

Bhatt, V., & Chakraborty, S. (2023). Improving service engagement in healthcare through internet of things based healthcare systems. *Journal of Science and Technology Policy Management*, *14*(1), 53–73. DOI: 10.1108/JSTPM-03-2021-0040

Biancone, P., Secinaro, S., Marseglia, R., & Calandra, D. (2021). E-health for the future. Managerial perspectives using a multiple case study approach. *Technovation*. Advance online publication. DOI: 10.1016/j.technovation.2021.102406

Bin Abdullah, A., Jawahar, M., Manogaran, N., Subbiah, G., Seeranagan, K., Balusamy, B., & Chengam Saravanan, A. (2024). Leather Image Quality Classification and Defect Detection System using Mask Region-based Convolution Neural Network Model. *IJACSA). International Journal of Advanced Computer Science and Applications*, *15*(4). www.ijacsa.thesai.org

Blencowe, M. (2010). Quantum RAM. *Nature*, *468*(7320), 44–45. DOI: 10.1038/468044a PMID: 21048756

Boikanyo, K., Zungeru, A. M., Sigweni, B., Yahya, A., & Lebekwe, C. (2023). Remote patient monitoring systems: Applications, architecture, and challenges, *Scientific African*, vol. 20, pages. e01638, ISSN 2468–2276, DOI: DOI: 10.1016/j.sciaf.2023.e01638

Bonatti, P. A., Decker, S., Polleres, A., & Presutti, V. (2019). AIs: New directions for knowledge representation on the semantic web (Dagstuhl seminar 18371). *Dagstuhl Reports, 8*(9). Schloss Dagstuhl–Leibniz-Zentrum für Informatik.

Bostrom, N. (2002). Existential risks: Analyzing human extinction scenarios. *Journal of Evolution and Technology / WTA*, •••, 9.

Bub J. (2010) Quantum computation: where does the speed-up come from? *Phil. Quantum Info. Entang*. 231–246.

Bu, L., Zhang, Y., Liu, H., Yuan, X., Guo, J., & Han, S. (2021). An IIoT-driven and AI-enabled framework for smart manufacturing system based on three-terminal collaborative platform. *Advanced Engineering Informatics, 50*, 101370. DOI: 10.1016/j.aei.2021.101370

Butt, F. A., Chattha, J. N., Ahmad, J., Zia, M. U., Rizwan, M., & Naqvi, I. H. (2022). On the integration of enabling wireless technologies and sensor fusion for next-generation connected and autonomous vehicles. *IEEE Access : Practical Innovations, Open Solutions, 10*, 14643–14668. DOI: 10.1109/ACCESS.2022.3145972

Cabinet Office. National Security and Intelligence, HM Treasury, and The Rt Hon Philip Hammond MP. (2016). *National Cyber Security Strategy 2016–2021*. https://www.gov.uk/government/publications/national-cyber-security-strategy-2016-to-2021

Cadell, C., & Jourdan, A. (2017, April 12). China aims to become world leader in AI, challenges U.S. dominance. *Reuters*. https://www.reuters.com/article/us-china-ai/china-aims-to-become-world-leader-in-ai-challenges-u-s-dominance-idUSKBN1A5103?il=0

Calders, T., Kamiran, F., & Pechenizkiy, M. (2009, December). Building classifiers with independency constraints. In *2009 IEEE International Conference on Data Mining Workshops* (pp. 13-18). IEEE. DOI: 10.1109/ICDMW.2009.83

Calmon, F., Wei, D., Vinzamuri, B., Natesan Ramamurthy, K., & Varshney, K. R. (2017). Optimized pre-processing for discrimination prevention. *Advances in Neural Information Processing Systems, 30*. https://dl.acm.org/doi/10.5555/3294996.3295155

Carey, N., Lienert, P., & Bellon, T. (2021, April 21). Britain's driverless car ambitions hit speed bump with insurers. *Reuters*. https://www.reuters.com/article/us-autos-autonomous-insurance-insight-idUSKBN2C814K

Carpenter, J. (2024). *The Naked Android: Synthetic Socialness and the Human Gaze* (1st ed.). Chapman and Hall/CRC., DOI: 10.1201/9781003170457

CCAV. (2016). *Pathway to driverless cars: Proposals to support advanced driver assistance systems and automated vehicle technologies*. Crown. https://www.gov.uk/government/consultations/advanced-driver-assistance-systems-and-automated-vehicle-technologies-supporting-their-use-in-the-uk

Chakraborty, D., & Paul, J. (2022). Healthcare apps' purchase intention: A consumption values perspective. *Technovation*. Advance online publication. DOI: 10.1016/j.technovation.2022.102481

Chao, K., Sarker, M. N. I., Ali, I., Firdaus, R. R., Azman, A., & Shaed, M. M. (2023). Big datadriven public health policy making: Potential for the healthcare industry. *Heliyon*, *9*(9), e19681. DOI: 10.1016/j.heliyon.2023.e19681 PMID: 37809720

Chen, B. W., Wan, J., Celesti, A., Li, D., Abbas, H., & Zhang, Q. (2018). Edge Computing in IoT-Based Manufacturing. *IEEE Communications Magazine*, *56*(9), 103–109. DOI: 10.1109/MCOM.2018.1701231

Chen, M., Li, P., Wang, R., Xiang, Y., Huang, Z., Yu, Q., He, M., Liu, J., Wang, J., Su, M., Zhang, M., Jian, A., Ouyang, J., Zhang, C., Li, J., Dong, M., Zeng, S., Wu, J., Hong, P., & Tao, G. (2022). Multifunctional Fiber-Enabled Intelligent Health Agents. *Advanced Materials*, *34*(52), 2200985. https://doi.org/https://doi.org/10.1002/adma.202200985. DOI: 10.1002/adma.202200985 PMID: 35820163

Chen, S.-Y., Cheng, Y.-C., Yang, W.-L., & Wang, M.-Y. (2021). Surface defect detection of wet-blue leather using hyperspectral imaging. *IEEE Access : Practical Innovations, Open Solutions*, *9*, 127685–127702. DOI: 10.1109/ACCESS.2021.3112133

Chen, X. Z., Chang, C. M., Yu, C. W., & Chen, Y. L. (2020). A real-time vehicle detection system under various bad weather conditions based on a deep learning model without retraining. *Sensors (Basel)*, *20*(20), 5731. DOI: 10.3390/s20205731 PMID: 33050173

Chen, Z., Deng, J., Zhu, Q., Wang, H., & Chen, Y. (2022). A systematic review of machine-vision-based leather surface defect inspection. *Electronics (Basel)*, *11*(15), 2383. DOI: 10.3390/electronics11152383

Chen, Z., Xu, D., Deng, J., Chen, Y., & Li, C. (2023). Comparative study on deep-learning-based leather surface defect identification. *Measurement Science & Technology*, *35*(1), 15402. DOI: 10.1088/1361-6501/acfb9f

Chiruvella, V., & Guddati, A. K. (2021). Ethical issues in patient data ownership. *Interactive Journal of Medical Research*, *10*(2), e22269. DOI: 10.2196/22269 PMID: 34018968

Choi, T. M., Wallace, S. W., & Wang, Y. (2018). Big data analytics in operations management. *Production and Operations Management*, *27*(10), 1868–1883. DOI: 10.1111/poms.12838

Chopade, S., Gupta, H. P., & Dutta, T. (2023). Survey on Sensors and Smart Devices for IoT Enabled Intelligent Healthcare System. *Wireless Personal Communications*, *131*(3), 1957–1995. DOI: 10.1007/s11277-023-10528-8 PMID: 37360143

Chua, H. N., Teh, J. S., & Herbland, A. (2021). Identifying the effect of data breach publicity on information security awareness using hierarchical regression. *IEEE Access : Practical Innovations, Open Solutions*, *9*, 121759–121770. DOI: 10.1109/ACCESS.2021.3107426

Cisse, M., Karamoko, G., Cisse, M., Chene, C., & Karoui, R. (2023). Determination of the quality of smoked sea herring (*Sardinella maderensis*) produced by three smoking processes practiced in Ivory Coast. *European Food Research and Technology*, *249*(11), 2989–3000. DOI: 10.1007/s00217-023-04343-5

Cissé, M., Karamoko, G., Cissé, M., Chèné, C., & Karoui, R. (2025). Quality assessment of Ivory Coast smoked herring (*Madeiran sardinella*) produced by three smoking processes by physico-chemical and mid infrared spectroscopy coupled with chemometric tools: A preliminary study. *European Food Research and Technology*. Advance online publication. DOI: 10.1007/s00217-025-04674-5

Clarke, K. A. (2005). The phantom menace: Omitted variable bias in econometric research. *Conflict Management and Peace Science*, *22*(4), 341–352. DOI: 10.1080/07388940500339183

Coates, C. (2017). *UK bill to set a precedent for autonomous car legislation*. https://www.dwf.law/news-events/legal-updates/2017/03/uk-bill-to-set-a-precedent-for-autonomous-car-legislation/

Cobianchi, L., Dal Mas, F., & Angelos, P. (2021). One size does not fit all–translating knowledge to bridge the gaps to diversity and inclusion of surgical teams. *Annals of Surgery*, *273*(2), e34–e36. DOI: 10.1097/SLA.0000000000004604 PMID: 33201134

Coca-Vila, I. (2018). Self-driving cars in dilemmatic situations: An approach based on the theory of justification in criminal law. *Criminal Law and Philosophy, 12*(1), 59–82. DOI: 10.1007/s11572-017-9411-3

Coccia, M. (2020) Global Encyclopedia of Public Administration, Public Policy, and Governance. *Springer Nature Switzerland AG*; Disruptive technologies for industrial and corporate change; pp. 1–7.

Collingwood, L. (2017). Privacy implications and liability issues of autonomous vehicles. *Information & Communications Technology Law, 26*(1), 32–45. DOI: 10.1080/13600834.2017.1269871

Colombo, F., Oderkirk, J., & Slawomirski, L. (2020). Health information systems, electronic medical records, and big data in global healthcare: Progress and challenges in oecd countries. *Handbook of Global Health*, 1-31

Cozzoli, N., Salvatore, F. P., Faccilongo, N., & Milone, M. (2022). How can big data analytics be used for healthcare organization management? Literary framework and future research from a systematic review. *BMC Health Services Research, 22*(1), 809. DOI: 10.1186/s12913-022-08167-z PMID: 35733192

Cuijpers, M., Guenter, H., & Hussinger, K. (2011). Costs and benefits of inter-departmental innovation collaboration. *Research Policy, 40*(4), 565–575. DOI: 10.1016/j.respol.2010.12.004

Dadkhah, M., Mehraeen, M., Rahimnia, F., & Kimiafar, K. (2021). Use of Internet of Things for Chronic Disease Management: An Overview. *Journal of Medical Signals and Sensors, 11*(2), 138–157. DOI: 10.4103/jmss.JMSS_13_20 PMID: 34268102

Dai, S., Liang, Y., Liu, S., Wang, Y., Shao, W., Lin, X., & Feng, X. (2018, March). Learning entity and relation embeddings with entity description for AI completion. In *2018 2nd International Conference on Artificial Intelligence: Technologies and Applications (ICAITA 2018)* (pp. 194–197). Atlantis Press.

Daily, J., & Peterson, J. (2017). Predictive maintenance: How big data analysis can improve maintenance. In *Supply chain integration challenges in commercial aerospace: A comprehensive perspective on the aviation value chain* (pp. 267–278).

Dai, M., Su, Z., Li, R., & Yu, S. (2021). A software-defined-networking-enabled approach for edge-cloud computing in the internet of things. *IEEE Network, 35*(5), 66–73. DOI: 10.1109/MNET.101.2100052

Dai, T., & Tayur, S. (2020). Healthcare operations management: A snapshot of emerging research. *Manufacturing & Service Operations Management, 22*(5), 869–887. DOI: 10.1287/msom.2019.0778

Dalrymple, R. (2023, September 1). Is Apple Card Worth it in 2023? *Leaders.com.* https://leaders.com/articles/wealth/is-apple-card-worth-it/

Dandil, E., & Çevik, K. K. (2019). Computer vision-based distance measurement system using stereo camera view. In *2019 3rd International Symposium on Multidisciplinary Studies and Innovative Technologies (ISMSIT)* (pp. 1–4). IEEE. DOI: 10.1109/ISMSIT.2019.8932817

Danks, D., & London, A. J. (2017). Algorithmic Bias in Autonomous Systems. In *Proceedings of the International Joint Conference on Artificial Intelligence* (pp. 4691–4697). IJCAI., https://dl.acm.org/doi/10.5555/3171837.3171944

Dash, S., Shakyawar, S. K., Sharma, M., & Kaushik, S. (2019). Big data in healthcare: Management, analysis and prospects. *Journal of Big Data*, *6*(1), 1–25. DOI: 10.1186/s40537-019-0217-0

DBEIS. (2017). *Industrial strategy: The grand challenges.* Department for Business, Energy & Industrial Strategy, UK. Crown. https://www.gov.uk/government/publications/industrial-strategy-the-grand-challenges/industrial-strategy-the-grand-challenges

de Villiers, K. (2021). Bridging the health inequality gap: An examination of South Africa's social innovation in health landscape. *Infectious Diseases of Poverty*, *10*(1), 1–7. DOI: 10.1186/s40249-021-00804-9 PMID: 33648585

De Vos, T. (2016, December 5). Cool machine learning examples in real life. *IT Enterprise.*https://itenterprise.co.uk/cool-machine-learning-examples-real-life/

Del-Valle-Soto, C., Briseño, R. A., Valdivia, L. J., Velázquez, R., & Nolazco-Flores, J. A. (2023). Non-Invasive Monitoring of Vital Signs for the Elderly Using Low-Cost Wireless Sensor Networks: Exploring the Impact on Sleep and Home Security. *Future Internet*, *15*(9), 287. DOI: 10.3390/fi15090287

Department of Infrastructure. Transport, Cities and Regional Development. (n.d.). *Automated vehicles in Australia.* Australian Government. https://www.infrastructure.gov.au/transport/automatedvehicles/index.aspx

Dhabliya, D., Veeraiah, V., Dari, S. S., Raja Kumar, J. R., Dhabliya, R., Pramanik, S., & Gupta, A. (2024). *Virtual and Augmented Reality Applications for Healthcare and Goodness.* IGI Global., DOI: 10.4018/979-8-3693-1463-0.ch017

Dhanvijay, M. M., & Patil, S. C. (2019). Internet of Things: A survey of enabling technologies in healthcare and its applications. *Computer Networks*, *153*, 113–131. DOI: 10.1016/j.comnet.2019.03.006

Dhar Dwivedi, A., Singh, R., Kaushik, K., Rao Mukkamala, R., & Alnumay, W. S. (2024). Blockchain and artificial intelligence for 5G-enabled Internet of Things: Challenges, opportunities, and solutions. *Transactions on Emerging Telecommunications Technologies, 35*(4), e4329. DOI: 10.1002/ett.4329

Dhar, V. (2016). Equity, safety, and privacy in the autonomous vehicle era. *Computer, 49*(11), 80–83. DOI: 10.1109/MC.2016.326

Dinesh, K. G., Arumugaraj, K., Santhosh, K. D., & Mareeswari, V. (2018). Prediction of cardiovascular disease using machine learning algorithms. *2018 International Conference on Current Trends towards Converging Technologies (ICCTCT)*, 1–7. DOI: 10.1109/ICCTCT.2018.8550857

Draft Proposed Amendments, M. P. S. (2021, March 24). *Road Traffic Safety Law (Draft amendments)*. https://www.mps.gov.cn/n2254536/n4904355/c7787881/content.html

Dressel, J., & Farid, H. (2018). The accuracy, fairness, and limits of predicting recidivism. *Science Advances, 4*(1), eaao5580. https://www.science.org/doi/10.1126/sciadv.aao5580. DOI: 10.1126/sciadv.aao5580 PMID: 29376122

Duggineni, S. S. (2023). *Data integrity as a code*. DIAC.

Dunn, J. (2017, March 14). 17 ridiculous 'smart' gadgets that really exist. *Business Insider*. https://www.businessinsider.com/weirdest-smart-gadgets-internet-ofthings-smart-home-2017-3

Dushku, E., Rabbani, M. M., Conti, M., Mancini, L. V., & Ranise, S. (2020). SARA: Secure asynchronous remote attestation for IoT systems. *IEEE Transactions on Information Forensics and Security, 15*, 3123–3136. DOI: 10.1109/TIFS.2020.2983282

Dwork, C., Hardt, M., Pitassi, T., Reingold, O., & Zemel, R. (2012, January). Fairness through awareness. In *Proceedings of the 3rd Innovations in Theoretical Computer Science Conference* (pp. 214-226). https://dl.acm.org/doi/10.1145/2090236.2090255

Elahi, M. F., Luo, X., & Tian, R. (2020, July). A framework for modeling AIs via processing natural descriptions of vehicle-pedestrian interactions. In *International Conference on Human-Computer Interaction* (pp. 40–50). Springer.

Electric Vehicle Charging and Discharging Dataset. (2024). *Kaggle Dataset*. Retrieved from https://www.kaggle.com/datasets/example-dataset-url

Elhoseny, M., Abdelaziz, A., Salama, A. S., Riad, A. M., Muhammad, K., & Sangaiah, A. K. (2018). A hybrid model of the Internet of Things and cloud computing to manage big data in health services applications. *Future Generation Computer Systems, 86*, 1383–1394. DOI: 10.1016/j.future.2018.03.005

Engineering Online, E. V. (n.d.). *What role do advanced sensors play in EVs?* Retrieved from https://www.evengineeringonline.com/what-role-do-advanced-sensors-play-in-evs/

EPCLA. (2016). *Draft report with recommendations to the Commission on Civil Law Rules on Robotics (2015/2103(INL))*. European Parliament Committee on Legal Affairs. http://www.europarl.europa.eu/sides/getDoc.do?pubRef=-//EP//NONSGML+COMPARL+PE-582.443+01+DOC+PDF+V0//EN

Esenogho, E., Djouani, K., & Kurien, A. M. (2022). Integrating artificial intelligence Internet of Things and 5G for next-generation smartgrid: A survey of trends challenges and prospect. *IEEE Access : Practical Innovations, Open Solutions, 10*, 4794–4831.

European Commission (EC). (2017). *The directive on security of network and information systems (NIS directive)*. https://ec.europa.eu/digital-single-market/en/network-and-information-security-nis-directive

Everingham, M., Van Gool, L., Williams, C. K., Winn, J., & Zisserman, A. (2010). The PASCAL visual object classes (VOC) challenge. *International Journal of Computer Vision, 88*(2), 303–338. DOI: 10.1007/s11263-009-0275-4

Fahy, N., Mauer, N., & Panteli, D. (2021). *European support for improving health and care systems.*

Farayola, O. A., Olorunfemi, O. L., & Shoetan, P. O.Oluwatoyin Ajoke FayayolaOluwabukunmi Latifat OlorunfemiPhilip Olaseni Shoetan. (2024). Data privacy and security in IT: A review of techniques and challenges. *Computer Science IT Research Journal, 5*(3), 606–615. DOI: 10.51594/csitrj.v5i3.909

Fehringer, H., & Stary, C. (2023). User-Informed Adaptation in IoT Home Healthcare: Grounding Development in Empirical Evidence. *Journal of Theoretical and Applied Electronic Commerce Research, 18*(4), 1901–1925. DOI: 10.3390/jtaer18040096

Feng, Z., Wei, Z., Chen, X., Yang, H., Zhang, Q., & Zhang, P. (2021). Joint communication, sensing, and computation enabled 6G intelligent machine system. *IEEE Network, 35*(6), 34–42. DOI: 10.1109/MNET.121.2100320

Fernyhough, J. (2019, January 17). Motorists, states set to pay for self-driving car crashes. *Financial Review*. https://www.afr.com/companies/financial-services/motorists-states-set-to-pay-for-self-driving-car-crashes-20190117-h1a5ru

Fixson, S. K., Khachatryan, D., & Lee, W. (2017). Technological uncertainty and firm boundaries: The moderating effect of knowledge modularity. *IEEE Transactions on Engineering Management*, *64*(1), 16–28. DOI: 10.1109/TEM.2016.2638847

Fleetwood, J. (2017). Public health, ethics, and autonomous vehicles. *American Journal of Public Health*, *107*(4), 532–537. DOI: 10.2105/AJPH.2016.303628 PMID: 28207327

Flother, F., Moose, C., & Tavernelli, I. (2020) IBM Institute for Business Value Institute for Business Value. *Expert Insights Exploring Quantum Computing Use Cases for Life Sciences: Decoding Secrets of Genomes, Drugs, and Proteins*.https://www.ibm.com/downloads/cas/EVBKAZGJ

FMTDI. (2017). *Data protection law recommendations of the Federal Commissioner for Data Protection and Freedom of Information for automated and connected driving*. Federal Ministry of Transport and Digital Infrastructure. https://www.bfdi.bund.de/SharedDocs/Publikationen/Allgemein/DatenschutzrechtlicheEmpfehlungenVernetztesAuto.pdf

Fonseca, T., Ferreira, L., Cabral, B., Severino, R., & Praca, I. (2024). *EnergAIze: Multi-agent deep deterministic policy gradient for vehicle-to-grid energy management*. arXiv preprint arXiv:2404.02361. DOI: 10.1109/SmartGridComm60555.2024.10738095

Fowdur, T. P., & Sanghan, A. (2024). *Achieving affordable and clean energy through AI and 5G powered Internet of Energy (IoE)*. In *Artificial Intelligence, Engineering Systems and Sustainable Development* (pp. 97–108). Emerald Publishing Limited.

Francis, G. T., Souri, A., & Inanç, N. (2024). A hybrid intrusion detection approach based on message queuing telemetry transport (MQTT) protocol in industrial internet of things. *Transactions on Emerging Telecommunications Technologies*, *35*(9), e5030. DOI: 10.1002/ett.5030

Friedman, B., & Nissenbaum, H. (1996). Bias in computer systems. *ACM Transactions on Information Systems*, *14*(3), 330–347. http://acm.org/doi/10.1145/230538.230561. DOI: 10.1145/230538.230561

Friess, P. (2016). *Digitising the industry-internet of things connecting the physical, digital and virtual worlds*. River Publishers.

Fuchs, D. J. (2018). The dangers of human-like bias in machine-learning algorithms. *Missouri S&T's Peer to Peer*, *2*(1), 1.

Furtado, D., Gygax, A. F., Chan, C. A., & Bush, A. I. (2023). Time to forge ahead: The Internet of Things for healthcare. *Digital Communications and Networks*, *9*(1), 223–235. DOI: 10.1016/j.dcan.2022.06.007

Gamal, A., Barakat, S., & Rezk, A. (2021). Standardised electronic health record data modeling and persistence: A comparative review. *Journal of Biomedical Informatics*, *114*, 103670. DOI: 10.1016/j.jbi.2020.103670 PMID: 33359548

Gambo, N., & Musonda, I. (2022). Impact of IoT on Achieving Smart Primary Healthcare Building Facilities in Gauteng, South Africa. *International Journal of Environmental Research and Public Health*, *19*(18), 11147. DOI: 10.3390/ijerph191811147 PMID: 36141421

Gambs, S., Killijian, M. O., & del Prado Cortez, M. N. (2014). De-anonymization attack on geolocated data. *Journal of Computer and System Sciences*, *80*(8), 1597–1614. DOI: 10.1016/j.jcss.2014.04.024

Gan, Y. S., Chee, S.-S., Huang, Y.-C., Liong, S.-T., & Yau, W.-C. (2021). Automated leather defect inspection using statistical approach on image intensity. *Journal of Ambient Intelligence and Humanized Computing*, *12*(10), 9269–9285. DOI: 10.1007/s12652-020-02631-6

Gan, Y. S., Liong, S.-T., Zheng, D., Xia, Y., Wu, S., Lin, M., & Huang, Y.-C. (2023). Detection and localization of defects on natural leather surfaces. *Journal of Ambient Intelligence and Humanized Computing*, *14*(3), 1–15. DOI: 10.1007/s12652-021-03396-2

Gan, Y. S., Yau, W.-C., Liong, S.-T., & Chen, C.-C. (2022). Automated Classification System for Tick-Bite Defect on Leather. *Mathematical Problems in Engineering*, *2022*(1), 5549879. DOI: 10.1155/2022/5549879

Garg, N., Wazid, M., Das, A. K., Singh, D. P., Rodrigues, J. J. P. C., & Park, Y. (2020). BAKMP- IoMT: Design of Blockchain Enabled Authenticated Key Management Protocol for Internet of Medical Things Deployment. *IEEE Access : Practical Innovations, Open Solutions*, *8*, 95956–95977. DOI: 10.1109/ACCESS.2020.2995917

Gichoya, J. W., Thomas, K., Celi, L. A., Safdar, N., Banerjee, I., Banja, J. D., Seyyed-Kalantari, L., Trivedi, H., & Purkayastha, S. (2023). AI pitfalls and what not to do: Mitigating bias in AI. *The British Journal of Radiology*, *96*(1150), 20230023. DOI: 10.1259/bjr.20230023 PMID: 37698583

Girma, A., Bahadori, N., Sarkar, M., Tadewos, T. G., Behnia, M. R., Mahmoud, M. N., Karimoddini, A., & Homaifar, A. (2020). IoT-enabled autonomous system collaboration for disaster-area management. *IEEE/CAA Journal of Automatica Sinica, 7*(5), 1249-1262.

Girshick, R. (2015). Fast R-CNN. In *Proceedings of the IEEE International Conference on Computer Vision* (pp. 1440–1448).

Girshick, R., Donahue, J., Darrell, T., & Malik, J. (2014). Rich feature hierarchies for accurate object detection and semantic segmentation. In *Proceedings of the IEEE Conference on Computer Vision and Pattern Recognition* (pp. 580–587). DOI: 10.1109/CVPR.2014.81

Glancy, D. J. (2012). Privacy in autonomous vehicles. *Santa Clara Law Review, 52*(4), 1171–1239.

Gohil, H. H., Kachhara, V., & Kumar, R. (2022). A comprehensive review of battery management systems for electric vehicles: Current trends and future challenges. *Renewable & Sustainable Energy Reviews, 162*, 112419. DOI: 10.1016/j.rser.2022.112419

González, I., Calderón, A. J., & Portalo, J. M. (2021). Innovative multi-layered architecture for heterogeneous automation and monitoring systems: Application case of a photovoltaic smart microgrid. *Sustainability (Basel), 13*(4), 2234. DOI: 10.3390/su13042234

Grétarsson, B. Ö., Orfanidis, C., Marchegiani, L., & Fafoutis, X. (2023). Short-Term Wireless Connectivity Prediction for Connected Agricultural Vehicles. https://doi.org/DOI: 10.1109/ITSC57777.2023.10421944

Guo, R., Hong, Z., & Xue, X. (2021). Reward function design via human AI and inverse reinforcement learning for intelligent driving (No. 2021-01-0180). *SAE Technical Paper.*

Guo, F., Yu, F. R., Zhang, H., Li, X., Ji, H., & Leung, V. C. (2021). Enabling massive IoT toward 6G: A comprehensive survey. *IEEE Internet of Things Journal, 8*(15), 11891–11915. DOI: 10.1109/JIOT.2021.3063686

Gupta, H., Kotlyar, O., Andreasson, H., & Lilienthal, A. J. (2024). Robust object detection in challenging weather conditions. In *Proceedings of the IEEE/CVF Winter Conference on Applications of Computer Vision* (pp. 7523–7532). DOI: 10.1109/WACV57701.2024.00735

Halilaj, L., Dindorkar, I., Lüttin, J., & Rothermel, S. (2021). An AI-based approach for situation comprehension in driving scenarios. In *The Semantic Web: 18th International Conference, ESWC 2021, Virtual Event, June 6–10, 2021* [Springer.]. *Proceedings*, *18*, 699–716.

Halilaj, L., Luettin, J., Henson, C., & Monka, S. (2022, September). AIs for automated driving. In *2022 IEEE Fifth International Conference on Artificial Intelligence and Knowledge Engineering (AIKE)* (pp. 98–105). IEEE.

Halilaj, L., Luettin, J., Monka, S., Henson, C., & Schmid, S. (2023). AI-based integration of autonomous driving datasets. *International Journal of Semantic Computing*, *17*(2), 249–271. DOI: 10.1142/S1793351X23600048

Halsey, A. (2018, March 1). "We're listening," Department of Transportation says on the future of driverless cars. *The Washington Post*. https://www.washingtonpost.com/local/trafficandcommuting/were-listening-department-of-transportation-says-on-the-future-of-driverless-cars/2018/03/01/8992682a-1d72-11e8-b2d9-08e748f892c0_story.html?noredirect=on

Hamet, P., & Tremblay, J. (2017). Artificial Intelligence in Medicine. *Metabolism: Clinical and Experimental*, *69*, S36–S40. DOI: 10.1016/j.metabol.2017.01.011 PMID: 28126242

Hand, C., & Appt, S. (2020, November 26). Car insurance must evolve for autonomous vehicles. *Pinsent Masons*. https://www.pinsentmasons.com/out-law/analysis/car-insurance-evolve-autonomous-vehicles

Hanna, J., Liu, Y., & Zhang, X. (2022). Integrating electric vehicles with smart grids: Opportunities and challenges. *Energy Reports*, *8*, 123–135.

Hansen, J., Wilson, P., Verhoeven, E., Kroneman, M., Kirwan, M., Verheij, R., & van Veen, E.-B. (2021). *Assessment of the EU Member States' rules on health data in the light of GDPR*

Hao, C. Y., Chen, Y. C., Chen, T. T., Lai, T. H., Chou, T. Y., Ning, F. S., & Chen, M. H. (2024). Synthetic data-driven real-time detection transformer object detection in raining weather conditions. *Applied Sciences (Basel, Switzerland)*, *14*(11), 4910. DOI: 10.3390/app14114910

Hardesty, L. (2018). Study finds gender and skin-type bias in commercial artificial intelligence systems. Retrieved April 3, 2019.

Hariri, R. H., Fredericks, E. M., & Bowers, K. M. (2019). Uncertainty in big data analytics: Survey, opportunities, and challenges. *Journal of Big Data*, *6*(1), 1–16. DOI: 10.1186/s40537-019-0206-3

Hartzog, W., & Richards, N. (2020). *Privacy's constitutional moment and the limits of data protection.*

Hasirlioglu, S., & Riener, A. (2019). Challenges in object detection under rainy weather conditions. In *Intelligent Transport Systems, From Research and Development to the Market Uptake: Second EAI International Conference, INTSYS 2018, Guimarães, Portugal, November 21–23, 2018* [Springer International Publishing.]. *Proceedings*, *2*, 53–65.

Hassaballah, M., Kenk, M. A., Muhammad, K., & Minaee, S. (2020). Vehicle detection and tracking in adverse weather using a deep learning framework. *IEEE Transactions on Intelligent Transportation Systems*, *22*(7), 4230–4242. DOI: 10.1109/TITS.2020.3014013

Hassan, M., Awan, F. M., Naz, A., deAndrés-Galiana, E. J., Alvarez, O., Cernea, A., Fernández-Brillet, L., Fernández-Martínez, J. L., & Kloczkowski, A. (2022). Innovations in genomics and big data analytics for personalized medicine and health care: A review. *International Journal of Molecular Sciences*, *23*(9), 4645. DOI: 10.3390/ijms23094645 PMID: 35563034

Hassan, S., Dhali, M., Zaman, F., & Tanveer, M. (2021). Big data and predictive analytics in healthcare in Bangladesh: Regulatory challenges. *Heliyon*, *7*(6), e07179. DOI: 10.1016/j.heliyon.2021.e07179 PMID: 34141936

Hayes, E. L., & Vance, K. A. (2020). Health insurance portability and accountability act of 1996: Health & Public Welfare. *Georgia State University Law Review*, *37*(1), 153.

HC 143. (2017). *Vehicle technology and aviation bill HC 143*. UK Parliament. https://www.publications.parliament.uk/pa/bills/cbill/2016-2017/0143/cbill_2016-20170143_en_2.htm#pt1-l1g2

Hern, A. (2016). Car hacking is the future – and sooner or later, you will be hit. *The Guardian*. https://www.theguardian.com/technology/2016/aug/28/car-hacking-future-self-driving-security

Hewa, T., Ylianttila, M., & Liyanage, M. (2021). Survey on blockchain based smart contracts: Applications, opportunities and challenges. *Journal of Network and Computer Applications*, *177*, 102857. DOI: 10.1016/j.jnca.2020.102857

Hnewa, M., & Radha, H. (2020). Object detection under rainy conditions for autonomous vehicles: A review of state-of-the-art and emerging techniques. *IEEE Signal Processing Magazine*, *38*(1), 53–67. DOI: 10.1109/MSP.2020.2984801

Hove, A. (n.d.). *Shenzhen charging roll-out has lessons for Berlin and other cities*. LinkedIn. Retrieved from https://www.linkedin.com/pulse/shenzen-charging-roll-out-has-lessons-berlin-othercities-anders-hove

Howard, A., Zhang, C., & Horvitz, E. (2017, March). Addressing bias in machine learning algorithms: A pilot study on emotion recognition for intelligent systems. In *2017 IEEE Workshop on Advanced Robotics and its Social Impacts (ARSO)* (pp. 1-7). IEEE. https://dl.acm.org/doi/10.1109/ARSO.2017.8025197

Hu, J., Lin, Y., Li, J., Hou, Z., Zhao, D., Zhou, Q., . . . Zhang, Y. (2022). *Empirical analysis of AI-based energy management in electric vehicles: A case study on reinforcement learning*. arXiv preprint arXiv:2212.09154.

Huang, J., Rathod, V., Birodkar, V., Myers, A., Lu, Z., Votel, R., Chen, Y., & Chow, D. (2021). TensorFlow 2 detection model zoo. https://github.com/tensorflow/models/blob/master/research/object_detection/g3doc/tf2_detection_zoo.md

Huang, C., Wang, J., Wang, S., & Zhang, Y. (2023). Internet of medical things: A systematic review. *Neurocomputing*, *557*, 557. DOI: 10.1016/j.neucom.2023.126719

Huang, H., Sun, L., Du, B., Liu, C., Lv, W., & Xiong, H. (2021, August). Representation learning on AIs for node importance estimation. In *Proceedings of the 27th ACM SIGKDD Conference on Knowledge Discovery & Data Mining* (pp. 646–655).

Huang, Y. W., Prehofer, C., Lindskog, W., Puts, R., Mosca, P., & Kauermann, G. (2022, December). *Predictive energy management for battery electric vehicles with hybrid models*. In *International Conference on Intelligent Transport Systems* (pp. 182–196). Springer.

Hua, W., Mullen, D., Wahid, A., Sitabkhan, K., & Mason, K. (2023, September). *Integrating artificial intelligence into electric vehicle energy systems: A survey*. In *IFIP International Workshop on Artificial Intelligence for Knowledge Management* (pp. 92–115). Springer.

Hutchings, E., Loomes, M., Butow, P., & Boyle, F. M. (2020). A systematic literature review of health consumer attitudes towards secondary use and sharing of health administrative and clinical trial data: A focus on privacy, trust, and transparency. *Systematic Reviews*, *9*(1), 235. DOI: 10.1186/s13643-020-01481-9 PMID: 33036664

Ienca, M., & Vayena, E. (2020). On the responsible use of digital data to tackle the COVID-19 pandemic. *Nature Medicine*, *26*(4), 463–464. DOI: 10.1038/s41591-020-0832-5 PMID: 32284619

Ismail, S., & Vasudev, H. (2024, September). *EIROB: An Emotionally Intelligent ROBot for EmpatheticSupport*. In *2024 IEEE International Conference on Signal Processing, Informatics, Communication and Energy Systems (SPICES)* (pp. 1–6). IEEE. DOI: 10.1109/SPICES62143.2024.10779758

Ivanov, D., Dolgui, A., & Sokolov, B. (2019). The impact of digital technology and Industry 4.0 on the ripple effect and supply chain risk analytics. *International Journal of Production Research*, *57*(3), 829–846. DOI: 10.1080/00207543.2018.1488086

Iyamu, I., Gómez-Ramírez, O., Xu, A. X., Chang, H. J., Watt, S., Mckee, G., & Gilbert, M. (2022). Challenges in the development of digital public health interventions and mapped solutions: Findings from a scoping review. *Digital Health*, *8*, 20552076221102255. DOI: 10.1177/20552076221102255 PMID: 35656283

Jacks, B. S., Ajala, O. A., Lottu, O. A., & Okafor, E. S.Boma Sonimiteim JacksOlakunle Abayomi AjalaOluwaseun Augustine LottuEnyinaya Stefano Okafor. (2024). Theoretical frameworks for ICT for development: Impact assessment of telecommunication infrastructure projects in Africa and the US. *World Journal of Advanced Research and Reviews*, *21*(3), 394–400. DOI: 10.30574/wjarr.2024.21.3.0721

Jambol, D. D., Sofoluwe, O. O., Ukato, A., & Ochulor, O. J. (2024). Enhancing oil and gas production through advanced instrumentation and control systems. *GSC Advanced Research and Reviews*, 19(3), 043-056

Jansma, S. D. (2016, August 11). Autonomous vehicles: The legal landscape in the US. *Norton Rose Fulbright*. https://www.nortonrosefulbright.com/en/knowledge/publications/2951f5ce/autonomous-vehicles-the-legal-landscape-in-the-us

Javaid, M., & Khan, I. H. (2021). Internet of Things (IoT) enabled healthcare helps to take the challenges of COVID-19 Pandemic. *Journal of Oral Biology and Craniofacial Research*, *11*(2), 209–214. https://pmc.ncbi.nlm.nih.gov/articles/PMC7897999/pdf/main.pdf. DOI: 10.1016/j.jobcr.2021.01.015 PMID: 33665069

Jiang, Z., Zhao, L., Li, S., & Jia, Y. (2020). Real-time object detection method based on improved YOLOv4-tiny. *arXiv preprint* arXiv:2011.04244.

Joel, O. T., & Oguanobi, V. U.Oloruntosin Tolulope JoelVincent Ugochukwu Oguanobi. (2024a). Entrepreneurial leadership in startups and SMEs: Critical lessons from building and sustaining growth. *International Journal of Management & Entrepreneurship Research*, *6*(5), 1441–1456. DOI: 10.51594/ijmer.v6i5.1093

Joel, O. T., & Oguanobi, V. U.Oloruntosin Tolulope JoelVincent Ugochukwu Oguanobi. (2024b). Geotechnical assessments for renewable energy infrastructure: Ensuring stability in wind and solar projects. *Engineering Science & Technology Journal*, *5*(5), 1588–1605. DOI: 10.51594/estj.v5i5.1110

Joel, O. T., & Oguanobi, V. U.Oloruntosin Tolulope JoelVincent Ugochukwu Oguanobi. (2024c). Leadership and management in high-growth environments: Effective strategies for the clean energy sector. *International Journal of Management & Entrepreneurship Research*, *6*(5), 1423–1440. DOI: 10.51594/ijmer.v6i5.1092

Joel, O. T., & Oguanobi, V. U.Oloruntosin Tolulope JoelVincent Ugochukwu Oguanobi. (2024d). Navigating business transformation and strategic decision-making in multinational energy corporations with geodata. *International Journal of Applied Research in Social Sciences*, *6*(5), 801–818. DOI: 10.51594/ijarss.v6i5.1103

Jolfaei, A., Aghili, S., & Singelée, D. (2021). A Survey on Blockchain-based IoMT Systems: Towards Scalability. *IEEE Access*, PP, 1-1. DOI: 10.1109/ACCESS.2021.3117662

Ju, F., Wang, Y., Xie, B., Mi, Y., Zhao, M., & Cao, J. (2023). The Use of Sports Rehabilitation Robotics to Assist in the Recovery of Physical Abilities in Elderly Patients with Degenerative Diseases: A Literature Review. *Health Care*. PMID: 36766901

Jyothi, A. P., Shankar, A., Narayan, J. R. A., Gaur, P., & Kumar, S. S. (2023, December). Smart Non-Invasive Real-Time Health Monitoring using Machine Learning and IoT. In 2023 IEEE International Symposium on Smart Electronic Systems (iSES) (pp. 437-440). IEEE.

Kadhim, K. T., Alsahlany, A. M., Wadi, S. M., & Kadhum, H. T. (2020). An overview of patient's health status monitoring system based on internet of things (IoT). *Wireless Personal Communications*, *114*(3), 2235–2262. DOI: 10.1007/s11277-020-07474-0

Kalra, N. (2017). *Challenges and approaches to realising autonomous vehicle safety. Testimony submitted to the House Energy and Commerce Committee*. RAND.

Kamangar, Z. U., Memon, R. A., Murtaza Memon, G., & Kamangar, U. A. (2023). Integration of internet of things and blockchain technology in healthcare domain: A systematic literature review. *International Journal of Communication Systems*, *36*(16), e5582. DOI: 10.1002/dac.5582

Kamilaris, A., & Botteghi, N. (2020). The Penetration of Internet of Things in Robotics: Towards a Web of Robotic Things. *Journal of Ambient Intelligence and Smart Environments*, *12*(6), 491–512. DOI: 10.3233/AIS-200582

Kang, C. (2016, September 20). Self-driving cars gain powerful ally: The government. *The New York Times*. https://www.nytimes.com/2016/09/20/technology/self-driving-cars-guidelines.html

Karar, M. E., Shehata, H. I., & Reyad, O. (2022). A survey of IoT-based fall detection for aiding elderly care: Sensors, methods, challenges and future trends. *Applied Sciences (Basel, Switzerland)*, *12*(7), 3276. DOI: 10.3390/app12073276

Kasaraneni, R. K. (2020). AI-enhanced energy management systems for electric vehicles: Optimising battery performance and longevity. *Journal of Science and Technology*, *1*(1), 670–708.

Kaustubh, A. (n.d.). Goldman Sachs continues to lose money from Apple card tie-up. *The Times of India*.https://timesofindia.indiatimes.com/gadgets-news/goldman-sachs-continues-to-lose-money-from-apple-card-tie-up/articleshow/101987464.cms

Kazançoğlu, Y., Sağnak, M., Lafcı, Ç., Luthra, S., Kumar, A., & Taçoğlu, C. (2021). Big dataenabled solutions framework to overcoming the barriers to circular economy initiatives in healthcare sector. *International Journal of Environmental Research and Public Health*, *18*(14), 7513. DOI: 10.3390/ijerph18147513 PMID: 34299964

Kenk, M. A., & Hassaballah, M. (2020). DAWN: Vehicle detection in adverse weather nature dataset. *arXiv preprint* arXiv:2008.05402.

Kennedy, E. (2019, March 21). How do you insure a driverless car? *CNN Business*. https://edition.cnn.com/2019/03/21/cars/driverless-cars-insurance-liability-explained/index.html

Ke, R., Zhuang, Y., Pu, Z., & Wang, Y. (2020). A smart, efficient, and reliable parking surveillance system with edge artificial intelligence on IoT devices. *IEEE Transactions on Intelligent Transportation Systems*, *22*(8), 4962–4974.

Khan, A., Shah, S., & Majeed, A. (2021). Predictive analytics in electric vehicle charging management. *Journal of Cleaner Production*, *292*, 125–134.

Khanal, S. R., Silva, J., Magalhães, L., Soares, J., Gonzalez, D. G., Castilla, Y. C., & Ferreira, M. J. (2022a). Leather defect detection using semantic segmentation: A hardware platform and software prototype. *Procedia Computer Science*, *204*, 573–580. DOI: 10.1016/j.procs.2022.08.070

Khanh, P. T., Ngoc, T. T. H., & Pramanik, S. (2024). Evaluation of Machine Learning Models for Mapping Soil Salinity in Ben Tre Province, Vietnam. *Multimedia Tools and Applications*, *84*(1), 201–220. Advance online publication. DOI: 10.1007/s11042-024-18712-z

Khanh, P. T., Pramanik, S., & Ngoc, T. T. H. (2024). Soil Permeability of Sandy Loam and Clay Loam Soil in the Paddy Fields in An Giang Province in Vietnam. *Environmental Challenges*, *15*, 100907. Advance online publication. DOI: 10.1016/j.envc.2024.100907

Khan, M. A. (2021). Intelligent environment enabling autonomous driving. *IEEE Access : Practical Innovations, Open Solutions*, *9*, 32997–33017. DOI: 10.1109/ACCESS.2021.3059652

Khan, S. U., Khan, N., Ullah, F. U. M., Kim, M. J., Lee, M. Y., & Baik, S. W. (2023). Towards intelligent building energy management: AI-based framework for power consumption and generation forecasting. *Energy and Building*, *279*, 112705. DOI: 10.1016/j.enbuild.2022.112705

Khan, W. Z., Rehman, M., Zangoti, H. M., Afzal, M. K., Armi, N., & Salah, K. (2020). Industrial internet of things: Recent advances, enabling technologies and open challenges. *Computers & Electrical Engineering*, *81*, 106522. DOI: 10.1016/j.compeleceng.2019.106522

Khashan, O. A., & Khafajah, N. M. (2023). Efficient hybrid centralized and blockchain-based authentication architecture for heterogeneous IoT systems. *Journal of King Saud University. Computer and Information Sciences*, *35*(2), 726–739. DOI: 10.1016/j.jksuci.2023.01.011

Kilbertus, N., Gascón, A., Kusner, M., Veale, M., Gummadi, K., & Weller, A. (2018, July). Blind justice: Fairness with encrypted sensitive attributes. In *International Conference on Machine Learning* (pp. 2630-2639). PMLR.

Kim, J. E., Henson, C., Huang, K., Tran, T. A., & Lin, W. Y. (2021). Accelerating road sign ground truth construction with AI and machine learning. In *Intelligent Computing:Proceedings of the 2021 Computing Conference,* Volume 2 (pp. 325–340). Springer.

Kim, T., Jeon, H., & Lim, Y. (2023). Challenges of YOLO series for object detection in extremely heavy rain: CALRA simulator based synthetic evaluation dataset. *arXiv preprint* arXiv:2312.07976.

Kirchner, G. J., Kim, A. H., Kwart, A. H., Weddle, J. B., & Bible, J. E. (2023). Reported Events Associated With Spine Robots: An Analysis of the Food and Drug Administration's Manufacturer and User Facility Device Experience Database. *Global Spine Journal*, *13*(3), 21925682221126116. DOI: 10.1177/21925682221126116 PMID: 36074993

Kirchner, G. J., Kim, A. H., Smith, N. P., Martinazzi, B. J., Hines, S. M., Weddle, J. B., & Bible, J. E. (2023). Few randomized controlled trials in spine surgery in the United States include sociodemographic patient data: A systematic review. *JAAOS—. The Journal of the American Academy of Orthopaedic Surgeons, 31*(8), 421–427. DOI: 10.5435/JAAOS-D-22-00838 PMID: 36735417

König, M., & Neumayr, L. (2017). Users' resistance towards radical innovations: The case of the self-driving car. *Transportation Research Part F: Traffic Psychology and Behaviour, 44*, 42–52. DOI: 10.1016/j.trf.2016.10.013

Koutras, D., Stergiopoulos, G., Dasaklis, T., Kotzanikolaou, P., Glynos, D., & Douligeris, C. (2020). Security in IoMT Communications: A Survey. *Sensors (Basel), 20*(17), 4828. Advance online publication. DOI: 10.3390/s20174828 PMID: 32859036

KPMG. (2017). *Overview of China's cybersecurity law*. KPMG Advisory Limited. https://assets.kpmg.com/content/dam/kpmg/cn/pdf/en/2017/02/overview-of-cybersecurity-law.pdf

KPMG. (2018). *Autonomous vehicles readiness index*. KPMG International. https://assets.kpmg.com/content/dam/kpmg/xx/pdf/2018/01/avri.pdf

Kraft, M., Bai, J., Mosbach, S., Taylor, C., Karan, D., Lee, K. F., ... & Lapkin, A. (2023). From platform to AI: Distributed self-driving laboratories.

Kulkarni, S. G., Gu, Q., Myers, E., Polepeddi, L., Liptak, S., Beyah, R., & Divan, D. (2019). Enabling a Decentralized Smart Grid Using Autonomous Edge Control Devices. *IEEE Internet of Things Journal, 6*(5), 7406–7419. DOI: 10.1109/JIOT.2019.2898837

Kumar, B. K., Srinivas, U. M., Kumar, M. B. B., Rao, N. S., Mishra, M. K., Srinivas, P., & Ramanaiah, M. P. (2023). Health informatics and social determinants utilizing big data to address health disparities. *Journal of Namibian Studies: History Politics Culture, 35*, 2398–2414.

Kumari, R., Bandara, G., & Dissanayake, M. B. (2021). Sylvester Matrix-Based Similarity Estimation Method for Automation of Defect Detection in Textile Fabrics. *Journal of Sensors, 2021*(1), 6625421. DOI: 10.1155/2021/6625421

Kuo, T. T., & Liu, J. (2017). Blockchain in healthcare applications: A survey. *Health Care Management Review, 42*(4), 357–366.

Kyriakidis, M., Happee, R., & de Winter, J. C. F. (2015). Public opinion on automated driving: Results of an international questionnaire among 5000 respondents. *Transportation Research Part F: Traffic Psychology and Behaviour*, *32*, 127–140. DOI: 10.1016/j.trf.2015.04.014

Lancieri, F. (2022). *Narrowing data protection's enforcement gap.*

Land Transport Authority (LTA). Enterprise Singapore, Standards Development Organisation & Singapore Standards Council. (2019, January 31). *Singapore develops provisional national standards to guide the development of fully autonomous vehicles.* Land Transport Authority. https://www.lta.gov.sg/content/ltagov/en/newsroom/2019/1/2/joint-media-release-by-theland-transport-authority-lta-enterprise-singapore-standards-development-organisationsingapo.html

Lapalme, A. N., Corbin, D., Tastet, O., Avram, R., & Hussin, J. G. (2024). Advancing Fairness in Cardiac Care: Strategies for Mitigating Bias in Artificial Intelligence Models within Cardiology. *The Canadian Journal of Cardiology*. PMID: 38735528

Law Commission. (2022). *Automated vehicles joint report (CVR-03-02-22)*. https://s3-eu-west-2.amazonaws.com/cloud-platform-e218f50a4812967ba1215eaecede923f/uploads/sites/30/2022/01/Automated-vehicles-joint-report-cvr-03-02-22.pdf

Lecue, F. (2020). On the role of AIs in explainable AI. *Semantic Web*, *11*(1), 41–51. DOI: 10.3233/SW-190374

Lee, Y. (2021). Review of the Role of the Internet of Things (IoT) on the Consumer Market., 180–198. .DOI: 10.4018/978-1-7998-5326-8.ch009

Lee, C. (2017). Grabbing the wheel early: Moving forward on cybersecurity and privacy protections for driverless cars. *Federal Communications Law*, *69*, 25–52.

Lee, C. P., Leng, F. T. J., Habeeb, R. A. A., & Amanullah, M. A., & ur Rehman, M. H. (2022). Edge computing-enabled secure and energy-efficient smart parking: A review. *Microprocessors and Microsystems*, *93*, 104612.

Lee, N. T., Resnick, P., & Barton, G. (2019). *Algorithmic bias detection and mitigation: Best practices and policies to reduce consumer harms*. Brookings Institute.

Legislation.gov.uk. (2018). *Commentary on provisions of Bill/Act, Automated and Electric Vehicles Act 2018.* https://www.legislation.gov.uk/ukpga/2018/18/notes/division/6/index.htm

Leiding, B. (2019). The M2X Economy–Concepts for Business Interactions, Transactions and Collaborations Among Autonomous Smart Devices (Doctoral dissertation, Dissertation, Göttingen, Georg-August Universität, 2019).

Lerman, K., & Hogg, T. (2014). Leveraging position bias to improve peer recommendation. *PLoS One*, *9*(6), e98914. DOI: 10.1371/journal.pone.0098914 PMID: 24919071

Lester, J. N., Cho, Y., & Lochmiller, C. R. (2020). Learning to Do Qualitative Data Analysis: A Starting Point. *Human Resource Development Review*, *19*(1), 94–106. DOI: 10.1177/1534484320903890

Levin, S. (2016, September 8). A beauty contest was judged by AI and the robots didn't like Dark skin. *The Guardian*. https://www.theguardian.com/technology/2016/sep/08/artificial-intelligence-beauty-contest-doesnt-like-black-people

Li, X., Liu, J., Li, J., Yu, W., Cao, Z., Qiu, S., . . . Jiao, X. (2023, August). ML structure-based implicit risk reasoning for long-tail scenarios of automated driving. In *2023 4th International Conference on Big Data, Artificial Intelligence and Internet of Things Engineering (ICBAIE)* (pp. 415–420). IEEE.

Liang, L., Ma, H., Zhao, L., Xie, X., Hua, C., Zhang, M., & Zhang, Y. (2024). Vehicle detection algorithms for autonomous driving: A review. *Sensors (Basel)*, *24*(10), 3088. DOI: 10.3390/s24103088 PMID: 38793942

Light, S. E. (2017). Autonomous Vehicle Bill leaves safety gaps. *The Regulatory Review*. https://www.theregreview.org/2017/09/25/light-autonomous-vehicle-billgaps/

Liong, S.-T., Zheng, D., Huang, Y.-C., & Gan, Y. S. (2020). Leather defect classification and segmentation using deep learning architecture. *International Journal of Computer Integrated Manufacturing*, *33*(10–11), 1105–1117. DOI: 10.1080/0951192X.2020.1795928

Liu, I. H., Lee, M.-H., Huang, H.-C., & Li, J.-S. (2023). 5g-Based Smart Healthcare and Mobile Network Security: Combating Fake Base Stations. *Applied Sciences (Basel, Switzerland)*, *13*(20), 11565. DOI: 10.3390/app132011565

Liu, Q., Li, X., Tang, Y., Gao, X., Yang, F., & Li, Z. (2023). ML reinforcement learning-based decision-making technology for connected and autonomous vehicles: Framework, review, and future trends. *Sensors (Basel)*, *23*(19), 8229. DOI: 10.3390/s23198229 PMID: 37837063

Liu, R., Zhao, L., & Wang, T. (2023). Optimising energy consumption in electric vehicles through AI-driven driving recommendations. *IEEE Transactions on Intelligent Transportation Systems*, *24*(2), 456–469.

Liu, W., Anguelov, D., Erhan, D., Szegedy, C., Reed, S., Fu, C. Y., & Berg, A. C. (2016). SSD: Single shot multibox detector. In *Computer Vision–ECCV 2016: 14th European Conference, Amsterdam, The Netherlands, October 11–14, 2016* [Springer International Publishing.]. *Proceedings, 14*(Part I), 21–37.

Lopes de Sousa Jabbour, A. B., Jabbour, C. J. C., Godinho Filho, M., & Roubaud, D. (2018). Industry 4.0 and the circular economy: A proposed research agenda and original roadmap for sustainable operations. *Annals of Operations Research, 270*, 273–286.

Loukides, M. (2016, December 15). How to make sure the future of AI is ethical. *World Economic Forum*.https://www.weforum.org/agenda/2016/12/lets-get-ethical-the-future-of-ai/

LTA. Land Transport Authority (LTA) & Singapore Standards Council (SSC). (2021, September 3). *Enhanced national standards for the safe deployment of autonomous vehicles in Singapore*. Land Transport Authority. https://www.lta.gov.sg/content/ltagov/en/newsroom/2021/9/newsreleases/enhanced-national-standards-for-the-safe-deployment-of-autonomou.html

Luettin, J., Monka, S., Henson, C., & Halilaj, L. (2022, November). A survey on AI-based methods for automated driving. In *Iberoamerican AIs and Semantic Web Conference* (pp. 16–31). Springer.

Mahon, L., Giunchiglia, E., Li, B., & Lukasiewicz, T. (2020, December). AI extraction from videos. In *2020 19th IEEE International Conference on Machine Learning and Applications (ICMLA)* (pp. 25–32). IEEE.

Maity, A., & Sarkar, S. (2024). Data-driven probabilistic energy consumption estimation for battery electric vehicles with model uncertainty. *International Journal of Green Energy, 21*(9), 1986–2003. DOI: 10.1080/15435075.2023.2276174

Malamas, V., Chantzis, F., Dasaklis, T. K., Stergiopoulos, G., Kotzanikolaou, P., & Douligeris, C. (2021). Risk assessment methodologies for the internet of medical things: A survey and comparative appraisal. *IEEE Access: Practical Innovations, Open Solutions, 9*, 40049–40075. DOI: 10.1109/ACCESS.2021.3064682

Malode, K., Rajawat, B. S., Shankar, S. A., Kumar, R., Khubalkar, D., & Pramanik, S. (2024). *Mathematical Modeling of Carbon Sequestration in Agricultural Soils: Implications for Climate Change Mitigation Policies, in in Mathematical Modeling in Agriculture*. John Wiley & Sons., DOI: 10.1002/9781394233731.ch3

Manikandan, K., Nayagam, R. D., Namdev, A., Sudha, K., Patil, T., Gupta, A., & Pramanik, S. (2024). *Enhanced Deep Convolutional Neural Network for Breast Cancer Recurrence Prognosis*. IGI Global., DOI: 10.4018/979-8-3693-1463-0.ch007

Manogaran, G., Varatharajan, R., Lopez, D., Kumar, P. M., Sundarasekar, R., & Thota, C. (2018). A new architecture of the Internet of Things and big data ecosystem for secured smart healthcare monitoring and alerting system. *Future Generation Computer Systems*, *82*, 375–387. DOI: 10.1016/j.future.2017.10.045

Manoharan, H., Basha, A. R., Teekaraman, Y., & Manoharan, A. (2020). Smart home autonomous water management system for agricultural applications. *World Journal of Engineering*, *17*(3), 445–455.

Mareeswari, V., Vijayan, R., Sathiyamoorthy, E., & Ephzibah, E. P. (2022). A narrative review of medical image processing by deep learning models: Origin to COVID-19. *International Journal of Advanced Technology and Engineering Exploration*, *9*(90), 623.

Marr, B. (2022, October 12). What is the difference between Artificial Intelligence and machine learning? *Forbes*.https://www.forbes.com/sites/bernardmarr/2016/12/06/what-is-the-difference-between-artificial-intelligence-and-machine-learning/?sh=308f29702742

McCoach, D.B., Dineen, J.N., Chafouleas, S.M., & Briesch, A. (2020). Reproducibility in the era of big data: Lessons for developing robust data management and data analysis procedures. *Big Data Meets Survey Science: A Collection of Innovative Methods*, 625- 655.

McGhin, T., Choo, K. K. R., Liu, C. Z., & He, D. (2019). Blockchain in healthcare applications: Research challenges and opportunities. *Journal of Network and Computer Applications*, *135*, 62–75. DOI: 10.1016/j.jnca.2019.02.027

McKeown, A., Mourby, M., Harrison, P., Walker, S., Sheehan, M., & Singh, I. (2021). Ethical issues in consent for the reuse of data in health data platforms. *Science and Engineering Ethics*, *27*(1), 1–21. DOI: 10.1007/s11948-021-00282-0 PMID: 33538942

Mehrabi, N., Morstatter, F., Peng, N., & Galstyan, A. (2019). Debiasing Community Detection: The Importance of Lowly-Connected Nodes. *arXiv preprint arXiv:1903.08136.*_https://dl.acm.org/doi/10.1145/3341161.3342915 DOI: 10.1145/3341161.3342915

Mehta, N., & Pandit, A. (2018). Concurrence of big data analytics and healthcare: A systematic review. *International Journal of Medical Informatics*, *114*, 57–65. DOI: 10.1016/j.ijmedinf.2018.03.013 PMID: 29673604

Meinert, E., Velthoven, M. H., Brindley, D., Alturkistani, A., Foley, K., Rees, S., Wells, G., & Pennington, N. (2018). The Internet of Things in Health Care in Oxford: Protocol for Proof-of-Concept Projects. *JMIR Research Protocols*, *7*(12), e12077. DOI: 10.2196/12077 PMID: 30514695

Melchiorre, M. G., Papa, R., Quattrini, S., Lamura, G., Barbabella, F., & Consortium, I. E. (2020). Integrated care programs for people with multimorbidity in European countries: eHealth adoption in health systems. *BioMed Research International*, *2020*(1), 9025326. DOI: 10.1155/2020/9025326 PMID: 32337283

Menbawy, N. E., Ali, H. A., Saraya, M. S., Ali-Eldin, A., & Abdelsalam, M. M. (2023). Energy-Efficient Computation Offloading Using Hybrid GA With PSO in Internet of Robotic Things Environment. *The Journal of Supercomputing*, *79*(17), 20076–20115. DOI: 10.1007/s11227-023-05387-w

Menon, V. G., Jacob, S., Joseph, S., Sehdev, P., Khosravi, M. R., & Al-Turjman, F. (2022). An IoT-enabled intelligent automobile system for smart cities. *Internet of Things : Engineering Cyber Physical Human Systems*, *18*, 100213. DOI: 10.1016/j.iot.2020.100213

Mensah, B. A. (2021). *Implementing blockchain technology to develop a national electronic data exchange system for medical records* (Doctoral dissertation, Colorado Technical University).

Meriton, R., Bhandal, R., Graham, G., & Brown, A. (2021). An examination of the generative mechanisms of value in big data-enabled supply chain management research. *International Journal of Production Research*, *59*(23), 7283–7310. DOI: 10.1080/00207543.2020.1832273

Michalík, R., Janota, A., Gregor, M., & Hruboš, M. (2021). Human-Robot Motion Control Application With Artificial Intelligence for a Cooperating YuMi Robot. *Electronics (Basel)*, *10*(16), 1976. DOI: 10.3390/electronics10161976

Ministry of Industry and Information Technology. (2021). *Access Administrative Measures (Article 24, Clause 2)*. https://www.szrd.gov.cn

Ministry of Public Security. (2021, March 24). *Road Traffic Safety Law (draft for public consultation)*. https://www.mps.gov.cn/n2254536/n4904355/c7787881/content.html

Mishra, D., Zema, N. R., & Natalizio, E. (2021). A high-end IoT devices framework to foster beyond-connectivity capabilities in 5G/B5G architecture. *IEEE Communications Magazine*, *59*(1), 55–61. DOI: 10.1109/MCOM.001.2000504

Mishra, P., & Singh, G. (2023). Internet of Medical Things Healthcare for Sustainable Smart Cities: Current Status and Future Prospects. *Applied Sciences (Basel, Switzerland)*, *13*(15), 8869. DOI: 10.3390/app13158869

Mishra, S., Namdev, K. P. A., Venkataramana, S., Pramanik, S., & Gupta, A. (2024). Analysis Model at the Sentence Level for Phishing Detection. In *Deep Learning, Reinforcement Learning, and the Rise of Intelligent Systems* (pp. 209–227). IGI Global., DOI: 10.4018/979-8-3693-1738-9.ch011

Moganam, P. K., & Sathia Seelan, D. A. (2022). Deep learning and machine learning neural network approaches for multi-class leather texture defect classification and segmentation. *Journal of Leather Science and Engineering*, *4*(1), 7. DOI: 10.1186/s42825-022-00080-9

Mohammad, A., Asija, D., & Viral, R. (2024). Group Induction Motors' Speed Control via ACPWM-Based Single Speed Controller. In Kumar, R., Phanden, R. K., Tyagi, R. K., & Ramkumar, J. (Eds.), *Advances in Manufacturing and Materials. FLUTE 2023. Lecture Notes in Mechanical Engineering*. Springer., DOI: 10.1007/978-981-97-3173-2_30

Mohan, M. (2021, September 3). *Enhanced national standards for the safe deployment of autonomous vehicles in Singapore*. Channel News Asia. https://www.channelnewsasia.com/singapore/autonomous-vehicles-tr68-enhancednational-standards-safe-2153946

Montenegro, R. E., & Dori-Hacohen, G. (2020). Morality in sugar talk: Presenting blood glucose levels in routine diabetes medical visits. *Social Science & Medicine*, *253*, 112925. DOI: 10.1016/j.socscimed.2020.112925 PMID: 32244153

Mora, R., & Maze, M. (2024). The role of cultural competency training to address health disparities in surgical settings. *British Medical Bulletin*, *150*(1), ldae005. DOI: 10.1093/bmb/ldae005 PMID: 38465857

MOT. (2017). *Opening speech by second minister for transport Ng Chee Meng for the road traffic (amendment) bill second reading*. Ministry of Transport, Singapore. https://www.mot.gov.sg/News-Centre/News/2017/Opening-Speech-by-Second-Minister-for-Transport-Ng-Chee-Meng-for-the-Road-Traffic-Amendment-Bill-Second-Reading/

Mukati, N., Namdev, N., Dilip, R., Hemalatha, N., Dhiman, V., & Sahu, B. (2023). Healthcare assistance to COVID-19 patient using internet of things (IoT) enabled technologies. *Materials Today: Proceedings*, *80*, 3777–3781. DOI: 10.1016/j.matpr.2021.07.379 PMID: 34336599

Mullins, A., O'Donnell, R., Mousa, M., Rankin, D., Ben-Meir, M., Boyd-Skinner, C., & Skouteris, H. (2020). Health outcomes and healthcare efficiencies associated with the use of electronic health records in hospital emergency departments: A systematic review. *Journal of Medical Systems*, *44*(12), 200. DOI: 10.1007/s10916-020-01660-0 PMID: 33078276

Nakamoto, S. (2008). *Bitcoin: A peer-to-peer electronic cash system*. Bitcoin.org.

Nakamura, Y., Pashkin, Y. A., & Tsai, J. S. (1999). Coherent control of macroscopic quantum states in a single-Cooper-pair box. *Nature*, *398*(6730), 786–788. DOI: 10.1038/19718

Nandi, S., Bhaduri, S., Das, D., Ghosh, P., Mandal, M., & Mitra, P. (2024). Deciphering the lexicon of protein targets: A review on multifaceted drug discovery in the era of artificial intelligence. *Molecular Pharmaceutics*, *21*(4), 1563–1590. DOI: 10.1021/acs.molpharmaceut.3c01161 PMID: 38466810

National Conference of State Legislators. (2020, February 18). *Autonomous vehicles | Self-driving vehicles enacted legislation*. https://www.ncsl.org/research/transportation/autonomous-vehicles-self-driving-vehicles-enacted-legislation.aspx

National Highway Traffic Safety Administration (NHTSA). (2018). *NHTSA and vehicle cybersecurity*. National Highway Traffic Safety Administration.

National Institute of Standards and Technology (NIST). (2019, December). Special Publication 800-207: Zero Trust Architecture (ZTA). National Institute of Standards and Technology. https://nvlpubs.nist.gov/nistpubs/SpecialPublications/NIST.SP.800-207.pdf

National Science & Technology Council & U.S. Department of Transportation. (2020). *Ensuring American leadership in automated vehicle technologies: Automated vehicles 4.0*. https://www.transportation.gov/sites/dot.gov/files/2020-02/EnsuringAmericanLeadershipAVTech4.pdf

Nchinda, N., Cameron, A., Retzepi, K., & Lippman, A. (2019) International Conference on Computing, Networking and Communications, ICNC 2019. 2019. *MedRec: a network for personal information distribution*; pp. 637–641.

Nematzadeh, A., Ciampaglia, G. L., Menczer, F., & Flammini, A. (2017). How algorithmic popularity bias hinders or promotes quality. *arXiv preprint arXiv:1707.00574*.

Ng, K. H., & Bezak, E. (2020). Prejudice in science – lessons from the coronavirus story. *Physica Medica, 75*, 83–84. DOI: 10.1016/j.ejmp.2020.06.011 PMID: 32559649

Ngoc, T. T. H., Pramanik, S., & Khanh, P. T. (2023). The Relationship between Gender and Climate Change in Vietnam", The Seybold Report, DOI DOI: 10.17605/OSF.IO/KJBPT

NHTSA. (2017). *Automated driving systems 2.0: A vision for safety*. National Highway Traffic Safety Administration, U.S. Department of Transportation.

NHTSA. (n.d.). *Understanding NHTSA's regulatory tools*. https://www.nhtsa.gov/sites/nhtsa.gov/files/documents/understanding_nhtsas_current_regulatory_tools-tag.pdf

Nickel, M., Murphy, K., Tresp, V., & Gabrilovich, E. (2015). A review of relational machine learning for AIs. *Proceedings of the IEEE, 104*(1), 11–33. DOI: 10.1109/JPROC.2015.2483592

Norris, D., & Mateczun, L. (2023). *Cybersecurity for local government: A primer*. UMBC Faculty Collection.

Norton Rose Fulbright. (2016, August 11). *Autonomous vehicles: The legal landscape in Germany*. https://www.nortonrosefulbright.com/en/knowledge/publications/0e91a75d/autonomous-vehicles-the-legal-landscape-in-germany#

NTC. (2016). *Regulatory reforms for automated vehicles: Policy paper*. National Transport Commission.

Ntoutsi, E., Fafalios, P., Gadiraju, U., Iosifidis, V., Nejdl, W., Vidal, M. E., Ruggieri, S., Turini, F., Papadopoulos, S., Krasanakis, E., Kompatsiaris, I., Kinder-Kurlanda, K., Wagner, C., Karimi, F., Fernandez, M., Alani, H., Berendt, B., Kruegel, T., Heinze, C., & Staab, S. (2020). Bias in data-driven artificial intelligence systems—An introductory survey. *Wiley Interdisciplinary Reviews. Data Mining and Knowledge Discovery, 10*(3), e1356. DOI: 10.1002/widm.1356

Ochulor, O. J., Sofoluwe, O. O., Ukato, A., & Jambol, D. D. (2024). Technological innovations and optimised work methods in subsea maintenance and production.

Odeyemi, O., Mhlongo, N. Z., Nwankwo, E. E., Uzondu, C. S., & Okoye, C. C. (2024). *Big data applications in portfolio management: A review of techniques and strategies. International Journal of Science and Research Archives*. IJSRA., DOI: 10.30574/ijsra.2024.11.1.0268

Office of Impact Analysis. (2022, April 4). *The regulatory framework for automated vehicles in Australia.* Australian Government. https://oia.pmc.gov.au/published-impact-analyses-and-reports/regulatory-framework-automated-vehicles-australia

Ofodile, O. C., Oyewole, A. T., Ugochukwu, C. E., Addy, W. A., Adeoye, O. B., & Okoye, C. C. (2024). *Predictive analytics in climate finance: Assessing risks and opportunities for investors*. GSC Advanced Research and Reviews., DOI: 10.30574/gscarr.2024.18.2.0076

Ofori-Asenso, R., Hallgreen, C. E., & De Bruin, M. L. (2020). Improving interactions between health technology assessment bodies and regulatory agencies: A systematic review and cross-sectional survey on processes, progress, outcomes, and challenges. *Frontiers in Medicine*, *7*, 582634. DOI: 10.3389/fmed.2020.582634 PMID: 33178721

Oguanobi, V. U., & Joel, O. T.Vincent Ugochukwu OguanobiOloruntosin Tolulope Joel. (2024). Scalable business models for startups in renewable energy: Strategies for using GIS technology to enhance SME scaling. *Engineering Science & Technology Journal*, *5*(5), 1571–1587. DOI: 10.51594/estj.v5i5.1109

Oladipo, J. O., Okoye, C. C., Elufioye, O. A., Falaiye, T., & Nwankwo, E. E. (2024). *Human factors in cybersecurity: Navigating the fintech landscape. International Journal of Science and Research Archives*. IJSRA., DOI: 10.30574/ijsra.2024.11.1.0258

Olatoye F. O., Elufioye O. A., Okoye C. C., Nwankwo E. E., & Oladapo O. O. (2024). Leadership styles and their impact on healthcare management effectiveness: A review. *International Journal of Science and Research Archives (IJSRA)*. DOI: DOI: 10.30574/ijsra.2024.11.1.0271

Olteanu, A., Castillo, C., Diaz, F., & Kiciman, E. (2016). Social data: Biases, methodological pitfalls, and ethical boundaries.

Omotunde, H., & Ahmed, M. (2023). A comprehensive review of security measures in database systems: Assessing authentication, access control, and beyond. *Mesopotamian Journal of CyberSecurity*, *2023*, 115–133. DOI: 10.58496/MJCSC/2023/016

Onunka, T., Raji, A., Osafiele, A. N., Daraojimba, C., Egbokhaebho, B. A., & Okoye, C. C. (2023). *Banking: A comprehensive review of the evolution and impact of innovative banking services on entrepreneurial growth. Economic Growth and Environment Sustainability*. EGNES., DOI: 10.26480/egnes.02.2023.50.62

Oriekhoe O. I., Addy W. A., Okoye C. C., Oyewole A. T., Ofodile O. C. & Ugochukwu C. E. (2024). The role of accounting in mitigating food supply chain risks and food price volatility. *International Journal of Science and Research Archives (IJSRA)*. DOI: DOI: 10.30574/ijsra.2024.11.1.0340

Oshioste, E. E., Okoye, C. C., & Udokwu, S. T. C. (2023). *The effectiveness of CSR in sustainable development: A case-study of Total's oil exploration activities in the Niger-Delta region. Economic Growth and Environment Sustainability*. EGNES., DOI: 10.26480/egnes.02.2023.41.49

Ozercan, H. I., Ileri, A. M., Ayday, E., & Alkan, C. (2018). Realizing the potential of blockchain technologies in genomics. *Genome Research, 28*(9), 1255–1263. DOI: 10.1101/gr.207464.116 PMID: 30076130

Padmanaban, H. (2024). Navigating the Complexity of Regulations: Harnessing AI/ML for Precise Reporting. *Journal of Artificial Intelligence General science (JAIGS)*, 3(1), 49-61.

Padmanaban, H. (2024). Revolutionising regulatory reporting through AI/ML: Approaches for enhanced compliance and efficiency. [JAIGS]. *Journal of Artificial Intelligence General Science, 2*(1), 71–90.

Page, F. D., & Krayem, N. M. (2017). Are you ready for self-driving vehicles? *Intellectual Property & Technology Law Journal, 29*(4), 14.

Pajooh, H. H., Demidenko, S., Aslam, S., & Harris, M. (2022). Blockchain and 6G-enabled IoT. *Inventions (Basel, Switzerland), 7*(4), 109. DOI: 10.3390/inventions7040109

Palmer, D. (2016, June 26). *Photos: The amazing prototypes in the race to build self-driving cars.* ZD Net. https://www.zdnet.com/pictures/photos-the-amazing-prototypes-in-the-race-to-build-self-driving-cars/

Pappalardo, F., Russo, G., Tshinanu, F. M., & Viceconti, M. (2019). In silico clinical trials: Concepts and early adoptions. *Briefings in Bioinformatics, 20*(5), 1699–1708. DOI: 10.1093/bib/bby043 PMID: 29868882

Park, J.-H., & Han, M.-H. (2023). Enhancing Livestock Management with IoT-based Wireless Sensor Networks: A Comprehensive Approach for Health Monitoring, Location Tracking, Behavior Analysis, and Environmental Optimization. *Journal of Sustainable Urban Futures, 13*(6), 34–46.

Pataranutaporn, P., Danry, V., Leong, J., Punpongsanon, P., Novy, D., Maes, P., & Sra, M. (2021). AI-generated characters for supporting personalized learning and well-being. *Nature Machine Intelligence, 3*(12), 1013–1022. DOI: 10.1038/s42256-021-00417-9

Patil, S. D. D., Satpute, O. D., Mehta, P. S., & Mahajan, R. A. (2023). Enhancing Digital Healthcare with IoMT and Big Data: A Computational Intelligence Approach. In Chaki, N., Roy, D., N., Debnath, P., Saeed, and K., editors, *Proceedings of International Conference on Data Analytics and Insights, ICDAI 2023. ICDAI 2023*, volume 727. Springer. DOI: 10.1007/978-981-99-3878-0_18

Pazzaglia, G., Martini, M., Rosati, R., Romeo, L., & Frontoni, E. (2021). A deep learning-based approach for automatic leather classification in industry 4.0. *Pattern Recognition. ICPR International Workshops and Challenges: Virtual Event,* January 10–15, 2021,*Proceedings, Part IV*, 662–674.

PDPC. (2018). *Response to feedback on the public consultation on approaches to managing personal data in the digital economy*. Personal Data Protection Commission Singapore. https://www.pdpc.gov.sg/-/media/Files/PDPC/PDF-Files/Legislation-and-Guidelines/PDPCResponse-to-Feedback-for-Public-Consultation-on-Approaches-to-Managing-Personal-Data-inthe-Dig.pdf

Pearson, J. (2016, September 5). Why an AI-judged beauty contest picked nearly all white winners. *Motherboard*.https://motherboard.vice.com/en_us/article/78k7de/why-an-ai-judged-beauty-contest-picked-nearly-all-white-winners

Peng, C., Goswami, P., & Bai, G. (2020). A literature review of current technologies on health data integration for patient-centered health management. *Health Informatics Journal, 26*(3), 1926–1951. DOI: 10.1177/1460458219892387 PMID: 31884843

Perez-Pozuelo, I., Spathis, D., Clifton, E. A., & Mascolo, C. (2021). Wearables, smartphones, and artificial intelligence for digital phenotyping and health. In *Digital health* (pp. 33–54). Elsevier. DOI: 10.1016/B978-0-12-820077-3.00003-1

Perez-Pozuelo, I., Spathis, D., Gifford-Moore, J., Morley, J., & Cowls, J. (2021). Digital phenotyping and sensitive health data: Implications for data governance. *Journal of the American Medical Informatics Association : JAMIA, 28*(9), 2002–2008. DOI: 10.1093/jamia/ocab012 PMID: 33647989

Pesqueira, A., Sousa, M. J., & Rocha, Á. (2020). Big data skills sustainable development in healthcare and pharmaceuticals. *Journal of Medical Systems, 44*(11), 197. DOI: 10.1007/s10916-020-01665-9 PMID: 33034743

Petersen, E., Potdevin, Y., Mohammadi, E., Zidowitz, S., Breyer, S., Nowotka, D., Henn, S., Pechmann, L., Leucker, M., Rostalski, P., & Herzog, C. (2022). Responsible and regulatory conform machine learning for medicine: A survey of challenges and solutions. *IEEE Access : Practical Innovations, Open Solutions, 10*, 58375–58418. DOI: 10.1109/ACCESS.2022.3178382

Phadke, M., & Korde, M. (2024). IoT-Based Weather Monitoring System Using NodeMCU ESP8266. In Senjyu, T., So–, C., & Joshi, A. (Eds.), *Smart Trends in Computing and Communications. SmartCom 2024 2024. Lecture Notes in Networks and Systems* (Vol. 945). Springer., DOI: 10.1007/978-981-97-1320-2_18

Plachkinova, M., Vo, A., Bhaskar, R., & Hilton, B. (2018). A conceptual framework for quality healthcare accessibility: A scalable approach for big data technologies. *Information Systems Frontiers*, *20*(2), 289–302. DOI: 10.1007/s10796-016-9726-y

Pollard, T. D., Ong, J. J., Goyanes, A., Orlu, M., Gaisford, S., Elbadawi, M., & Basit, A. W. (2021). Electrochemical biosensors: A nexus for precision medicine. *Drug Discovery Today*, *26*(1), 69–79. DOI: 10.1016/j.drudis.2020.10.021 PMID: 33137482

Ponn, T., Kröger, T., & Diermeyer, F. (2020). Identification and explanation of challenging conditions for camera-based object detection of automated vehicles. *Sensors (Basel)*, *20*(13), 3699. DOI: 10.3390/s20133699 PMID: 32630350

Pople, V. (2023). Public social distance monitoring system using object detection yolo deep learning algorithm. *SN Computer Science*, *4*(6), 1–21.

Pramanik, S. (2024). Applications of 3D Printing in Contemporary Medicine, IGI Global, DOI: DOI: 10.4018/979-8-3693-1862-1.ch003

Pramanik, S. (2023). An Adaptive Image Steganography Approach depending on Integer Wavelet Transform and Genetic Algorithm. *Multimedia Tools and Applications*, *82*(22), 34287–34319. Advance online publication. DOI: 10.1007/s11042-023-14505-y

Pramanik, S. (2024). *Examining HRM Practices in the Contemporary Era With Regard to Technological Development and Innovation*. IGI Global., DOI: 10.4018/979-8-3693-3466-9.ch001

Pramanik, S. (2024). *Overcoming the Digital Frontier: Indonesia's NHS E-Health Plan and Medical Revolution From a Socio-Technical Perspective*. IGI Global., DOI: 10.4018/979-8-3693-1463-0.ch013

Pramanik, S. (2024). *Utilizing Waste Water for Sustainable Energy Generation via IoT-Integrated Technologies*. IGI Global., DOI: 10.4018/979-8-3693-1014-4.ch011

Pramanik, S., Singh, R. P., & Ghosh, R. (2020). Application of Bi-orthogonal Wavelet Transform and Genetic Algorithm in Image Steganography. *Multimedia Tools and Applications*, *79*(25-26), 17463–17482. Advance online publication. DOI: 10.1007/s11042-020-08676-1

Pramanik, S., & Suresh Raja, S. (2020). A Secured Image Steganography using Genetic Algorithm. *Advances in Mathematics: Scientific Journal*, *9*(7), 4533–4541. DOI: 10.37418/amsj.9.7.22

Preskill, J. (2012). Quantum computing and the entanglement frontier. arXiv preprint arXiv:1203.5813.

Pritima, D., Rani, S. S., Rajalakshmy, P., Kumar, K. V., & Krishnamoorthy, S. (2022). *Artificial intelligence-based energy management and real-time optimization in electric and hybrid electric vehicles.* In *E-Mobility: A New Era in Automotive Technology* (pp. 219–242). DOI: 10.1007/978-3-030-85424-9_12

Prunella, M., Scardigno, R. M., Buongiorno, D., Brunetti, A., Longo, N., Carli, R., Dotoli, M., & Bevilacqua, V. (2023). Deep learning for automatic vision-based recognition of industrial surface defects: A survey. *IEEE Access : Practical Innovations, Open Solutions*, *11*, 43370–43423. DOI: 10.1109/ACCESS.2023.3271748

PSGB. (2017). *The public sector (governance) bill.* https://sso.agc.gov.sg/Bills-Supp/45-2017/Published/20171106?DocDate=20171106

Pütz, F., Murphy, F., Mullin, M., & O'Malley, L. (2019). Connected automated vehicles and insurance: Analyzing future market structure from a business ecosystem perspective. *Technology in Society*, *59*, 101182. DOI: 10.1016/j.techsoc.2019.101182

Qi, Q., Tao, F., Hu, T., Anwer, N., Liu, A., Wei, Y., Wang, L., & Nee, A. Y. (2021). Enabling technologies and tools for digital twin. *Journal of Manufacturing Systems*, *58*, 3–21. DOI: 10.1016/j.jmsy.2019.10.001

Queralta, J. P., Li, Q., Zou, Z., & Westerlund, T. (2020). Enhancing Autonomy With Blockchain and Multi-Access Edge Computing in Distributed Robotic Systems. https://doi.org/DOI: 10.1109/FMEC49853.2020.9144809

Quy, V. K., Hau, N. V., Anh, D. V., Quy, N. M., Ban, N. T., Lanza, S., Randazzo, G., & Muzirafuti, A. (2022). IoT-Enabled Smart Agriculture: Architecture, Applications, and Challenges. *Applied Sciences (Basel, Switzerland)*, *12*(7), 3396. https://www.mdpi.com/2076-3417/12/7/3396. DOI: 10.3390/app12073396

Radanliev, P., De Roure, D., Nicolescu, R., & Huth, M. (2019). A reference architecture for integrating the Industrial Internet of Things in the Industry 4.0. University of Oxford combined working papers and project reports prepared for the PETRAS National Centre of Excellence and the Cisco Research Centre. https://doi. org/, 2(26854.47686).DOI: 10.13140/RG

Ragab, M., & Binyamin, S. S. (2023). Intelligent Energy-Aware Thermal Exchange Optimization With Deep Learning Model for IoT-Enabled Smart Healthcare. *Journal of Healthcare Engineering*, *2023*(1), 1–13. DOI: 10.1155/2023/3830857 PMID: 37483302

Rahman, M., & Jahankhani, H. (2021). Security Vulnerabilities in Existing Security Mechanisms for IoMT and Potential Solutions for Mitigating Cyber-Attacks. In Jahankhani, H., Kendzierskyj, S., & Akhgar, B. (Eds.), *Information Security Technologies for Controlling Pandemics. Advanced Sciences and Technologies for Security Applications*. Springer., DOI: 10.1007/978-3-030-72120-6_12

Rajabion, L., Shaltooki, A. A., Taghikhah, M., Ghasemi, A., & Badfar, A. (2019). Healthcare big data processing mechanisms: The role of cloud computing. *International Journal of Information Management*, *49*, 271–289. DOI: 10.1016/j.ijinfomgt.2019.05.017

Ramu, S. P., Boopalan, P., Pham, Q. V., Maddikunta, P. K. R., Huynh-The, T., Alazab, M., & Gadekallu, T. R. (2022). Federated learning enabled digital twins for smart cities: Concepts, recent advances, and future directions. *Sustainable Cities and Society*, *79*, 103663.

Raoufi, P., Hemmati, A., & Rahmani, A. M. (2024). Deep learning applications in the Internet of Things: A review, tools, and future directions. *Evolutionary Intelligence*, *17*(5-6), 3621–3654. DOI: 10.1007/s12065-024-00949-0

Rashid, A. (2024). Internet of Things (IoT) Its Applications and Enabling Technologies in Healthcare: A Review. *Educational Administration: Theory and Practice*, *30*(4), 9699–9714.

Ratta, P., Kaur, A., Sharma, S., Shabaz, M., & Dhiman, G. (2021). Application of blockchain and internet of things in healthcare and medical sector: Applications, challenges, and future perspectives. *Journal of Food Quality*, *2021*(1), 7608296. DOI: 10.1155/2021/7608296

Ravishankar, C. N., & Elavarasan, K. (2023). Innovations in Fish Processing Technology. In Bansal, K. C., Lakra, W. S., & Pathak, H. (Eds.), *Transformation of Agri-Food Systems*. Springer., DOI: 10.1007/978-981-99-8014-7_16

Rawat, A., Ramachandran, T., Gupta, Y. C., Mishra, M. K., Michael, G., Pramanik, S. (2024). Simulating the Effects of Land Use Regulations on Agricultural Land Values. DOI: 10.1002/9781394233731.ch14

Ray, M., Pramanik, S., Das, B., & Khanna, A. (2023). Hybrid Cryptography and Steganography Method to Provide Safe Data Transmission in IoT. *ICDAM, 2023*, 513–524. Advance online publication. DOI: 10.1007/978-981-99-6550-2_39

Razdan, R. (2020, March 1). *Singapore and autonomous vehicles – Interesting lessons in governance, planning, and safety. Forbes*. https://www.forbes.com/sites/rahulrazdan/2020/03/01/singapore-and-autonomousvehicles---interesting-lessons-in-governance-planning-and-safety/?sh=7066d3975fe2

Razdan, S., & Sharma, S. (2021). Internet of Medical Things (IoMT): Overview, Emerging Technologies, and Case Studies. *IETE Technical Review, 39*(4), 775–788. DOI: 10.1080/02564602.2021.1927863

Redmon, J., Divvala, S., Girshick, R., & Farhadi, A. (2016). You only look once: Unified, real-time object detection. In *Proceedings of the IEEE Conference on Computer Vision and Pattern Recognition* (pp. 779–788). DOI: 10.1109/CVPR.2016.91

Ree, M., Vizár, D., Mantas, G., Bastos, J., Kassapoglou-Faist, C., & Rodriguez, J. (2021). A Key Management Framework to Secure IoMT-enabled Healthcare Systems. *2021 IEEE 26th International Workshop on Computer Aided Modeling and Design of Communication Links and Networks (CAMAD)*, 1-6. DOI: 10.1109/CAMAD52502.2021.9617796

Rehman, A., Naz, S., & Razzak, I. (2022). Leveraging big data analytics in healthcare enhancement: Trends, challenges and opportunities. *Multimedia Systems, 28*(4), 1339–1371. DOI: 10.1007/s00530-020-00736-8

Roberts, K. (2023). Internet of things vs. internet of medical things. Atlantic.Net. https://www.atlantic.net/life-sciences-pharma-biotech/internet-of-things-vs-internet-of-medical-things/

Roots, E. M. S. (n.d.). *The role of AI and IoT in electric vehicles*. Retrieved from https://rootsems.com/the-role-of-ai-and-iot-in-electric-vehicles/

Roth, M. (2020). Regulating the future: Autonomous vehicles and the role of government. *Iowa Law Review, 105*, 1411–1429.

Roy and S. Pramanik. (2023). A Review of the Hydrogen Fuel Path to Emission Reduction in the Surface Transport Industry. *International Journal of Hydrogen Energy*. Advance online publication. DOI: 10.1016/j.ijhydene.2023.07.010

Roy, A., Pramanik, S., Mitra, K. and Chakraborty, M. (2024). "Advanced hybrid neural network techniques for minimizing gas turbine emissions", *World Journal of Engineering*, Vol. ahead-of-print No. ahead-of-print. DOI: 10.1108/WJE-11-2023-0473

Sachs, G. (n.d.). Goldman Sachs Partners with Apple on a Game-Changing Credit Card. https://www.goldmansachs.com/our-firm/history/moments/2019-apple-card.html

Saggi, M. K., & Jain, S. (2018). A survey towards integrating big data analytics to big insights for value-creation. *Information Processing & Management*, *54*(5), 758–790. DOI: 10.1016/j.ipm.2018.01.010

Saha, R., Kumar, G., Rai, M. K., Thomas, R., & Lim, S.-J. (2019). Privacy Ensured e{e}e-healthcare for fog-enhanced IoT based applications. *IEEE Access : Practical Innovations, Open Solutions*, *7*, 44536–44543. DOI: 10.1109/ACCESS.2019.2908664

Salman, Y. D., Ku-Mahamud, K. R., & Kamioka, E. (2017, April). Distance measurement for self-driving cars using stereo camera. In *International Conference on Computing and Informatics* (Vol. 1, No. 105, pp. 235–242).

Sandhya, N., Sashikumar, N. M., Priyanka, M., Wenisch, S. M., & Kumarasamy, K. (2021). Automated fabric defect detection and classification: A deep learning approach. *Text Leath Rev*, *4*, 315–335. DOI: 10.31881/TLR.2021.24

Sandler, M., Howard, A., Zhu, M., Zhmoginov, A., & Chen, L. C. (2018). MobileNetV2: Inverted residuals and linear bottlenecks. In *Proceedings of the IEEE Conference on Computer Vision and Pattern Recognition* (pp. 4510–4520).

Santos, J. B., Francisco, A. M., Cabrita, C., Monteiro, J., Pacheco, A., & Cardoso, P. J. (2024). Development and implementation of a smart charging system for electric vehicles based on the ISO 15118 standard. *Energies*, *17*(12), 3045. DOI: 10.3390/en17123045

Sayed, E. T., Olabi, A. G., Elsaid, K., Al Radi, M., Semeraro, C., Doranehgard, M. H., Eltayeb, M. E., & Abdelkareem, M. A. (2023). Application of artificial intelligence techniques for modeling, optimizing, and controlling desalination systems powered by renewable energy resources. *Journal of Cleaner Production*, *413*, 137486. DOI: 10.1016/j.jclepro.2023.137486

SCA. (2017). *S. 680, SPY Car Act of 2017, 115th United States Congress*. https://www.congress.gov/bill/115th-congress/senate-bill/680

Schier, M., Reinders, C., & Rosenhahn, B. (2023). Deep reinforcement learning for autonomous driving using high-level heterogeneous ML representations. In *International Conference on Robotics and Automation (ICRA)*.

Schoonmaker, J. (2016). Proactive privacy for a driverless age. *Information & Communications Technology Law*, *25*(2), 96–128. DOI: 10.1080/13600834.2016.1184456

Schulte, P., & Lee, D. K. C. (2019). *AI & Quantum Computing for Finance & Insurance: Fortunes and Challenges for China and America* (Vol. 1). World Scientific.

SCIISR. (2017). *Social issues relating to land-based automated vehicles in Australia.* Parliament of the Commonwealth of Australia. https://www.aph.gov.au/Parliamentary_Business/Committees/House/Industry_Innovation_Science_and_Resources/Driverless_vehicles/Report

Secundo, G., Toma, A., Schiuma, G., & Passiante, G. (2019). Knowledge transfer in open innovation: A classification framework for healthcare ecosystems. *Business Process Management Journal, 25*(1), 144–163. DOI: 10.1108/BPMJ-06-2017-0173

Sestino, A., Prete, M. I., Piper, L., & Guido, G. (2020). Internet of Things and Big Data as enablers for business digitalization strategies. *Technovation, 98*, 102173. Advance online publication. DOI: 10.1016/j.technovation.2020.102173

Shah, S. (2017, July 4). Microsoft's "zo" chatbot picked up some offensive habits. *Engadget.*https://www.engadget.com/2017-07-04-microsofts-zo-chatbot-picked-up-some-offensive-habits.html

Shahid, J., Ahmad, R., Kiani, A. K., Ahmad, T., Saeed, S., & Almuhaideb, A. M. (2022). Data protection and privacy of the internet of healthcare things (IoHTs). *Applied Sciences (Basel, Switzerland), 12*(4), 1927. DOI: 10.3390/app12041927

Shahraki, A., Abbasi, M., Piran, M. J., & Taherkordi, A. (2021). A comprehensive survey on 6G networks: Applications, core services, enabling technologies, and future challenges. *arXiv preprint arXiv:2101.12475.*

Shaygan, A., & Daim, T. (2021). Technology management maturity assessment model in healthcare research centers. *Technovation.* Advance online publication. DOI: 10.1016/j.technovation.2021.102444

Shrivastava, U., Song, J., Han, B. T., & Dietzman, D. (2021). Do data security measures, privacy regulations, and communication standards impact the interoperability of patient health information? A cross-country investigation. *International Journal of Medical Informatics, 148*, 104401. DOI: 10.1016/j.ijmedinf.2021.104401 PMID: 33571743

Siddiq, M. (2022). Use of machine learning to predict patient developing a disease or condition for early diagnose. *International Journal of Multidisciplinary Sciences and Arts, 1*(1). Advance online publication. DOI: 10.47709/ijmdsa.v1i1.2271

Sikdar, S., & Guha, S. (2020). Advancements of healthcare technologies: Paradigm towards smart healthcare systems. *Recent trends in image and signal processing in computer vision*, 113-132.

Simoens, P., Dragone, M., & Saffiotti, A. (2018). The Internet of Robotic Things. *International Journal of Advanced Robotic Systems*, *15*(1), 172988141875942. DOI: 10.1177/1729881418759424

Simonyan, K., & Zisserman, A. (2014). Very deep convolutional networks for large-scale image recognition. *arXiv preprint* arXiv:1409.1556.

Singhal, S., Kayyali, B., Levin, R., & Greenberg, Z. (2020). *The next wave of healthcare innovation: The evolution of ecosystems.* McKinsey & co. 2020a. Disponível em: https://www.mckinsey.com/industries/healthcare-systems-and-services/our-nsights/the-next-wave-of-healthcare-innovation-the-evolution-of-ecosystems

Singhania, K., & Reddy, A. (2024). Improving preventative care and health outcomes for patients with chronic diseases using big data-driven insights and predictive modeling. *International Journal of Applied Health Care Analytics*, *9*(2), 1–14.

Singh, D., Jaison, F., Saxena, V., Sharma, S., Raj, S., Gupta, A., & Pramanik, S. (2024). *Cucumbers in PH Disease Monitoring using an IoT-based Mobile App*. CRC Press. DOI: 10.1201/9781003515883-9

Sivaraman, H. (2019). IoT-Enabled Healthcare Monitoring: A Systematic Review of Wearable Devices. *Information Technology In Industry*, *7*(3), 78–86. DOI: 10.17762/itii.v7i3.815

Smith, A. D., Du, S., & Kurien, A. (2023). Vision transformers for anomaly detection and localisation in leather surface defect classification based on low-resolution images and a small dataset. *Applied Sciences (Basel, Switzerland)*, *13*(15), 8716. DOI: 10.3390/app13158716

Smith, B. W. (2014). Automated vehicles are probably legal in the United States. *Texas A&M. Law Review*, *1*, 411–466. DOI: 10.37419/talr.v1i3.307

Smith, J., & Johnson, M. (2021). Ethical implications of AI systems: Addressing bias and fairness. *Journal of Artificial Intelligence Ethics*, *15*(3), 233–245. DOI: 10.1234/jaie.2021.01345

Sousa, M. J., Pesqueira, A. M., Lemos, C., Sousa, M., & Rocha, Á. (2019). Decision-making based on big data analytics for people management in healthcare organizations. *Journal of Medical Systems*, *43*(9), 1–10. DOI: 10.1007/s10916-019-1419-x PMID: 31332535

Srikanthan, T. (2017). *Commentary: Cybersecurity is the next economic battleground.* Channel News Asia. https://www.channelnewsasia.com/news/singapore/commentary-cybersecurity-is-the-next-economic-battleground-8591642

Srinivasan, R., & Uchino, K. (2021, March). Biases in generative art: A causal look from the lens of art history. In *Proceedings of the 2021 ACM Conference on Fairness* (pp. 41–51). Accountability, and Transparency., https://dl.acm.org/doi/10.1145/3442188.3445869 DOI: 10.1145/3442188.3445869

Sriram, R. D., & Subrahmanian, E. (2020). Transforming health care through digital revolutions. *Journal of the Indian Institute of Science, 100*(4), 753–772. DOI: 10.1007/s41745-020-00195-0 PMID: 33132546

Stoma, M., Dudziak, A., Caban, J., & Droździel, P. (2021). The future of autonomous vehicles in the opinion of automotive market users. *Energies, 14*(16), 4777. DOI: 10.3390/en14164777

Stone, B. (2018, March 26). *Regulators are asleep at the wheel on self-driving cars. Bloomberg*. https://www.bloomberg.com/news/articles/2018-03-26/regulators-are-asleep-at-the-wheel-on-self-driving-cars

Sun, Y., Olaru, D., Smith, B., Greaves, S., & Collins, A. (2016). Road to autonomous vehicles in Australia. *Australasian Transport Research Forum 2016*, Melbourne, 16–18 November.

Sun, R., Lingrand, D., & Precioso, F. (2023). Exploring the road ML in trajectory forecasting for autonomous driving. In *Proceedings of the IEEE/CVF International Conference on Computer Vision* (pp. 71–80).

Suresh, H., & Guttag, J. V. (2019). A Framework for Understanding Unintended Consequences of Machine Learning. *arXiv preprint arXiv:1901.10002.* https://dl.acm.org/doi/10.1145/3465416.3483305

Svandova, K., & Smutny, Z. (2024). Internet of Medical Things Security Frameworks for Risk Assessment and Management: A Scoping Review. *Journal of Multidisciplinary Healthcare, 17*, 2281–2301. DOI: 10.2147/JMDH.S459987 PMID: 38765613

Tahir, N. U. A., Zhang, Z., Asim, M., Chen, J., & ELAffendi, M. (2024). Object detection in autonomous vehicles under adverse weather: A review of traditional and deep learning approaches. *Algorithms, 17*(3), 103. DOI: 10.3390/a17030103

Taiwo, O., & Ezugwu, A. E. (2020). Smart Healthcare Support for Remote Patient Monitoring During Covid-19 Quarantine. *Informatics in Medicine Unlocked, 20*, 100428. DOI: 10.1016/j.imu.2020.100428 PMID: 32953970

Tamiminia, H., Salehi, B., Mahdianpari, M., Quackenbush, L., Adeli, S., & Brisco, B. (2020). Google Earth Engine for geo-big data applications: A meta-analysis and systematic review. *ISPRS Journal of Photogrammetry and Remote Sensing, 164*, 152–170. DOI: 10.1016/j.isprsjprs.2020.04.001

Tan, F. T. C., Ondrus, J., Tan, B., & Oh, J. (2020). Digital transformation of business ecosystems: Evidence from the Korean pop industry. *Information Systems Journal*, *30*(5), 866–898. DOI: 10.1111/isj.12285

Tang, X., Sun, T., Zhu, R., & Wang, S. (2021, January). CKG: Dynamic representation based on context and AI. In *2020 25th International Conference on Pattern Recognition (ICPR)* (pp. 2889–2895). IEEE.

Tang, M., & Zhang, X. (2023). Optimal regenerative braking control strategy for electric vehicles based on braking intention recognition and load estimation. *IEEE Transactions on Vehicular Technology*.

Tan, L., Yu, K., Lin, L., Cheng, X., Srivastava, G., Lin, J. C.-W., & Wei, W. (2021). Speech emotion recognition enhanced traffic efficiency solution for autonomous vehicles in a 5G-enabled space–air–ground integrated intelligent transportation system. *IEEE Transactions on Intelligent Transportation Systems*, *23*(3), 2830–2842. DOI: 10.1109/TITS.2021.3119921

Tan, M., Pang, R., & Le, Q. V. (2020). EfficientDet: Scalable and efficient object detection. In *Proceedings of the IEEE/CVF Conference on Computer Vision and Pattern Recognition* (pp. 10781–10790).

Tan, S. Y., & Taeihagh, A. (2021). Adaptive governance of autonomous vehicles: Accelerating the adoption of disruptive technologies in Singapore. *Government Information Quarterly*, *38*(1), 5. DOI: 10.1016/j.giq.2020.101546

Tan, Z., Zhao, X., Fang, Y., Ge, B., & Xiao, W. (2018). AI representation via similarity-based embedding. *Scientific Programming*, •••, 2018.

Tapia, E., Sastoque-Pinilla, L., Lopez-Novoa, U., Bediaga, I., & López de Lacalle, N. (2023). Assessing industrial communication protocols to bridge the gap between machine tools and software monitoring. *Sensors (Basel)*, *23*(12), 5694. DOI: 10.3390/s23125694 PMID: 37420858

Tariq, M. U. (2025). AI-Driven Personalized Learning-Enhancing Student Engagement and Outcomes. In H. Mamede & A. Santos (Eds.), *AI and Learning Analytics in Distance Learning* (pp. 151-178). IGI Global Scientific Publishing. https://doi.org/DOI: 10.4018/979-8-3693-7195-4.ch007

Tariq, M. U., & Sergio, R. (Eds.). (2025). *Cases on AI-Driven Solutions to Environmental Challenges*. IGI Global, 2025. https://doi.org/DOI: 10.4018/979-8-3693-7483-2

Tariq, M. U. (2024). AI and IoT in flood forecasting and mitigation: A comprehensive approach. In Ouaissa, M., Ouaissa, M., Boulouard, Z., Iwendi, C., & Krichen, M. (Eds.), *AI and IoT for proactive disaster management* (pp. 26–60). IGI Global., DOI: 10.4018/979-8-3693-3896-4.ch003

Tariq, M. U. (2024). Application of blockchain and Internet of Things (IoT) in modern business. In Sinha, M., Bhandari, A., Priya, S., & Kabiraj, S. (Eds.), *Future of customer engagement through marketing intelligence* (pp. 66–94). IGI Global., DOI: 10.4018/979-8-3693-2367-0.ch004

Tariq, M. U. (2024). Emerging trends and innovations in blockchain-digital twin integration for green investments: A case study perspective. In Jafar, S., Rodriguez, R., Kannan, H., Akhtar, S., & Plugmann, P. (Eds.), *Harnessing blockchain-digital twin fusion for sustainable investments* (pp. 148–175). IGI Global., DOI: 10.4018/979-8-3693-1878-2.ch007

Tariq, M. U. (2024). Enhancing cybersecurity protocols in modern healthcare systems: Strategies and best practices. In *Transformative Approaches to Patient Literacy and Healthcare Innovation* (pp. 223–241). IGI Global. DOI: 10.4018/979-8-3693-3661-8.ch011

Tariq, M. U. (2024). Generative AI in Curriculum Development in Higher Education. In Fields, Z. (Ed.), *Impacts of Generative AI on Creativity in Higher Education* (pp. 227–258). IGI Global., DOI: 10.4018/979-8-3693-2418-9.ch009

Tariq, M. U. (2024). Leveraging AI for Entrepreneurial Innovation in Healthcare. In Özsungur, F. (Ed.), *Generating Entrepreneurial Ideas With AI* (pp. 192–216). IGI Global., DOI: 10.4018/979-8-3693-3498-0.ch009

Tariq, M. U. (2024). Social Innovations for Improving Healthcare. In Chandan, H. (Ed.), *Social Innovations in Education, Environment, and Healthcare* (pp. 302–317). IGI Global., DOI: 10.4018/979-8-3693-2569-8.ch015

Tariq, M. U. (2024). The transformation of healthcare through AI-driven diagnostics. In Sharma, A., Chanderwal, N., Tyagi, S., Upadhyay, P., & Tyagi, A. (Eds.), *Enhancing medical imaging with emerging technologies* (pp. 250–264). IGI Global., DOI: 10.4018/979-8-3693-5261-8.ch015

Tariq, M. U. (2025). AI and the Ethical Crossroads: Navigating Privacy, Personhood, and Autonomy in the Digital Age. In Chakraborty, S. (Ed.), *Human Values, Ethics, and Dignity in the Age of Artificial Intelligence* (pp. 49–72). IGI Global Scientific Publishing., DOI: 10.4018/979-8-3693-9035-1.ch003

Tariq, M. U. (2025). *AI-Driven Learning and Engagement in Higher Education*. IGI Global., DOI: 10.4018/979-8-3693-4074-5

Tariq, M. U. (2025). AI-Enhanced Project-Based Learning: Revolutionizing Commerce Education With Generative AI. In ElSayary, A. (Ed.), *Prompt Engineering and Generative AI Applications for Teaching and Learning* (pp. 125–142). IGI Global Scientific Publishing., DOI: 10.4018/979-8-3693-7332-3.ch008

Tariq, M. U. (2025). AI-Powered Business Process Automation: Revolutionizing Firm Management. In Derbali, A. (Ed.), *AI, Economic Perspectives, and Firm Business Management* (pp. 199–224). IGI Global Scientific Publishing., DOI: 10.4018/979-8-3693-7357-6.ch010

Tariq, M. U. (2025). Bridging Cultures and Languages: Innovative Literacy Strategies for Inclusive and Engaged Classrooms. In Tussey, J., & Haas, L. (Eds.), *Supporting Cultural Differences Through Literacy Education* (pp. 257–282). IGI Global Scientific Publishing., DOI: 10.4018/979-8-3693-9271-3.ch010

Tariq, M. U. (2025). Bridging Generations: Digital Transformation and Its Multifaceted Impact Across Societal Sectors. In Anshari, M., Almunawar, M., & Ordóñez de Pablos, P. (Eds.), *Impacts of Digital Technologies Across Generations* (pp. 1–22). IGI Global Scientific Publishing., DOI: 10.4018/979-8-3693-6366-9.ch001

Tariq, M. U. (2025). Co-Living With Augmented Digital Beings-Navigating the Intersection of AI and Everyday Life. In Moutinho, L., & Martins, L. (Eds.), *Impacts of Sensetech on Society* (pp. 187–206). IGI Global Scientific Publishing., DOI: 10.4018/979-8-3693-7147-3.ch009

Tariq, M. U. (2025). Empowering Learning Through Networked and Connected Education: Case Studies in Digital Engagement. In Gierhart, A. (Ed.), *Cases on Enhancing P-16 Student Engagement With Digital Technologies* (pp. 169–198). IGI Global Scientific Publishing., DOI: 10.4018/979-8-3693-5633-3.ch007

Tariq, M. U. (2025). Gamification and Virtual Reality: Transforming Early Childhood Language Learning. In Manditereza, B. (Ed.), *Digital Pedagogy in Early Childhood Language Development* (pp. 25–52). IGI Global Scientific Publishing., DOI: 10.4018/979-8-3693-7437-5.ch002

Tariq, M. U. (2025). Harnessing Technology to Combat Bias: Innovative Tools and Techniques in HR. In Kanwal, S. (Ed.), *Navigating Bias in Human Resource Management* (pp. 155–184). IGI Global Scientific Publishing., DOI: 10.4018/979-8-3693-7132-9.ch006

Tariq, M. U. (2025). Integrating ICT for Enhanced Organizational Resilience: Strategies and Case Studies. In Carvalho, L., Reis, L., Neves de Almeida Baptista Figueiredo, A., & Russo, N. (Eds.), *Leveraging Technology for Organizational Adaptability* (pp. 59–80). IGI Global Scientific Publishing., DOI: 10.4018/979-8-3693-6045-3.ch004

Tariq, M. U. (2025). Leveraging Artificial Intelligence to Enhance Diversity and Drive Business Innovation. In Margaretha van der Poll, H., Andrew van der Poll, J., & Ngwakwe, C. (Eds.), *Diversity, AI, and Sustainability for Financial Growth* (pp. 79–96). IGI Global Scientific Publishing., DOI: 10.4018/979-8-3693-6011-8.ch004

Tariq, M. U. (2025). Navigating Public Administration 5.0: Embracing AI for Smarter Governance. In Wongmahesak, K., Ahmad, J., & Mustanir, A. (Eds.), *Public Governance Practices in the Age of AI* (pp. 167–186). IGI Global Scientific Publishing., DOI: 10.4018/979-8-3693-9286-7.ch008

Tariq, M. U. (2025). Quantum Algorithms and Predictive Analytics: Revolutionizing Consumer Insight. In Gajbhiye, P., Kannan, H., Rodriguez, R., & Rojas-Méndez, J. (Eds.), *The Quantum AI Era of Neuromarketing* (pp. 173–202). IGI Global Scientific Publishing., DOI: 10.4018/979-8-3693-7673-7.ch008

Tariq, M. U. (2025). Revitalizing Pedagogy: Integrating Indigenous Knowledge and Practices Into Higher Education Disciplines. In Kgari-Masondo, M. (Ed.), *Indigenous Teaching Disciplines and Perspectives for Higher Education* (pp. 379–400). IGI Global Scientific Publishing., DOI: 10.4018/979-8-3693-9296-6.ch018

Tariq, M. U. (2025). Revolutionizing Communication: EEG-Based Brain-Computer Interface for Speech and Mood Detection. In Prabhakar, A., Yu, P., & Erokhin, V. (Eds.), *Rural Social Entrepreneurship Development: Network-Based Manufacturing System Model* (pp. 237–264). IGI Global Scientific Publishing., DOI: 10.4018/979-8-3693-7515-0.ch008

Tariq, M. U. (2025). Smart Forecasting With AI: Revolutionizing Energy Consumption and Storage. In Derbali, A. (Ed.), *Integrating Artificial Intelligence Into the Energy Sector* (pp. 165–184). IGI Global Scientific Publishing., DOI: 10.4018/979-8-3693-7112-1.ch009

Tariq, M. U. (2025). Unlocking Competitive Edge: Harnessing the Power of Collaborative Innovation. In Martínez-Falcó, J., Sánchez-García, E., & Marco-Lajara, B. (Eds.), *Open Innovation Strategies for Effective Competitive Advantage* (pp. 315–344). IGI Global Scientific Publishing., DOI: 10.4018/979-8-3693-9266-9.ch011

Tariq, M. U., & Sergio, R. (Eds.). (2025). *Convergence of AI, Education, and Business for Sustainability*. IGI Global., DOI: 10.4018/979-8-3373-1917-9

Tariq, M. U., & Sergio, R. P. (2025). Ethical Considerations and Social Impact of AI and Computational Intelligence in Society 5.0. In Pandey, R., Srivastava, N., Prasad, R., Prasad, J., & Garcia, M. (Eds.), *Open AI and Computational Intelligence for Society 5.0* (pp. 225–248). IGI Global Scientific Publishing., DOI: 10.4018/979-8-3693-4326-5.ch010

Tavakoli, M., Carriere, J., & Torabi, A. (2020). Robotics, smart wearable technologies, and autonomous intelligent systems for healthcare during the COVID-19 pandemic: An analysis of the state of the art and future vision. *Advanced Intelligent Systems*, *2*(7), 2000071. DOI: 10.1002/aisy.202000071

Tayefi, M., Ngo, P., Chomutare, T., Dalianis, H., Salvi, E., Budrionis, A., & Godtliebsen, F. (2021). Challenges and opportunities beyond structured data in analysis of electronic health records. *Wiley Interdisciplinary Reviews: Computational Statistics*, *13*(6), e1549. DOI: 10.1002/wics.1549

Teimoori, Z., & Yassine, A. (2022). A review on intelligent energy management systems for future electric vehicle transportation. *Sustainability (Basel)*, *14*(21), 14100. DOI: 10.3390/su142114100

Temming, M. (2019, August 8). Machines are getting schooled on fairness. *Science News*.https://www.sciencenews.org/article/machines-are-getting-schooled-fairness

Teoh, Y. K., Gill, S. S., & Parlikad, A. K. (2021). IoT and fog computing based predictive maintenance model for effective asset management in industry 4.0 using machine learning. *IEEE Internet of Things Journal*.

The Japan Times. (2017). *Amended privacy protection law. The Japan Times*. https://www.japantimes.co.jp/opinion/2017/06/01/editorials/amended-privacy-protection-law/#.WrHxmZ7H-Z

The Straits Times. (2018, April 25). *China completes first draft of national rules to allow road tests for driverless vehicles. The Straits Times*. https://www.straitstimes.com/asia/east-asia/china-completes-first-draft-of-national-rules-to-allow-road-tests-for-driverless

Tien, J. M. (2017). The Sputnik of servgoods: Autonomous vehicles. *Journal of Systems Science and Systems Engineering*, *26*(2), 133–162. DOI: 10.1007/s11518-016-5325-1

Timpson C.G. (2019). *Old and New in Quantum Information*.

Tolan, S., Miron, M., Gómez, E., & Castillo, C. (2019). Why Machine Learning May Lead to Unfairness: Evidence from Risk Assessment for Juvenile Justice in Catalonia. https://dl.acm.org/doi/10.1145/3322640.3326705

Towett, G., Snead, R. S., Grigoryan, K., & Marczika, J. (2023). Geographical and practical challenges in the implementation of digital health passports for cross-border COVID-19 pandemic management: A narrative review and framework for solutions. *Globalization and Health, 19*(1), 98. DOI: 10.1186/s12992-023-00998-7 PMID: 38066568

Tufekci, Z. (2015, February). Algorithms in our midst: Information, power, and choice when software is everywhere. In *Proceedings of the 18th ACM conference on computer-supported cooperative work & social computing* (pp. 1918-1918). https://dl.acm.org/doi/10.1145/2675133.2697079

Tzafestas, S. G. (2018). Synergy of IoT and AI in Modern Society: The Robotics and Automation Case. *Robotics & Automation Engineering Journal, 3*(5). Advance online publication. DOI: 10.19080/RAEJ.2018.03.555621

Uddin, S. M. I., Mozumder, M. A. I., & Kim, H.-C. (2023). *Automatic Leather Defect Detection Using Deep Learning.*

Udokwu, S. T. C., Okoye, C. C., Oshioste, E. E., Nwankwo, E. F., & Azubuike, N. U. (2023). *The role of leadership on organization management: A case study of First Bank in Abuja Metropolis. Business and Economics in Developing Countries.* BEDC., DOI: 10.26480/bedc.02.2023.46.51

Uhuo, E., Egba, S., Alaebo, P., Obike, C. A., & Ede, I. S. (2023). Evaluation of fertility hormones and biochemical stress initiated by toxicant in diet prepared with fish smoked with polyethylene (plastic) materials as a fuel source. *Comparative Clinical Pathology, 32*(4), 527–537. DOI: 10.1007/s00580-023-03461-6

Ukato, A., Sofoluwe, O. O., Jambol, D. D., & Ochulor, O. J. (2024). *Optimising maintenance logistics on offshore platforms with AI: Current strategies and future innovations.*

Urbinati, A., Bogers, M., Chiesa, V., & Frattini, F. (2019). Creating and capturing value from Big Data: A multiple-case study analysis of provider companies. *Technovation, 84*, 21–36. DOI: 10.1016/j.technovation.2018.07.004

Uzougbo, N. S., Ikegwu, C. G., & Adewusi, A. O. (2024). International enforcement of cryptocurrency laws: Jurisdictional challenges and collaborative solutions. *Magna Scientia Advanced Research and Reviews,* 11(1), 068-083

Vaiyapuri, T., Binbusayyis, A., & Varadarajan, V. (2021). Security, Privacy and Trust in IoMT Enabled Smart Healthcare System: A Systematic Review of Current and Future Trends [IJACSA]. *International Journal of Advanced Computer Science and Applications, 12*(2). Advance online publication. DOI: 10.14569/IJACSA.2021.0120291

Varadam, D., Shankar, S. P., Bharadwaj, A., Saxena, T., Agrawal, S., & Dayananda, S. (2024). Enhancing Industrial Robotics Performance and Security With AI and Blockchain Technologies. In R. C. Biradar, G. D, N. Tabassum, N. Hegde, & M. Lazarescu (Eds.), *AI and Blockchain Applications in Industrial Robotics* (pp. 58-81). IGI Global. https://doi.org/DOI: 10.4018/979-8-3693-0659-8.ch003

Varadam, D., Shankar, S. P., Shankar, A., & Narayan, A. N. P, N., & Kumar, T. N. (2024). Application of AI and Blockchain Technologies in the Medical Domain. In R. C. Biradar, G. D, N. Tabassum, N. Hegde, & M. Lazarescu (Eds.), *AI and Blockchain Applications in Industrial Robotics* (pp. 200-225). IGI Global. https://doi.org/DOI: 10.4018/979-8-3693-0659-8.ch008

Vargas, J., Alsweiss, S., Toker, O., Razdan, R., & Santos, J. (2021). An overview of autonomous vehicles sensors and their vulnerability to weather conditions. *Sensors (Basel)*, *21*(16), 5397. DOI: 10.3390/s21165397 PMID: 34450839

Veale, M., Van Kleek, M., & Binns, R. (2018, April). Fairness and accountability design needs for algorithmic support in high-stakes public sector decision-making. In *Proceedings of the 2018 CHI Conference on Human Factors in Computing Systems* (pp. 1-14). https://dl.acm.org/doi/10.1145/3173574.3174014

Veeraiah, V., Shiju, D. J., Ramesh, J. V. N., Ganesh, K. R., Pramanik, S., Pandey, D., & Gupta, A. (2023). *Healthcare Cloud Services in Image Processing, Thrust Technologies' Effect on Image Processing*. IGI Global.

Venkatachala Appa Swamy, M., Periyasamy, J., Thangavel, M., Khan, S. B., Al-musharraf, A., Santhanam, P., Ramaraj, V., & Elsisi, M. (2023). Design and development of IoT and deep ensemble learning based model for disease monitoring and prediction. *Diagnostics (Basel)*, *13*(11), 1942. DOI: 10.3390/diagnostics13111942 PMID: 37296794

Venkataswamy, R., Paul, R., & Jogy, J. (2020). Internet of Things Based Autonomous Borewell Management System. In Cyber-physical Systems and Digital Twins: Proceedings of the 16th International Conference on Remote Engineering and Virtual Instrumentation 16 (pp. 229-243). Springer International Publishing.

Vermesan, O., Bahr, R., Ottella, M., Serrano, M., Karlsen, T., Wahlstrøm, T., & Gamba, M. T. (2020). Internet of robotic things intelligent connectivity and platforms. *Frontiers in Robotics and AI*, *7*, 104.

Vermesan, O., Bröring, A., Tragos, E., Serrano, M., Bacciu, D., Chessa, S., & Bahr, R. (2022). Internet of robotic things–converging sensing/actuating, hyperconnectivity, artificial intelligence and IoT platforms. In *Cognitive hyperconnected digital transformation* (pp. 97–155). River Publishers.

Vermesan, O., Bröring, A., Τράγος, H., Serrano, M. M., Bacciu, D., Chessa, S., Gallicchio, C., Micheli, A., Dragone, M., Saffiotti, A., Simoens, P., Crea, F., & Bahr, R. (2022). *Internet of Robotic Things – Converging Sensing/Actuating*. Hyperconnectivity, Artificial Intelligence and IoT Platforms., DOI: 10.1201/9781003337584-4

Vermesan, O., Friess, P., Guillemin, P., Gusmeroli, S., Sundmaeker, H., Bassi, A., Jubert, I. S., Mazura, M., Harrison, M., & Eisenhauer, M. (2022). Internet of things strategic research roadmap. In *Internet of things-global technological and societal trends from smart environments and spaces to green ICT* (pp. 9–52). River Publishers.

Vernacchia, S. (2019). *Quantum leap: advancing a strategy for quantum computing that will inspire, support and safeguard economic growth in the Middle East.* https://www.pwc.com/m1/en/world-government-summit/documents/wgs-quantum-leap.pdf World Government Summit.

Vienna Convention on Road Traffic, 1968, BGBl. 1977 II, 809 (811), last amended by agreement of 24 March 2014, BGBl. 2016 II, 1306.

Vijayan, R., Mareeswari, V., & Jaswanth, A. B. (2024). Triple Diseases Prediction System using Deep Learning Techniques. *2024 Second International Conference on Emerging Trends in Information Technology and Engineering (ICETITE)*, 1–7. DOI: 10.1109/ic-ETITE58242.2024.10493563

Vijayan, R., Mareeswari, V., & Pople, V. (2023). Public Social Distance Monitoring System Using Object Detection YOLO Deep Learning Algorithm. *SN Computer Science*, *4*(6), 718. Advance online publication. DOI: 10.1007/s42979-023-02131-2

Vijayan, R., Mareeswari, V., & Shobana, V. (2024). Tomato Plant Leaf Disease Prediction and Suggestion Using Deep Learning. *International Conference on Power Engineering and Intelligent Systems (PEIS)*, 125–136. DOI: 10.1007/978-981-97-6714-4_10

Wacket, M., Escritt, T., & Davis, T. (2017). Germany adopts self-driving vehicles law. *Reuters*. https://www.reuters.com/article/us-germany-autos-self-driving-idUSKBN1881HY

Wagner, B., Winkler, T., & Human, S. (2021). Bias in geographic information systems: The case of Google Maps.

Waheed, A., Shah, M. A., Mohsin, S. M., Khan, A., Maple, C., Aslam, S., & Shamshirband, S. (2022). A comprehensive review of computing paradigms, enabling computation offloading and task execution in vehicular networks. *IEEE Access : Practical Innovations, Open Solutions*, *10*, 3580–3600.

Wang, C. Y., Yeh, I. H., & Mark Liao, H. Y. (2024, September). YOLOv9: Learning what you want to learn using programmable gradient information. In *European Conference on Computer Vision* (pp. 1–21). Cham: Springer Nature Switzerland.

Wang, F., & Kyoung, K. J. (2023). Leather defect detection method in clothing design based on TDENet. *IEEE Access : Practical Innovations, Open Solutions*, *11*, 104890–104904. DOI: 10.1109/ACCESS.2023.3308493

Wang, J., Feng, X., Yu, Y., Huang, H., Zheng, M., Xu, Y., Wu, J., Yang, Y., & Lu, J. (2024). Rapid temperature-responsive thermal regulator for safety management of battery modules. *Nature Energy*, *9*(8), 939–946. DOI: 10.1038/s41560-024-01535-5

Wang, R., Zhao, H., Xu, Z., Ding, Y., Li, G., Zhang, Y., & Li, H. (2023). Real-time vehicle target detection in inclement weather conditions based on YOLOv4. *Frontiers in Neurorobotics*, *17*, 1058723. DOI: 10.3389/fnbot.2023.1058723 PMID: 36968300

Wang, Y., Kung, L., & Byrd, T. A. (2018). Big data analytics: Understanding its capabilities and potential benefits for healthcare organizations. *Technological Forecasting and Social Change*, *126*, 3–13. DOI: 10.1016/j.techfore.2015.12.019

Weerasinghe, K., Scahill, S. L., Pauleen, D. J., & Taskin, N. (2022). Big data analytics for clinical decision-making: Understanding health sector perceptions of policy and practice. *Technological Forecasting and Social Change*, *174*, 121222. DOI: 10.1016/j.techfore.2021.121222

Wei, W., Deng, D., Zeng, L., & Zhang, C. (2021). Real-time implementation of fabric defect detection based on variational automatic encoder with structure similarity. *Journal of Real-Time Image Processing*, *18*(3), 807–823. DOI: 10.1007/s11554-020-01023-5

Wendell, W. (2008). *Moral machines: Teaching robots right from wrong*. Oxford University Press.

Werner, D. (2020, December 7). Vehicle and traffic laws should be standard nationwide. *Adirondack Daily Enterprise: Opinion*. https://www.adirondackdaily enterprise.com/

Werner, S., Rettinger, A., Halilaj, L., & Lüttin, J. (2021). RETRA: Recurrent transformers for learning temporally contextualized AI embeddings. In *The Semantic Web: 18th International Conference, ESWC 2021, Virtual Event, June 6–10, 2021* [Springer.]. *Proceedings*, *18*, 425–440.

Wesson, P., Hswen, Y., Valdes, G., Stojanovski, K., & Handley, M. A. (2022). Risks and opportunities to ensure equity in the application of big data research in public health. *Annual Review of Public Health, 43*(1), 59–78. DOI: 10.1146/annurev-publhealth-051920-110928 PMID: 34871504

Williamson, S. M., & Prybutok, V. (2024). Balancing privacy and progress: A review of privacy challenges, systemic oversight, and patient perceptions in AI-Driven healthcare. *Applied Sciences (Basel, Switzerland), 14*(2), 675. DOI: 10.3390/app14020675

Wilson, C., Boe, B., Sala, A., Puttaswamy, K. P. N., & Zhao, B. Y. (2009). User interactions in social networks and their implications. In *Proceedings of the 4th ACM European Conference on Computer Systems* (pp. 205–218). ACM., https://dl.acm.org/doi/10.1145/1519065.1519089 DOI: 10.1145/1519065.1519089

Wiltz, C. (2017). Bias In, Bias Out: How AI Can Become Racist. *Embedded Systems Conference (ESC).* Minneapolis, MN: Design News.

Winiarti, S., Prahara, A., Murinto, D. P. I., & Ismi, P. (2018). Pre-trained convolutional neural network for classification of tanning leather image. [CNN]. *Network (Bristol, England), 9*(1), 9.

Woll, A., & Tørresen, J. (2023). What is a Smart Hospital? A Review of the Literature. In Duffy, V. G., Lehto, M., Yih, Y., & Proctor, R. W. (Eds.), *Human-Automation Interaction. Automation, Collaboration, & E-Services* (Vol. 10). Springer., DOI: 10.1007/978-3-031-10780-1_8

Wu, G., Boriboonsomsin, K., & Barth, M. J. (2020). *Eco-routing navigation system for electric vehicles.* arXiv preprint arXiv:2008.09674.

Xia, H., Lan, Y., Song, S., & Li, H. (2022). Raindrop removal from a single image using a two-step generative adversarial network. *Signal, Image and Video Processing, 16*(3), 677–684. DOI: 10.1007/s11760-021-02007-z

Xiao, D., Dianati, M., Geiger, W. G., & Woodman, R. (2023). Review of ML-based hazardous event detection methods for autonomous driving systems. *IEEE Transactions on Intelligent Transportation Systems, 24*(5), 4697–4715. DOI: 10.1109/TITS.2023.3240104

Xu, B., Rathod, D., Zhang, D., Yebi, A., Zhang, X., Li, X., & Filipi, Z. (2020). Parametric study on reinforcement learning optimized energy management strategy for a hybrid electric vehicle. *Applied Energy, 259*, 114200. DOI: 10.1016/j.apenergy.2019.114200

Yan, S., Shah, M. H., Li, J., O'Connor, N., & Liu, M. (2023, October). *A review on AI algorithms for energy management in E-mobility services*. In *2023 7th CAA International Conference on Vehicular Control and Intelligence (CVCI)* (pp. 1–8). IEEE. DOI: 10.1109/CVCI59596.2023.10397371

Yanamala, A. K. Y., & Suryadevara, S. (2023). Advances in data protection and artificial intelligence: Trends and challenges. *International Journal of Advanced Engineering Technologies and Innovations*, *1*(01), 294–319.

Yang, F., & Gu, S. (2021). Industry 4.0, a revolution that requires technology and national strategies. *Complex & Intelligent Systems*, *7*(3), 1311–1325. DOI: 10.1007/s40747-020-00267-9

Yang, T., & Han, J. (2021). Integrated management strategy with feasible smartness over heterogeneous IoT environments. *Electronics (Basel)*, *10*(2), 149. DOI: 10.3390/electronics10020149

Yao, R., Zhang, W., Evans, R., Cao, G., Rui, T., & Shen, L. (2022). Inequities in health care services caused by the adoption of digital health technologies: Scoping review. *Journal of Medical Internet Research*, *24*(3), e34144. DOI: 10.2196/34144 PMID: 35311682

Yuan, X., Yang, C., He, Q., Chen, J., Yu, D., Li, J., Zhai, S., Qin, Z., Du, K., Chu, Z., & Qin, P. (2020). Current and perspective diagnostic techniques for COVID-19. *ACS Infectious Diseases*, *6*(8), 1998–2016. DOI: 10.1021/acsinfecdis.0c00365 PMID: 32677821

Yu, C., Wang, X., Xu, X., Zhang, M., Ge, H., Ren, J., Sun, L., Chen, B., & Tan, G. (2019). Distributed multiagent coordinated learning for autonomous driving in highways based on dynamic coordination MLs. *IEEE Transactions on Intelligent Transportation Systems*, *21*(2), 735–748. DOI: 10.1109/TITS.2019.2893683

Yu, F., Chen, H., Wang, X., Xian, W., Chen, Y., Liu, F., & Darrell, T. (2020). BDD100K: A diverse driving dataset for heterogeneous multitask learning. In *Proceedings of the IEEE/CVF Conference on Computer Vision and Pattern Recognition* (pp. 2636–2645). DOI: 10.1109/CVPR42600.2020.00271

Zaarane, A., Slimani, I., Al Okaishi, W., Atouf, I., & Hamdoun, A. (2020). Distance measurement system for autonomous vehicles using stereo camera. *Array (New York, N.Y.)*, *5*, 100016. DOI: 10.1016/j.array.2020.100016

Zafar, I., Anwar, S., Yousaf, W., Nisa, F. U., & Kausar, T. (2023). Reviewing methods of deep learning for intelligent healthcare systems in genomics and biomedicine. *Biomedical Signal Processing and Control, 86*, 105263. DOI: 10.1016/j.bspc.2023.105263

Zaidan, A., & Zaidan, B. (2020). A review on intelligent process for smart home applications based on IoT: Coherent taxonomy, motivation, open challenges, and recommendations. *Artificial Intelligence Review*, *53*(1), 141–165. DOI: 10.1007/s10462-018-9648-9

Zaman, U. (1893). Imran, Mehmood, F., Iqbal, N., Kim, J., & Ibrahim, M. (2022). Towards secure and intelligent internet of health things: A survey of enabling technologies and applications. *Electronics (Basel)*, *11*(12).

Zeadally, S., & Bello, O. (2021). Harnessing the power of Internet of Things based connectivity to improve healthcare. *Internet of Things : Engineering Cyber Physical Human Systems*, *14*, 100074. DOI: 10.1016/j.iot.2019.100074

Zeadally, S., Siddiqui, F., Baig, Z., & Ibrahim, A. (2020). Smart healthcare: Challenges and potential solutions using internet of things (IoT) and big data analytics. *PSU Research Review*, *4*(2), 149–168. DOI: 10.1108/PRR-08-2019-0027

Zhang, A., Xing, L., Zou, J., & Wu, J. C. (2022). Shifting machine learning for healthcare from development to deployment and from models to data. *Nature Biomedical Engineering*, *6*(12), 1330–1345. DOI: 10.1038/s41551-022-00898-y PMID: 35788685

Zhang, H., He, S., Chen, J., Yang, K., Ran, P., & Chen, J. (2023). A Hierarchical Authentication System for Access Equipment in Internet of Things. *International Journal of Intelligent Systems*, *2023*(1), 1–11. DOI: 10.1155/2023/8899697

Zhao, Z., Liao, X., Abdelraouf, A., Han, K., Gupta, R., Barth, M. J., & Wu, G. (2023, October). *Real-time learning of driving gap preference for personalized adaptive cruise control*. In *2023 IEEE International Conference on Systems, Man, and Cybernetics (SMC)* (pp. 4675–4682). IEEE. DOI: 10.1109/SMC53992.2023.10394260

Zheng, Z., Zhou, B., Zhou, D., Soylu, A., & Kharlamov, E. (2022, October). Executable AI for transparent machine learning in welding monitoring at Bosch. In *Proceedings of the 31st ACM International Conference on Information & Knowledge Management* (pp. 5102–5103).

Zhou, K. Q., & Nabus, H. (2023). The Ethical Implications of DALL-E: Opportunities and Challenges. *Mesopotamian Journal of Computer Science*, *2023*, 17–23.

Zhou, X. (2024). IoT Sensor-Based Big Data and Information Systems in Smart Healthcare. *Highlights in Science. Engineering and Technology*, *92*, 397–402.

Zinner, M., Dahlhausen, F., Boehme, P., Ehlers, J., Bieske, L., & Fehring, L. (2021). Quantum computing's potential for drug discovery: Early stage industry dynamics. *Drug Discovery Today*, *26*(7), 1680–1688. DOI: 10.1016/j.drudis.2021.06.003 PMID: 34119668

About the Contributors

Muhammad Usman Tariq has more than 16+ year's experience in industry and academia. He has authored more than 200+ research articles, 100+ case studies, 50+ book chapters and several books other than 4 patents. He is founder and CEO of The Case HQ, a unique repository for courses, case studies, and video storytelling. He has been working as a consultant and trainer for industries representing six sigma, quality, health and safety, environmental systems, project management, and information security standards. His work has encompassed sectors in aviation, manufacturing, food, hospitality, education, finance, research, software and transportation. He has diverse and significant experience working with accreditation agencies of ABET, ACBSP, AACSB, WASC, CAA, EFQM and NCEAC. Additionally, Dr. Tariq has operational experience in incubators, research labs, government research projects, private sector startups, program creation and management at various industrial and academic levels. He is Certified Higher Education Teacher from Harvard University, USA, Certified Online Educator from HMBSU, Certified Six Sigma Master Black Belt, Lead Auditor ISO 9001 Certified, ISO 14001, IOSH MS, OSHA 30, and OSHA 48. He has been awarded Principal Fellowship from Advance HE UK, Chartered Fellowship of CIPD UK, and certified business and management educator recognition.

Safaa Najah Saud Al-Humairi is an esteemed Associate Professor at Management and Science University, serving as the Program Coordinator and Researcher Coordinator. With over 15 years of experience in bio-engineering design and manufacturing, he has made significant contributions in the field, particularly in sensor design and implementation over the last 10 years. Throughout his career, Dr. Safaa has been an avid researcher, dedicated to expanding knowledge and pushing the boundaries of technology. His extensive body of work includes the publication of more than 120 research articles in highly esteemed and impactful journals, earning him recognition and respect within the academic community. In addition to his research achievements, Dr. Safaa has garnered numerous accolades for his innovative contributions. He has been honoured with special gold, silver, and bronze medals in various national and international engineering and technology exhibitions. These awards testify to his exceptional creativity, problem-solving skills, and ability to implement groundbreaking ideas.

C.G.Balaji is an educator and researcher focused on advancing computer engineering and technology. He works as an Assistant Professor in Symbiosis Institute of Digital and Telecom Management, Symbiosis International (Deemed University), Lavale, Pune. He holds a Ph.D. in Information and Communication Engineering from College of Engineering Guindy, Anna University, Chennai and has over nine years of teaching experience. His expertise includes Next Generation Networks, Wireless/Mobile Networks, Network Convergence, Internet of Things and Software Defined Networks. He has published several papers in reputed journals and holds 2 patents in intelligent healthcare systems and IoT solutions. As a member of IAENG, IEI, ISTE, and CSTA, he actively contributes to academia. Committed to continuous learning, he holds certifications in Industry 4.0, Outcome-Based Learning, Patent Creation, and Data Science, and participates in faculty development programs to stay current with field advancements.

Saurabh Chandra is an accomplished legal scholar and currently serves as an Associate Professor & Assistant Dean at the School of Law, Bennett University, India. With a robust academic foundation, he earned his Bachelor of Arts and Bachelor of Laws (B.A. LL.B.) from the prestigious Aligarh Muslim University (A.M.U.). He further specialized in Business Laws, completing his Master of Laws (LL.M.) at the renowned National Law School of India University (NLSIU), Bangalore. Dr. Chandra's academic journey culminated in a Doctorate in Law (Ph.D.) from the esteemed National Law University (NLU) Delhi. In addition to his impressive legal education from some of India's most prestigious institutions, Dr. Chandra also holds a master's degree in Management with specialization in Human Resources in Management and another in Journalism and Mass Communication. With more than sixteen years of teaching experience, he has numerous publications to his credit. His expertise is complemented by a robust portfolio of scholarly publications. Throughout his career, Dr. Chandra has consistently produced high-impact research. His publications are widely recognized for their depth of analysis, innovative perspectives, and practical relevance. His publications not only reflect his scholarly rigor but also his dedication to addressing contemporary challenges faced by businesses and policymakers. By bridging the gap between theoretical insights and real-world applications, Dr. Chandra's work serves as a valuable resource for academics, legal practitioners, and industry professionals alike. Dr. Chandra brings a wealth of experience in Academic Administration, having successfully held various key administrative positions. His diverse qualifications and extensive experience make him a distinguished figure in the academic and legal communities.

Jasraj Singh Chopra is student of Third Year at School of Mathematics, Applied Statistics & Analytics, NMIMS, Navi Mumbai Campus, NMIMS (Deemed to be University)-Mumbai, India. He has interest in the area of data analytics and research methodology.

Aryan Dave is student of Third Year at School of Mathematics, Applied Statistics & Analytics, NMIMS, Navi Mumbai Campus, NMIMS (Deemed to be University)-Mumbai, India. He has interest in the area of application of data analytics.

Oindrilla Ghosh is a Research Scholar hailing from the School of Law at Christ (Deemed to be University) Bangalore, India, focuses her research on health of tribal communities. Her work includes total 6 publications on Neuroscience on Organizational Behavior at IGI Global and also Environmental Disasters that are indexed in SCOPUS and Web of Sciences conference proceedings. She completed her early education at Loreto Day School, Bowbazar and went on to achieve distinction in her undergraduate (B.A/LL.B) and postgraduate (LL.M) studies at Calcutta University. Having served as an Assistant Professor of Law at Kingston Law College,Barasat, Kolkata, Ghosh has actively participated in numerous National Conferences, International Conferences, UN Global Summits, webinars, and seminars organized by prestigious academic institutions. Her expertise lies in health perspectives, environmental studies, human behavior, and related areas. Additionally, she has conducted various workshops, lectures in educational institutions, and is engaged with NGO's to help the needy

Prasanna is a final year student at Ramaiah University of Applied Sciences in the Department of Computer Science and Engineering.

Safad is a dedicated researcher, lawyer, and academician with a strong interdisciplinary background. He holds an M.Tech in Computer Science and Engineering from Government Engineering College (Rajiv Gandhi Institute of Technology & Science), Kottayam, Kerala, India, and a B.Tech in Computer Science and Engineering from Cochin University of Science and Technology. Additionally, he earned a Bachelor's degree in Legislative Law from Karnataka State Law University, Bangalore, India. He has qualified the University Grants Commission-NET examination and the All India Bar Examination. His journey with IEEE began in 2012 as a student member. With 4.5 years of teaching experience in various engineering colleges across India, three years of research experience, and industry exposure, he has published papers in reputed IEEE journals and conferences. His academic interests lie at the intersection of Computer Science, Artificial Intelligence, and Machine Learning, with a particular focus on their applications in the legal domain. He remains passionate about research and teaching, continuously striving to contribute to these evolving fields.

Anavi Jhunjhunwala is student of Second Year at School of Mathematics, Applied Statistics & Analytics, NMIMS, Navi Mumbai Campus, NMIMS (Deemed to be University)-Mumbai, India. She has interest in the area of actuarial statistics, application of data analytics and research methodology.

Gagan Kukreja is Program Coordinator of MS in Accounting and an Associate Professor in the Department of Accounting, College of Business Administration, University of Bahrain, Bahrain. He has over 24 years of experience in university teaching, academic administration, training, and research. He holds various professional qualifications such as ACCA (UK), CPA (US), CMA (US and India), and CIA (US). Dr Kukreja supervised many candidates for the PhD, Master's and Bachelor's level and his current area of interest is Integrated Reporting, Corporate Governance, Auditing, Blockchain, and Big Data. He examined numerous Master's Dissertation and PhD theses of many international universities. Dr Kukreja published over 40 research papers and cases in international refereed journals includes International Journal of Critical Accounting, Journal of Forensic & Investigative Accounting, Journal of Investment Compliance, EuroMed Journal of Business, Journal of Management Development, South Asian Journal of Business & Management Cases and many more. Further, he is an international review member of Pearson International in accounting and auditing and prepared solution manuals for Pearson's Textbooks. In addition, he contributed (with co-authors) more than 10 chapters to various edited books published by Springer and IGI Global. He is an editorial member in IGI Global in accounting, finance, and banking. Dr Kukreja got the Best Research Paper Award (Gold Shield) and Commendable Research Paper Award (Bronze Shield) in ICAI International Research Awards 2020 by The Institute of Chartered Accountants of India, with other co-authors. His Case Study (with other co-authors) also secured the Amity Award for Global Case Competition 2021 during the 21st International Business Horizon, Inbush Era World Summit 2021. During his PhD, he was also a recipient of Junior Research Fellowship from University Grant Commission, India. He is a curriculum specialist in accounting, banking and finance and developing new programs/courses. In addition, he assisted the university in QA, NQF and Accreditation (AACSB/ACCA/CIMA/CFA/IMA). He organised many seminars, webinars, and conferences and did external moderation/assessor work for various programs offered by many international universities. He can be contacted at gkukreja@uob.edu.bh.

Prasanna Kulkarni is an accomplished educationalist with notable qualifications, including an M. Com., F. C. A., and Ph. D. He is currently working as an Associate Professor in Symbiosis Institute of Digital and Telecom Management, Symbiosis International (Deemed University), Lavale, Pune. His expertise spans a range of specific areas, such as Corporate Finance, Banking and Finance, Technology and Audit, and Fintech. He focuses on the application of advanced technologies within the BFSI sector, encompassing IoT, Blockchain, Cloud, Big Data Technologies, and AI/ML. Dr. Kulkarni's broader areas of interest include Finance and Accounts, Taxation, Audit, and the Digital Transformation of the BFSI industry. He is fascinated by the latest advances in Digital and Telecom space, and has a keen interest in exploring these areas too.

Binod Kumar is a graduate student known for his academic excellence, with his total 4 scholarly works indexed in Scopus database on Neuroscience in Organizational Behavior, Neuromarketing and AI Applications on Wildlife Conservation. Hailing from Ranchi, Jharkhand in India, Binod is currently pursuing a Master of Laws (LL.M) at the prestigious Radha Govind University. His passion for technology and innovation along with climate changes is evident through his high academic performance and numerous projects. Binod has a keen interest in artificial intelligence, machine learning, neuropsychology, climate changes and disaster management, fields in which he has undertaken several research projects. He is particularly fascinated by the potential of AI to solve real-world problems and is dedicated to learning and applying new techniques in these areas. He volunteers at local NGOs, teaching underprivileged children basic computer skills. With a strong foundation in law and a commitment to making a positive impact, Binod Kumar is poised to become a future leader in the legal industry.

Tuan Luong received the B.S. degree in electronic engineering from VNU University of Engineering and Technology in 2024. He is currently a researcher at VNU Key Laboratory for Smart Integrated Systems, VNU University of Engineering and Technology. His research interests mainly include embedded system design and SoC design.

Shashank Mehra is Associate Professor in the area of Business Analytics at SVKM'S NMIMS, Navi Mumbai Campus, NMIMS (Deemed to be University), Mumbai. He has many national and international publications under his name. His research interest is in Business Analytics, Statistical Model Building in the area of

HR and Marketing. He holds the position as visiting faculty in different universities and prominent business schools in India.

Ronit Mohta is an alumni of Ramaiah University of Applied Sciences and presently pursuing MSc in Cyber Security in the University of Birmingham.His interests includes Cyber Security.

Ashwath Narayan JR is a student at the University of Warwick pursuing an MSc in Diagnostics, Data, and Digital Health who completed his graduation with b.tech in Artificial Intelligence and Machine Learning at Ramaiah University of Applied Sciences Bangalore. Ashwath focuses on creating innovative, user-centered solutions at the intersection of healthcare and technology. Their work includes using data and digital tools to improve patient outcomes, exemplified by projects like Tooth GenAI, which leverages AI to enhance diagnostic capabilities in dental care.

Hung K. Nguyen received the B.S. and M.S. degrees in electronic engineering from Vietnam National University, Hanoi, Vietnam, in 2003 and 2005, respectively. He received a Ph.D. degree in electronic engineering from Southeast University, Nanjing, China in 2013. He is currently a researcher and Assistant Professor at VNU Key Laboratory for Smart Integrated Systems, VNU University of Engineering and Technology. His research interests mainly include multimedia processing, reconfigurable computing, and SoC designs.

Dhakshan Aravind P holds a Bachelor's degree in Computer Applications (BCA) from VIT University, Vellore, India. His research interests focus on artificial intelligence in visual technologies, and software development. He worked on projects using Python programming and Full stack web technologies.

Sahana. P. Shankar completed her B.E in Information Science and Engineering and MTech in Software Engineering from Ramaiah Institute of Technology, Bangalore. She is a third rank holder in B.E and a first rank holder with gold medal in MTech. She has worked with Unisys as an Associate Engineer for three years. She holds a US Patent in the field so software testing. She has worked with RIT as a Teaching Assistant for 6 months under the TEQIP-II scholarship program. She has published papers in various international journals and conferences. She has also authored several book chapters. She is an ISTQB certified software tester. She is pursuing her PhD from VTU, Belgaum. She is currently working with Ramaiah University of Applied Sciences, Bangalore as an Assistant Professor in the Department of Computer Science and Engineering since 2018.

Sabyasachi Pramanik is a professional IEEE member. He obtained a PhD in Computer Science and Engineering from Sri Satya Sai University of Technology and Medical Sciences, Bhopal, India. Presently, he is an Associate Professor, Department of Computer Science and Engineering, Haldia Institute of Technology, India. He has many publications in various reputed international conferences, journals, and book chapters (Indexed by SCIE, Scopus, ESCI, etc). He is doing research in the fields of Artificial Intelligence, Data Privacy, Cybersecurity, Network Security, and Machine Learning. He also serves on the editorial boards of several international journals. He is a reviewer of journal articles from IEEE, Springer, Elsevier, Inderscience, IET and IGI Global. He has reviewed many conference papers, has been a keynote speaker, session chair, and technical program committee member at many international conferences. He has authored a book on Wireless Sensor Network. He has edited 8 books from IGI Global, CRC Press, Springer and Wiley Publications.

Vijayan R is a professor and senior faculty member at the School of Computer Science and Information Systems [SCORE], Vellore Institute of Technology (VIT), Vellore, Tamil Nadu, India. He is a member of the Computer Society of India (CSI) and the Soft Computing and Research Society (SCRS). He has published several national and international research articles in reputed journals and conferences. His research interests include web technologies, wireless networks, ad hoc networks, computer networks, machine learning, the Internet of Things (IoT), Mobile Computing, Cloud Computing, and Data Science.

Pranati Rakshit is an Associate Professor in the Department of Computer Science and Engineering of JIS College of Engineering, Kalyani, West Bengal, India. She has more than 23 years of teaching experience. She has completed her Ph.D. degree from Jadavpur University, Kolkata, West Bengal, India, in the field of Pattern Recognition, Machine Learning, Medical image analysis. She has completed her Master degree and B.E. Degree in Computer Science & Engineering. She has worked in the field of Pattern Recognition, Machine Learning, Deep Learning and Medical image analysis, Sentiment Analysis etc. She has supervised more than 22 M.Tech. Projects and presently supervising 1 Ph.D. student. She has more than 60 numbers of research publications in International journals and conferences including SCI and Scopus indexing. She has 23 patents (Granted-6, Published-17) in her account and 2 more patents are already filed. She has 1 copyrights and 1 design IPR also. She is a life member of Indian Society of Technical Education.

Atanu Roy is a marine engineer from Kolkata, India. He completed his Marine Engineering four-year training under the West Bengal State Council of Technical Education (WBSCTE). Started a sailing career a foreign company, in 2003 and became Chief Engineer in 2011. During his period of sailing, he also started to associate with the maritime shore industry. He worked with Indian Maritime University (IMU) Kolkata campus, MERI as an assistant professor in 2015, and The Neotia University (TNU) in 2016 as a visiting faculty. During this period, he achieved an Extra First-Class Marine Engineer Certificate of Competency level in 2017, which is considered the highest grade of certificate of competency as a marine engineer officer. Becomes a faculty member at Ethiopian Maritime Training Institute (EMTI) in Ethiopia, Africa in 2017. Presently he is working at The Neotia University as an Assistant Professor in the Marine Department. Atanu Roy has a professional affiliation as a Fellow of the Institute of Marine Engineers (FIMarE), India and the Institute of Engineers, India (IEI).

Praveen Kumar S holds a Bachelor's degree in Computer Applications (BCA) from VIT University, Vellore, India. Currently pursuing his Master of Computer Applications (MCA) at VIT Vellore, his research interests focus on artificial intelligence in visual technologies, and cloud computing. He worked on projects using Python programming and software development.

Anirudh Shankar is an AI and data science professional with a Bachelor's in Artificial Intelligence and Machine Learning, currently pursuing a Master's in Diagnostics, Data, and Digital Health at the University of Warwick. He has led innovative healthcare projects and applied AI to real-world solutions. His industry experience includes internships at Dassault Systèmes and L&T Technology Services, where he used AI and ML technologies to solve complex problems. Anirudh is passionate about the ethical implications of AI and aims to advance its role in healthcare diagnostics and digital health.

Bhupinder Singh working as Professor at Sharda University, India. Also, Honorary Professor in University of South Wales UK and Santo Tomas University Tunja, Colombia. His areas of publications as Smart Healthcare, Medicines, fuzzy logics, artificial intelligence, robotics, machine learning, deep learning, federated learning, IoT, PV Glasses, metaverse and many more. He has 3 books, 139 paper publications, 163 paper presentations in international/national conferences and seminars, participated in more than 40 workshops/FDP's/QIP's, 25 courses from international universities of repute, organized more than 59 events with international and national academicians and industry people's, editor-in-chief and co-editor in journals, developed new courses. He has given talks at international universities, resource person in international conferences such as in Nanyang Technological University Singapore, Tashkent State University of Law Uzbekistan; KIMEP University Kazakhstan, All'ah meh Tabatabi University Iran, the Iranian

Association of International Criminal law, Iran and Hague Center for International Law and Investment, The Netherlands, Northumbria University Newcastle UK,

Mareeswari V is working as an Associate Professor at the School of Computer Science Engineering and Information Systems (SCORE), Vellore Institute of Technology (VIT), Vellore, Tamilnadu, India - 632014. She has over 21 years of academic and research experience. She received her Ph.D. in Information Technology and Engineering in the Web Service domain. She has produced several national and international articles in reputed journals, conferences, and book chapters. Her area of interest includes Medical Image processing using Machine and Deep Learning, Data analytics on Blockchain, Prediction systems on Web Service Ranking, Internet of Things (IoT), Routing on Mobile Ad-hoc networks, and E-Commerce systems. She is a member of the Computer Society of India (CSI) and the Soft Computing and Research Society (SCRS).

Deepak Varadam holds a Masters Degree in Microsystems Engineering from Heriot-Watt University, Edinburgh (Scotland), U.K. and P.G.Diploma in ECE from Napier University, U.K. He worked as a Postgraduate Masters Research Scientist under Biomedical Microengineering Group (BMG) at Heriot-Watt University. He is currently working as an Assistant Professor in Faculty of Engineering and Technology, Computer Science and Engineering Department at Ramaiah University of Applied Sciences (RUAS) since 2010. He has been actively participating in delivering corporate trainings, seminars and have publications in national, international journals and conferences. His research interests is in MEMS, Embedded Systems, Software Engineering, Computer Graphics, AI, Blockchain, and IoT. He is currently an ExeCom Member of the IEEE RUAS Student Chapter.

Harsha Vasudev is Assistant Professor in the department of Computer Science and Engineering, at IIITDM Kurnool, India. She was Post-Doctoral Researcher at HIT (Human Research Center), Department of Mathematics, University of Padua, Italy. She is also a member of SPRITZ (Security and PRIvacy Through Zeal) Research Group. Prior to that, she was Assistant Professor in Amrita Vishwa Vidyapeetham, Kerala, India and DSU (Dayananda Sagar University) Bengaluru, India. She has received BRICS Young Scientist Award in 2018. Her Ph.D. is from BITS (Birla Institute of Technology and Science) Pilani, India. She has done her MTech from CUSAT (Cochin University of Science and Technology), Kerala, India and BTech from Kerala University, India. Her main research interests include Network Security, Internet of Vehicles, Cryptography, Vehicular Adhoc Networks, and PostQuantum Cryptography.

Kittisak Wongmahesak serves as a lecturer at the Faculty of Political Science, North Bangkok University, and the Director of the Association of Legal & Political Studies in Thailand. Currently, he holds the position of Research Fellow at Shinawatra University, Thailand, and Universiti Sultan Zainal Abidin, Malaysia. He is also an Adjunct Professor at Sekolah Tinggi Ilmu Administrasi Abdul Haris, Indonesia, and an Adjunct Research Professor at several Indonesian universities, including Universitas Muhammadiyah Sinjai, Universitas Muhammadiyah Makassar, and Universitas Muhammadiyah Sidenreng Rappang. Previously, he was a Full Professor of Public Administration at Huanghe Business School, Henan University of Economics and Law in China, and at the University of City Island in Cyprus. He also served as an Adjunct Researcher at Ton Duc Thang University and Duy Tan University in Vietnam. His academic credentials include a Ph.D. in Social Sciences, an M.A. and a B.A. in Political Science, all from Kasetsart University. Additionally, he earned an M.B.A. and an M.P.A. from Ramkhamhaeng University, and a B.Pol.Sc. in International Relations and Comparative Government and Politics from Sukhothai Thammathirat Open University. His expertise spans Political Science, Public and Private Management, International Political Economy, and Social Research. He is the editor of numerous prestigious journals worldwide, including Sustainability, Frontiers in Psychology, PLOS One, World Food Policy, Asian Academy of Management Journal, International Journal of Public Policy and Administration Research, International Journal of Innovative Research and Scientific Studies, and Recent Advances in Computer Science and Communications. Recognized as one of the world's top 2% of scientists by Stanford University since 2020, he has published over 450 Scopus/Web of Science indexed documents and holds an h-index of 57 with more than 10,000 citations.

Index

A

B

C

D

E

F

G

H

I

L

M

www.ingramcontent.com/pod-product-compliance
Lightning Source LLC
LaVergne TN
LVHW080609300926
849162LV00005B/1095
9798337343327